Introduction to
Environment, Biodiversity and
Climate Change

Introduction to
Environment, Biodiversity and Climate Change

Navale Pandharinath

CRC Press
Taylor & Francis Group
Boca Raton   London   New York

CRC Press is an imprint of the
Taylor & Francis Group, an **informa** business

BSP **BS Publications**
A unit of **BSP Books Pvt. Ltd.**
4-4-309/316, Giriraj Lane,
Sultan Bazar, Hyderabad - 500 095

First published 2022
by CRC Press
2 Park Square, Milton Park, Abingdon, Oxon, OX14 4RN

and by CRC Press
6000 Broken Sound Parkway NW, Suite 300, Boca Raton, FL 33487-2742

© 2022 BS Publications

CRC Press is an imprint of Informa UK Limited

Trademark notice: Product or corporate names may be trademarks or registered trademarks, and are used only for identification and explanation without intent to infringe.

Print edition not for sale in South Asia (India, Sri Lanka, Nepal, Bangladesh, Pakistan or Bhutan).

British Library Cataloguing-in-Publication Data
A catalogue record for this book is available from the British Library

Library of Congress Cataloging-in-Publication Data
A catalog record has been requested

ISBN: 978-1-032-13798-8 (hbk)
ISBN: 978-1-003-23086-1 (ebK)

DOI: 10.1201/9781003230861

Dedicated to,

My Parents

Late Smt. Narasubai R. Navale
Late Shri. Rukmaji Rao Narsoji Rao Navale
Resident of Kalyani, Tq. Yellareddy,
Nizamabad Dist. Telangana

and

To all my Teachers

from Primary School Kalyani, Middle School Yellareddy,
High School Mahabubnagar and City College,
Nizam College and Osmania University Hyderabad

Navale Pandharinath

PREFACE

United Nations Conference on Human Environment was held at StockHolm to take appropriate steps for the protection and improvement, because the state of environment declined greatly in quality and threatening to effect the available natural resources for the present and future generations, besides causing great damage to the health of living species and extinction of some species of life.

Environment includes air, water, land, biosphere and inter-relationship among and between air, water, land and humanbeings, other living creatures, plants, micro-organisms and property.

Biological diversity means the variability among living organisms from all sources and the ecological complexes including diversity within species or between species and ecosystems.

The physical state of the atmosphere at any place/location rarely exhibits a steady state even during short intervals of time. The ever changing physical state of atmosphere constitutes the weather, which is described by the values of meteorological elements on surface of the earth and in the atmosphere such as temperature, pressure, humidity, wind (speed and direction), state of sky (clouds), precipitation etc. The average conditions of weather at a place over a long period (more than 30 years) together with its extremes (including probabilities of its occurrence) constitute the climate of the place. Weather and climate has been changing right from the origin of atmosphere and life on earth and reached a balanced state that is suitable for life on earth. Climate never fits into rigid demarcation. It changes from one type to another, one generation to the next, from one century to the next and from one ice age to the next. Climate normal's are average values of weather elements over a place for about 10 days or one month, averaged over long period about 30 to 100 years of records. Climate normals are used as yardstick to describe its behaviors or variance of weather/climate.

From the dawn of civilization man has been trying to use the natural resources for making life easier and comfortable. Science helped him to understand the nature of animate and inanimate objects around him.

Acquisition of the knowledge helped him to understand the physico-chemical laws/principles which govern the behavior of living and non-living things and their inter-relationship. The beneficial application of science to mankind is called technology. Application of technology in industry greatly modified the life of man in twentieth century.

It is now known that all forms of life is made up of protoplasm and carry identical processes but they live in totally different surroundings/ environment. Environment of an organism is its surrounding media. Ecology is the relationship of an organism to its environment. Ecosystem is the functioning of living and non-living components of environment. The primary physical factors of an environment are soil condition, temperature, inclination of the sun (light), water, atmospheric conditions and topography.

Biological resources means plants, animals and micro organisms, their genetic material byproducts with actual or potential use/value. Forests are the natural renewable resources and forestry is a specialized science. Forests are the important resource of economy of a country world over. Forests occupy about 27 % of worlds land area, of which about 10% is occupied by trees. According to forest report 1991 (statement the total area of India 3287263 square kilometers, of which forest area 770078 square kilometers (about 23.4 % of total area).

Earth environment effects the wellbeing of man, animal and plants world over. As compared to other species of life, man is more advanced in intellect and hence it is the duty of man to protect the environment from undesired pollution, which is injurious to all life. For many reasons man is mostly responsible for the pollutions of air, water, soil and in turn responsible for global warming, climate change and health hazards, dwindling of forests, water resources and undesirable changes in ecological balance of the biosphere, destruction/dwindling of natural resources. It is the fundamental duty of every citizen (by virtue of constitution) to protect and improve the natural environment including air, forests, lakes, rivers and wild life. It is in this context the following government protection Acts came into existence.

1. The Environmental Protection Act, 1986
2. The Air Pollution Act 1981
3. The Water (protection and control) Pollutions Act 1974
4. The India Forest Act 1927 and Forest (protection) Conservation Act 1980
5. The Wildlife (biodiversity) Protection Act 1972 and
6. The Public Liability Insurance Act 1991.

The stress on the biological diversity is being brought by human dominance and not nature. The world population explosion is of great concern. During last few centuries human beings wholly dominated over other species of life. Equally responsible are the climate changes. Weather pattern changes with rise in extreme weather events. Global warming causing decline of polar bears along with polar ice melt.

During 1990-2015, 177 mammal species lost, at least 30% of the geographical area (lost) that they inhabited. The National Academy of sciences mapped 27600 species by population and range (about half of the terrestrial vertebrate species). Scientist naming this loss biological annihilation or sixth mass extinction. Since 1970, biological diversity shows down trend. India is the home for about one third of known life forms in the world. According to survey of India, Department of Science and Technology, these are more than 500 species of mammals, and 206 forms of species of birds which are truly Indian. The twentieth century witnessed the loss of a few magnificient wildlife. India has 1256 orchids of which 388 species, which are endemic to India.

Marine Biosphere: The part of the earth which is capable sustaining life (both plant & animal) is called biosphere. The three main sub-division or habitats termed bio-cycles. They are the terrestrial, the marine & the fresh-water bio-cycles. Each cycle has ecological features and associated plants and animals. Some animal species may migrate freely one type to another.

It is known the oceans cover about 71% of the earth's surface. The marine environment provides about 300 times the inhabitable space compared to terrestrial and fresh water biocycles together. Terrestrial environment provides habitable biosphere in a shallow zone of about a few feet, while marine habitat provides habitat (at least to some form of life) to the abyssal depth of several kilometers. The fresh water bio-cycle has a small fraction of the other two bio-cycles. The aerial portion of the globe not considered as separate bio-cycle, because this portion is a temporary journey of birds and insects. Marine environment broadly divided as primary biotic and secondary biotic divisions.

The two primary divisions of the sea are called the Benthic and the Pelagic. The benthic includes all life on the ocean floor while the pelagic includes the life in whole mass of the sea/ocean water. Marine environment population divided as Benthos, Nekton and Plankton. Benthos belong to benthic region and the other two belong to pelagic region.

In Greek Benthos meaning deep/deeper sea. Benthos comprises, (i) sessile animals (like sponges, oysters, corals, hydroids, bryozoa, sea weeds and eelgrasses, diatoms (ii) creeping forms-lobsters, certain copepods, amphipods, crustacea, many protozoa, snails, fishes and (iii) burrowing forms-most of the elams and worms, some crustacea and echinoderms.

The Nekton (in Greek-swimming) is composed of swimming animals found in the pelagic division, which includes squids, fishes and whales all of marine animals which migrate freely over long distance. There are no plants in this group. In the Plankton (in Greek-wanderer) include all of the floating or drifting life of the Pelagic divisions of the sea. The Plankton is divided as the phytoplankton (floating plants) and Zooplankton (myriads of animals in floating state, larvae & eggs of the animal benthos and nekton).

The populations of the sea are–plant groups of the sea and the animal population of the sea as such deserves study in brief is being dealt.

In addition to these the following briefly dealt–forest fires, fossil fuels, minerals, Natural gas, soils, Desert soils and disasters.

For the study of field study some portable instruments have been described which may be important for the students and teachers.

In the appendix a brief description of some statistical methods given with a practical example of climate change of Hyderabad. Sustainable development, objectives of the UNEF and the components of biological diversity preambles given.

Authors hope that this introductory book will be a welcome by students & teachers as it covers main syllabus of UGC. Authors gladly welcome any suggestion for the improvement of the book with thanks.

In writing this book relevant material collected from various books, published papers, a few given in references. We express our sincere thanks with indebtedness to M/S Nikhil Shah, Anil Shah B.S publications for patronage and their staff Shri Naresh, Smt. Sandhya and others.

-Authors

CONTENTS

Preface ....(vii)

Abbreviations ....(xix)

About the Author ....(xiii)

Chapter 1| Environment

1.1   Man and Environment 3

1.2   Environments in Nature.. 4

1.3   Racial Development Theories 5

1.4   World Environmental Day (WED).. 6

        1.4.1   Environmental Main Issues 7

1.5   Environmental Protection Act.. 7

1.6   Ecosystems 8

1.7   Autotrophs and Heterotrophs 9

1.8   Components of an Ecosystem 10

1.9   Ecology.. 10

        1.9.1   Types of Ecosystems 12

1.10   Types of Climax.. 12

1.11   Food Chain 13

1.12   Trophic Levels 14

1.13   Food Webs 14

1.14   Ecological Pyramids 15

Chapter 2| Biological Environment of Sea

2.1   Marine Environment 19

2.2   Marine Biosphere 20

2.3   Marine Environment Classifications 20

2.4   Sea Life 23

2.5   Populations of the Sea 23

2.6   Animal Populations of the Sea 24

2.7   B – Vertebrates 26

2.8   Phytoplankton in Relation to Marine Environment 27

2.9   Factors of Phytoplankton Reproduction 28

2.10   Marine Organisms 30

2.11   The Nitrogen Cycle 32

2.12   Ecological Groups of Marine Animals 34

2.13   Organic Production in the Sea 34

2.14   Plant Nutrient Consumption as an
       Index of Organic Production 35

2.15   Oxygen Production and Consumption
       as an Index of Organic Production 35

2.16   Zooplankton Production 36

2.17   Marine Biosphere Atmosphere Interaction 36

Chapter 3| Natural Resources

3.1   Natural Resources-(A) Land Resources 39

       3.1.1 Soil   40

3.2   Minerals 44

3.3   Soil Survey and Mapping 46

3.4   Soil Testing (Practical) 49

3.5   Land Utilization 50

3.6   Agriculture 52

3.7   Definition of Terms used in Land Utilization Statistics 52

3.8   Soil Erosion 54

3.9   Soil Fertility 55

3.10   Natural Resources – (B) Water Resources 57

       3.10.1 Hydrology 58

3.11   World's Water 60

3.12   Surface Water Resources Estimates for India 65

3.13   Groundwater Resources 67

3.14   Utilization of Water Resources 67

3.15   Cryosphere or Ice Sphere 68

3.16   Water Uses 69

Chapter 4| Environmental Degradation and Protection

4.1   Disaster Management 72

    4.1.1   Mitigation 74

    4.1.2   Climate Disasters 76

4.2   The National Environment Appellate Authority Act 1997 77

4.3   World Environmental Day 5th June 78

4.4   Environmental Issues 78

4.5   Environmental Degradation 79

    4.5.1   Air Pollution Control Boards 80

4.6   Environmental Hazards 81

    4.6.1   Types of Hazards/Disasters
Natural or Man-made 82

    4.6.2   The Chemical Weapons and Weapons of
Mass Destruction 83

    4.6.3   Prohibition of Chemical Weapons 83

    4.6.4   Prohibition Relating to Weapons of
Mass Destruction 84

    4.6.5   IPCC (Inter Governmental Panel on
Climate Change) 85

4.7   Industrial Chemical Wastes and Toxic Chemicals 86

    4.7.1   Some Toxic Chemicals 86

    4.7.2   National Authority 88

Chapter 5| The Atmosphere

5.1   Composition of the Atmosphere 91

5.2   Atmospheric Heat Process 92

5.3   The Vertical Structure of the
Atmosphere based on Temperature 93

5.4   The Sun 97

Chapter 6| The Oceans

6.1   Composition of Sea Water 103

6.2   The Exchange of Earth Materials between
Sea and Atmosphere 104

6.3   The Cryosphere 106

    6.3.1   Sea Icing 107

6.4   Some Physical Properties of Sea Water 107

6.5   Energy Exchange Processes between
Sea and Atmosphere Interface 108

6.6   Ocean Currents 109

6.7   Motion of the Sea and the Effect of
       Wind on Ocean Currents 110

6.8   Ocean Currents caused by Density Differences 111

6.9   Some Ocean Currents 113

6.10   Indian Ocean Currents 114

6.11   Some Definitions and Explanation of Terms 114

6.12   Geophysical Properties of Oceans and Earth 115

6.13   Major Oceans 117

6.14   Sea Floor Features 118

       6.14.1 Basin 119

Chapter 7| Soils

7.1   Soil Erosion Types 121

       7.1.1 Mechanism of Erosion 122

       7.1.2 Factors Influencing Erosion 123

7.2   Soil and Water Conservation Measures 123

       7.2.1 Soil and Water Conservation on
              Agricultural Land 124

7.3   Erosion Control-Mechanical Measures 125

       7.3.1 Soil Types 126

       7.3.2 Desert and Desertification 127

7.4   Interior Structure of the Earth 130

7.5   Plate Tectonics 132

7.6   Seismicity of India 144

Chapter 8| Environmental Pollution

8.1   Air Pollution 154

8.2   pH Value 159

8.3   Pollutions Standard Index 160

       8.3.1 Hazards of Nuclear Fall-out 167

       8.3.2 Nuclear Energy and Radiation Hazards 170

8.4   Carbon and its Compounds 171

8.5   Sulphur (S) 173

8.6   Atmospheric Nitrogen and its Compounds 173

8.7   Ozone (O_3) 175

8.8   Elementary Ways of Pollution Control 177

8.9   Acid Rain 178

8.10   BAPMoN.. 179

8.11   SMOG 181

8.12   Toxic Air Pollutants... 182

8.13   Dispersal of Air Pollutants 183

8.14   Methods of Estimation of
        Particulate Matter in Air....................................... 186

8.15   Global Warming Effects.. 188

8.16   Solid Wastes 198

8.17   Fire Services (Prevention, Control and Forecasting).................. 199

Chapter 9| Water and its Pollutions

9.1   The Water Pollution Act 1974 207

9.2   Global and India Water uses.................................... 209

9.3   Functions of Central/State Water
        Pollution Control Boards 210

9.4   Prohibition on use of Stream or Well for
        Disposal of Polluting Substances/Matter...................... 211

9.5   Emergency Measures in Case of
        Pollutions of Stream or Well................................... 211

9.6   Rivers in India.. 212

9.7   Hydrological Cycle... 212

9.8   Water Resources.. 214

9.9   Hardness of Water 217

9.10   Drinking Water.. 220

Chapter 10| Forests

10.1   Definition 225

10.2   Forests in India... 226

10.3   Forest Policy.. 226

10.4   Forests and Man 227

10.5   Indian Forest Organization...................................... 227
        10.5.1 Main Forest Products 228

10.6   ICFRE (Indian Council of Forestry Research and Education)
        HQ Dehradun 228

10.7   Forest Fires 229

10.8   Forest 229

10.9   Coral Reefs 232

10.10   The Indian Forest Act 1927 232
10.11   Reserved Forest 233
          10.11.1 Acts Prohibited in Reserve Forest Area 233
          10.11.2 Forest Conservation Act 1980 233
10.12   Forest Products 234
10.13   Danger due to Deforestation 237

Chapter 11| The Biodiversity

11.1   The Bio-Diversity Act 2002 240
11.2   The National Green Tribunal Act 2010 241
11.3   Biodiversity 241
11.4   Coastal and Marine Biodiversity 242
11.5   Wild Life 244
          11.5.1 The Variety of Fauna in India 245
          11.5.2 Human Population Growth 246
          11.5.3 The Present Environmental Issues 248
          11.5.4 Biological Diversity Under Stress 249
          11.5.5 Variety of Fauna in India 251
          11.5.6 Coastal and Marine Bio-diversity 256
          11.5.7 The Wild Life (Protection) Act, 1972 258
          11.5.8 The National Board for Wild Life 259
          11.5.9 State Boards for Wild Life 259
11.6   Biogeographic Zones in India 260
11.7   The Public Liability Insurance Act, 1991 261
11.8   Sustainable Development 262

Chapter 12| The Climate-Role of Ocean in Climate Fluctuations

12.1   Worlds Land Utilization (1990) 270
          12.1.1 Climate 271
12.2   Climatology 274
12.3   Climate of India 277
12.4   The Role of Ocean in Climate Fluctuation 284
          12.4.1 Role Factors (Importance) of
                    Ocean in Climate Variability 286
12.5   Present and Past Radioactivity of the Earth 287
12.6   Climatic Controls 289
12.7   Components of Climate System 291

12.8   Climate System or Earth System 293
        12.8.1 Extinction of Dinosaurs on the Earth........................ 293
        12.8.2 Climate Change.. 293
        12.8.3 Disaster (Measurement) Management
                System-Instruments 295
        12.8.4 Environ Automatic Weather Station 296
12.9   Role Factors (Importance) of Ocean in Climate Variability 301
12.10   The Earth-Atmospheric System 302
12.11   Climate over Earth History 305
        12.11.1 Annual Carbon Fluxes in the
                 Earth Atmosphere System 306
12.12   The Evolution of the Earth's Atmosphere 306
12.13   Climate during Geological Cretaceous
        Period of Earthwarm Epoch 135-180 MY Ago 309
12.14   Geological Time Scale 310
12.15   Milankovitch Cycles.. 313
12.16   Impact of Climate Change on Agriculture 315
        12.16.1 Agroclimatology.. 317
        12.16.2 Topoclimatology 318

Chapter 13| International Geosphere-Biosphere Programme IGBP-1989

13.1   The Linkages of General Circulation Models
        (GCM) with Ecosystem Models................................... 321
13.2   Global Changes of the Past 322
13.3   Law of the Sea 324
        13.3.1 The Role of Ocean in Climate Variability 325

Chapter 14| Modern Aids of Communication and Detection

14.1   Radio Spectrum... 318
14.2   Satellite Communication 320
        14.2.1 Radar... 324
14.3   Satellite Meteorology.. 325
        14.3.1 Satellite Pay Loads, TV Camera 328
        14.3.2 Remote Sensing.. 329
        14.3.3 Geostationary Satellites 331
        14.3.4 IMDPS (INSAT Meteorological
                Data Processing System)............................... 332

14.4   Global Positioning System 332
        14.4.1 Indian Regional Navigation
                Satellite System (IRNSS) 335
        14.4.2 GAGAN (GPS Aided geo Augmented Navigation) ... 336
        14.4.3 IRS Applications 337
        14.4.4 Geostationary Operational
                Environmental Satellite (GOES)............................ 337
        14.4.5 Satellite Measurements of
                Ocean Parameters Routinely 338
        14.4.6 GSAT-17 Communication Satellite 339
        14.4.7 Cartostat-series 339
        14.4.8 Satellite Navigation (Sat Nav) 340
        14.4.9 UN Indirect Satellite Imagery for
                Disaster Risk Reduction (UNISDR) Model................ 343
14.5   Geophysical Information System (GIS)........................... 343
        14.5.1 Some Applications of GIS 344

Chapter 15| Energy

15.1   Different Forms of Energy 348
15.2   Energy Resources... 358
15.3   Fossil Fuels 360
15.4   Petroleum Refining... 361
        15.4.1 Petroleum Refining Products 362
        15.4.2 Petroleum Resources................................... 363
15.5   Mineral Wealth of India 364
15.6   Metallurgy... 366
15.7   Metal Extraction Furnaces 366

Index 369

ABBREVIATIONS

APT	–	Automatic Picture Transmission
ART	–	Accident Relief Trains
AWS	–	Automatic Weather Station
AGCM	–	Atmospheric General Circulation Model
BOD	–	Biological Oxygen Demand
CRIDA	–	Central Research Institute for Dry land Agriculture
CRZ	–	Coastal Regulation Zone
CBD	–	Convention on Biological Diversity
CBRI	–	Central Building Research Institute
CBRN	–	Chemical, Biological, Radiological and Nuclear
CPU	–	Central Processing Unit
CWC	–	Central Water Commission
CPCB	–	Central Pollution Control Bond
COP	–	Conference of Parties
DCP	–	Data Collection Platform
DCT	–	Data Collection Transponder
DRT	–	Data Relay Transponders
DOD	–	Department of Ocean Development
GCM	–	General Circulation Model
EPA	–	Environmental Pollution Act
ESSA	–	Environmental Science Services Administration
ECMWF	–	European Centre Medium Range Weather Forest
FSI	–	Forest Surveys of India
GDAS	–	Grass Data Assimilation System
GDP	–	Global Domestic Product
GIS	–	Geographical Information System/Geophysical Information System
GPS	–	Global Positioning System
GOES	–	Geostationary Operational Environment Satellite
HWRF	–	Hurricane Weather Research Forecast

GMOs	–	Genetically Modified Organisms
IMD	–	India Meteorological Department
INSAT	–	Indian National Satellite
IPCC	–	Inter governmental Panel on Climate Change
ISRO	–	Indian Space Research Organisation
IITM	–	Indian Institute of Tropical Meteorology
IDNDR	–	Internal Decade for Natural Disaster Reduction
IGBP	–	International Geosphere Biosphere Programme
LPA	–	Long Period Average
Monex	–	Monsoon Experiment
LMOs	–	Living Modified Organisms
LGM	–	Last Glacial Maximum
NCC	–	National Climate Centre Pune
NAAQS	–	National Ambient Air Quality Standards
NASA	–	National Aeronautical Space Administration
NDMA	–	National Disaster Management Authority
NDRF	–	National Disaster Response Force
NGRI	–	National Geophysical Research Institute
NOAA	–	National Oceanic and Atmospheric Administration
NWP	–	Numerical Weather Prediction
NMC	–	National Meteorological Centre-Washington
NOC	–	No Objection Certificate
OLR	–	Outgoing Long wave Radiation
PPM	–	Parts Per Million
PCB	–	Polychlorinated Bicarbon / Pollution Control Board
PRA	–	Participatory Rural Appraisal
PSI	–	Pollutions Standard Index
SPCB	–	State Pollution Control Board
SASE	–	Snow and Avalanche Study Establishment
SDMA	–	State Disaster Management Authority
SR	–	Scanning Radiometer
SST	–	Sea Surface Temperature
TERI	–	The Energy Resources Institute
TIROS	–	Television Infrared Observational Satellite
TSPM	–	Total Suspended Particulate Matter
UN	–	United Nations
UNEP	–	United Nations Environmental Programme
UNDP	–	United Nations Development Programme

UNCED	–	United Nations Conference of Environment and Development
VHF	–	Very High Frequency
VHRR	–	Very High Resolution Radiometer
VTPR	–	Vertical Temperature Profile Radiometer
WHO	–	World Health Organization
WMO	–	World Meteorological Organization
WV	–	Water Vapour

ABOUT THE AUTHOR

Navale Pandharinath, M.Sc., (Maths), M.Sc. (Statistics) is a retired Director of India Meteorological Dept. He has undergone departmental advanced training in Meteorology at Pune. He has rich experience in Aviation, Non-Aviation weather forecasting, cyclone warnings, flood met-work and Agri-met-work. He has published about forty papers in Mausam and Vayumandal departmental journals and in National and International seminars and symposium. He has published five books–A Course in Dynamic Meteorology, The Science of Weather and Environment, Aviation Meteorology, Earth and Atmospheric Disasters Management and An Introduction to Ocean Dynamics. After retirement he worked as Ground Instructor Meteorology in A.P. Aviation Academy and Flytech Aviation Academy Hyderabad. He has worked as a Guest Faculty (2013-2016) University Centre for Earth and Space Science at University of Hyderabad.

CHAPTER - 1
Environment

INTRODUCTION

According to the constitution of India (Article 51A, g), it shall be one of fundamental duty of every citizen of India to protect and improve the natural environment including forests, lakes, rivers, wildlife and to have compassion for living creatures.

The state of environment observed to be declined in quality world over since 1960 due to population growth, loss of vegetation, loss of biodiversity and climate change UN Conference on the Human Environment held Stockholm in June 1972.

UN Conference on the Human Environment and Development (UNCED) held in Rio de Janeiro in 1990, where all members agreed of protect the environment from further degradation of air, soil, water and biosphere and to improve social, economic development to ensure sustainable development. In 2002, at world summit on sustainable development in Johannesburg world members of UN reaffirmed their commitment (a collective effort necessary at local, national, regional and global level) to ensure economic, social development and environmental protection.

Biological environment includes parts of atmosphere, hydrosphere and lithosphere. Environment thus includes air, water, land and human beings other living creatures, plants, micro-organisms and property. Environmental pollution means presence of solid. Liquid or gaseous substances in such concentration as may be injurious or tend to be injurious to environment.

Environment may be defined as the place in which an organism lives and also includes the conditions under which it survives. Environment includes water, air and land and the inter-relationship which exists among and between water, air and land and human beings, other living creatures, plants, micro-organisms and property. From the dawn of civilization man has been trying to use the natural resources for making life easier and comfortable. Science has helped him to understand the nature of animate and inanimate objects around him. Acquisition of knowledge helped him to understand the physic-chemical laws/principles which govern the behaviors of living and non-living things and their inter-relations ship. The beneficial application of science to mankind is called technology. Application of technology in industry greatly modified the life style of man in twentieth century.

It is now known that all forms of life is made up of protoplasm and carry identical processes but they live in totally different surroundings or environments. There exists a critical relationship between livings things and its physical environment. Biosphere is the space (area) near the earth's surface which encompasses all living organisms. This region includes parts of atmosphere, hydrosphere and lithosphere. *Environment* of an organism is its surrounding media. *Ecology* is the relationship of an organism to its environment. *Ecosystem* is the functioning of living and non-living components of environment.

The principal physical factors of an environment are: soil conditions, temperature, inclination of the sun (light), water, atmospheric conditions and topography.

The application of technology in twentieth century not only brought benefits but also brought decline in the environmental quality and threatening to effect the available natural resources for the present and future generations besides causing damage to the health of living being and extinction of some species of life. Thus technology proved to be both Boon and Bane to life on earth. The UN Conference on Human Environment was held at Stockholm in June 1972 to take appropriate steps for the protection and improvement of human environment. This step is required because the state of environment has substantially declined in quality due to increasing pollution, loss of vegetation cover and loss of biological diversity. There has been excessive concentration of harmful chemicals in the ambient atmosphere and in food chains. Environmental accidents risk increased and threatened the life supporting systems. The first World Environment Day Observed in 1973.

All living organisms are composed of carbon, Oxygen, Nitrogen, Hydrogen, which are also the basic elements of water and air-shells of the earth. A large part of the living matter contained in green plants, which entrap solar energy and make complex compounds by photosynthesis process. The main sources of plant feeding are carbon dioxide and water. Plants use about 2% of insolation (incoming solar radiation) for photosynthesis process, about $7w/m^2$ is consumed by plants.

Plant kingdom provides about 10^{17} kg of biomass annually and an equal amount of oxygen. An average size of tree supplies about 3500kg of oxygen per year which is sufficient for three people. Because of this plant kingdom is also called the lungs of the earth. On an average a man requires about 3.13 kg oxygen or 15 kg of air per day.

The total mass of CO_2 in the atmosphere is about 0.23×10^{16} kg while the mass of CO_2 in the ocean water is about 1.4×10^{17} kg, which is more than 60 times the mass of CO_2 in the atmosphere. Sea water plays a peculiar role in respect of dissolved natural gases of N, O_2, CO_2 and H_2S. these gases are closely related to living matter on the land and in the sea. Carbon dioxide enters into the atmosphere by human and animal breathing decay and burning of materials containing carbon and volcanic activity. More than 90% of the earth CO_2 is dissolved in sea waters. The stability of CO_2 in sea changes with temperature.

A jet airliner consumes 6-9 tons of oxygen in one hour, 25-50 thousand hectares of forest expels (gives) 50-75 tons of O_2 in 8 hours.

1.1   MAN AND ENVIRONMENT

Earth environment affects the well being of man, animal and plants world over. Compared to other species of life, man is more advanced in intellect and hence it is the duty of man to protect the environment from undesired pollution, which is injurious to all life. For many reasons man is mostly responsible for the pollution of air, water, soil and in turn responsible for global warning, climate change and health hazards, dwindling of forests, water resources and undesirable changes in ecological balance of the biosphere, destruction/dwindling of natural resources. It is the fundamental duty of every citizen (by virtue of constitution) to protect and improve the natural environment including air, forests, lakes, rivers and wildlife. It is in this context, the following government protection Acts came into existence.

1. The Environmental Protections Act, 1986
2. The Air pollution Act, 1981
3. The water (protections and control) Pollution Act, 1974
4. The Indian Forest Act 1927 and Forest (protection) Conservation Act, 1980
5. The Wildlife (Biodiversity) Protection Act, 1972 and
6. The Public Liability Insurance Act, 1991

Man is the highest form of life on the earth, but he is dependent on other forms of life and environment. Man has the ability to modify or destroy an environment or burn a whole forest and cause damage to the ecosystem, yet he is very closely dependent on the nature-environmental system, like atmospheric oxygen for breathing, and animal, plant kingdom for his food.

According to UN Report, at present mankind is persisting with thoughtless and extravagant consumption of natural resources and damaging the natural resources in an unprecedented manner. UNEP (United Nations Environment Programme) outlook-4 report (25 October 2007) says, at the present consumption rate it requires 21.9 hectares per person while earths capacity (on an average) 15.7 hectares per person. As a result of this 116550 square kilometers of forest area being lost across the world each year. 60% of the world's major rivers have been damaged or diverted and fresh fish population declined by 50% during last 20 year. About 30% amphibians, 23% of mammals and 12% of birds are under threat of extinction due to human activity, while 10% of the world's large rivers are running dry every year before reaching in to sea. The US's consumption of energy increased by 20% over the last two decades. Of the total GHGs (greenhouse gases) increased in atmosphere by human activity, 70% accounts to energy sector. The IPCC further noted that due to global warming there would be glacier retreat in the Himalayas, sea level rise, production of wheat, rice, maize would drop in India and China and more than one billion people may face fresh water shortage by 2020. International Renewable Energy Agency promotes the adaptation of renewable energy worldwide and aims to provide a definite policy advice and facilitates capacity building and technology transfer.

1.2   ENVIRONMENTS IN NATURE

It is common observation plants and animals live under variety of conditions. A wood, a field, a deep ravine or a marshy land all create totally different environments from each other. All they have a direct effect on the organisms under their influence. As said earlier each environment made of many factors. The physical factors include soil condition, temperature, sunlight, water, atmospheric conditions and changes in earth. Equally important are the biological factors or the living surroundings of organisms and non-living surroundings. A plant growth depends on soil, temperature, water, sunlight and atmospheric conditions. The plants in turn create environment for animals. Almost any set of environmental conditions are suitable for certain plants and animals. Because of this we find living organisms on land, under water bodies and mountain tops to mountain valleys and all atmospheric conditions from north pole to south pole.

BIOLOGICAL FACTORS

A plant or animal lives under influence of other living things together with non-living matter. An individual organism is a part of a large society of life, which is bound to the non living surroundings. The earth offers a wide variety of conditions for life (which is biodiversity). The 5 environmental factors are given below.

Plants and animals have varied requirements.

1. **Soil:** Soil is a basic factor of environment. The main types of soils-clay loam, sandy loam. Soils may be acidic or alkaline.

2. **Temperature:** It controls environment. All organisms withstand the day and night temperatures. Animals and birds face the problems of seasonal temperature changes. During severe winter animals and birds migrate or move to warmer places. During warm summers they seek cool places. Some undergo into hibernations like frog, tortoise etc.

3. **Light:** Solar energy is a critical factor in the environment of living things.

4. **Water:** Offers profound bearing on the environment living things. Deserts, droughts, evergreen trees, tropical forests etc. are all the effects of water or rainfall.

5. **Atmosphere:** All living things require oxygen hence atmosphere is a factor of environment. Plants and animals which live in deep sea and in the soil seek closer surface to water and soil surface for oxygen supply. Topography/land formation/physical features of the earth have great effect on living things. As the earth changes, plants and animals migrate to new favorable places and new species develop in the new environment, that is, as environment changes living things change with it.

Tropism: The automatic response of a plant or any of its parts toward or away from a stimulus is called tropism

Sensitivity: It is the response of protoplasm to its surroundings

Based on responses, tropism is of six types.

1. Chemotropism – response to chemicals (like soil minerals)
2. Geotropism – response to gravity
3. Hydrotropism – response to water
4. Photo-tropism – response to sunlight/light
5. Thermo-tropism – response to heat

1.3   RACIAL DEVELOPMENT THEORIES

Lamarck's theories: The first racial development theory was presented by French biologist Jean Baptiste Lamarck, in 1801. His three theories are:

1. Theory of need - The production of a new organ or part of a plant or animal results from a need.
2. Theory of use and disuse – Organs remains active as long as they are used and they disappear gradually with disuse.
3. Theory of inheritance – All that has been acquired or modified structure of individual during their life is passed on by heredity to the next generation. Lamarck's theory of use and disuse has little scientific basis.

Darwin's Theory of Natural Selection (1859):

Charles Darwin, an English scientist published his origin of Species by Natural selections. Broadly his theory accepted is given below.

The main factors that accounts for the development of new species from common ancestry are:

(i) Over production of individuals.

(ii) Struggle for existence,

(iii) Variation among individuals

(iv) Survival of the fittest

(v) Inheritance of favorable characteristics and

(vi) New forms better adapted to survive, "Naturally selected" as new species.

Scientists say "God is Nature and Nature is God". This bespeaks volumes about environment and climate change.

The life on earth undergoes gradual changes with changes in environment and climate is termed Racial Development. Fossils are foot prints of past ages, preserved remains or mineral replacements of past ages, preserved remains or minerals replacements of living things of previous ages. The racial evidence has been collected through the study of fossils, homologues organs vestigial organs, embryology, geographical distribution experiments in genetics.

1.4 WORLD ENVIRONMENTAL DAY (WED)

Every year on 5 June, world environmental day is observed to create awareness all over the world about environment. The aim of observation of WED is to focus the environmental issues, to empower people for sustainable and equitable development. To encourage the people to look into environmental issues and to act for sustainable development. To advocate participation of all nations for protection of environment, in order to create safety for future generations. The awareness can be created by street processions competitions in schools, planting trees, rainwater harvesting, recycling of water, cleaning up campaign like swatch Bharat etc.

It stimulates worldwide awareness about Environment and increases political attention and action. WED aims to give human face to environmental issues, empower people as active agents of sustainable and equitable development. It Encourages communications to change attitude towards environmental issues. It advocates partnership for all nations to enjoy secure future. The day is observed by street processions, competitions in schools and colleges, Tree planting, recycling and cleaning up campaigns etc.

The earth charter seeks to inspire all people a sense of global interdependence and shared responsibility for the well being of humanity.

1.4.1 ENVIRONMENTAL MAIN ISSUES

Global warming, climate change, deforestation, energy crisis, air, water, soil pollution, waste material management, control oil spills, population control, use of nuclear energy, nuclear hazard protection, preservation of natural resources for future generation.

During last 150 years, all species of life on earth are affected by environmental issues, which are invading and threatening the air, water and soil pollutions. The main issues are given below.

1. Global warming and climate change
2. Deforestation, weakening the lungs of the earth
3. Energy shortage crisis
4. Environmental pollution
5. Harmful toxic wastes/Radioactive waste
6. Oil spills which harms/extinct some marine life. Plastic dumping in lakes and seas will have adverse effect on marine life
7. Depletion of natural resources
8. World population explosion, 74 million people per year. 2010 world population estimated to be 6909×10^6 and it would be 8012×10^6 by 2025
9. *Nuclear Issues:* Nuclear weapons-dangerous to environmental issues
10. Strengthening world greenery or lungs of the earth-Go Green
11. Globally, phytoplankton absorbs as much as CO_2 as tropical rainforest. It is therefore, very important to understand their response to global warming.

1.5 ENVIRONMENTAL PROTECTION ACT

To prevent decline of environment, biodiversity world community decided to protect environment. The UN Conference on environment held in Stockholm (in June 1972), recommended to enact a comprehensive law to take action for environmental protection. As a result Government of India introduced Environmental protection Bill and passed in 1986, and came into force on 19-11-1986 vide and GSR 1198 E, 12-11-1986 (GSR = Gazette Statute Rules). The environmental act aims at the protection and improvement of Environment, prevention of hazards to human beings and other living creatures, plants and property.

According to IPCC (Inter Governmental Panel on Climate Change) Report 2007, global average temperature rose by $[0.74 \pm 0.18]$ °C during 1906-2005. Widespread ice melt caused average sea level rise at the rate 1.8mm per year during 1960-2003 and Arctic sea ice reduced by 2.7% per decade since 1978. Over the period 1901-2009, the mean annual temperature seen to be increasing trend of 0.56°C/100 year and increase in mean temperature of the world + 0.74°C, while in India it was recorded a rise of + 0.5°C during the same period. In the changed state the effects on biodiversity is unimaginable. University Leeds, UK predicted that 15-37% of plant and animal species over the world face extinction by the year 2050.

Note: Geotropism means the response of plants to gravity

Gene: A determiner of heridity, located in a chromosome.

Tropism: It is the involuntary response of an organism to a stimulus

Hydrotropism: It is the response of roots to water

Omnivores Organism: It is one which eats both plant and animal substance.

Predator: Any animal which preys on other animals.

Parasite: An organism which gets its food entirely from another living organism

Phylum: one of the large divisions in the classification of plants.

Protoplasm: The living substance which is the physical basis of all life.

Movement of Energy in Ecosystem

Ecosystem is the totality of organisms in a particular place (or region) and the environment in which they live. The living organisms interact with each other and with environment (in which they live)

Ecosystem may be viewed as energy processing units

The complex nature of ecosystem definition

1.6   ECOSYSTEMS

Soils, vegetation and climate together form environments/ecosystems, whereas a group of specially adopted plants and animals coexist, each depending on the other for their survival. As long as the ecosystem remains undisturbed the plants and animals remain almost constant. Disturbance imbalance may be disastrous. For example, goats are voracious animals which were responsible for the destruction of original coniferous forest in Mediterranean lands. The present day climax vegetation is actually secondary plant cover.

The sun is the main energy source for all living matter. Green plants and some bacteria derive their energy (food) from the sun through photosynthesis process. This energy is stored in plant tissues and converted into mechanical and heat forms in metabolic activities.

The chemical equation of photosynthesis is given below:

$6CO_2$ (atmospheric CO_2) + $6H_2O$ (water) + sunlight energy $\xrightarrow{\text{photosynthesis}}$ $C_6H_{12}O_6$ (Sugar or glucose) + $6O_2$ (oxygen)

In respiration process, the sugar produced in plants is broken into energy by plant organells. This energy is used by plant for its growth, repair and reproduction. This respiration is a reverse process of photosynthesis . The respiration chemical equation is given below:

$$C_6H_{12}O_6 \text{ (sugar)} + 6O_2 \xrightarrow[\text{(REVERSEPHOTOSYNTHESIS)}]{\text{Resperation}} 6CO_2 + 6H_2O + \text{energy}$$

1.7   AUTOTROPHS AND HETEROTROPHS

Plants and bacteria which use insolation for the production of their food are called Autotrophs while other life forms which depends on autotrophs for their food (life energy) are called Hetrotrophs. In this biological process the energy flows from sun to plants (autotrophs) then to all heterotrophic organisms (like microbes, animals, human beings). This energy flow is represented as below:

Sun → Plants or producers (metabolism)
(a) Consumers → heat energy
(b) Chemical energy → decompose (Metabolism) → heat energy
(c) Heat energy (digestion)

Energy produced by mechanical motion is called kinetic energy while that is stored by virtue of its position is called potential energy.

Biological activity of an ecosystem is dynamic and required energy utilization.

All living organisms store potential energy by way of chemical energy of food. Oxidation of food gives out energy which is used to do work (that is chemical energy is transformed into mechanical energy)

Few autotrophs use energy released by oxidation process for the synthesis of organic food.

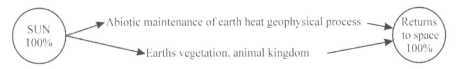

Fig. 1.1 Flow of energy on the earth

Chloroplasts: Only green organs (like leaves) plants are capable of assimilating atmospheric CO_2. The green pigment is called chloroplast (a plastid containing chlorophyll)

Metabolism: Refers to all energy transformations in living matter (that is, the physiological and chemical processes of the body)

1.8   COMPONENTS OF AN ECOSYSTEM

Definition: According to Odum (1983)

Ecosystem has six components, three biotic and three abiotic

The biotic components are:

1. Producers – all autotrophs
2. Macro consumers – animals which eat or ingest other organisms
3. Micro consumers – (Saprobes or decomposers) which include hetrotrophs decomposing dead organic matter.

The three abiotic components are:

1. Inorganic substances, like Carbon, Nitrogen, Hydrogen
2. Organic compounds like proteins, carbohydrates, and
3. Climatic factors like rainfall, temperature wind, light etc.

Note: *Ecology.* The specialized term for the environmental biology is ecology [derived from the Greek roots "oikos" meaning "house"]

1.9   ECOLOGY

It is the study of inter relationship between living things and their surroundings (or environment)

Ecological Succession: Plants and animals continually move on earth. Plant population of an area gradually changes. Along with this change animals find new living homes (habitats). This movement of living things is termed Succession. An initial open field gradually turns to meadow, then it will have shrubs, then turns into forest like area and changes the environment and then into dense forest with big stable trees with Climax plants.

Open field → Meadows → Sun loving trees and shrubs → Forest like area → Environmental change → dense forest with tall tress (like maple, beech) that control the area, which are climax plants.

Grasses are climax plants in the Great plains

Climax species may be destroyed by fire, wind or cleared by man. The area is left to the nature. In this area succession starts again and eventually will be taken over by climax plants again.

The changes in the physical environment and sequential changes lead to establishment of stable community. At any location (area) due to changes in physical environment one community replaces by another. This establishes the pattern of ecological succession with the following parameters.

Succession is an orderly process, community changes in certain direction. Changes in the physical environment by community results in succession. Because it is a biological process on a site (location) it leads to stable ecosystem with succession climaxes.

All living organisms are under the constant influence of physical factors of the environment, like soil, temperature, water, light and atmospheric conditions. These factors determine growth of plants and in turn provide an environment for animals. A set of conditions are favourable for development of certain plants and animals. As a consequence living organisms occupy the land, water bodies and atmosphere from North Pole the South Pole, mountain tops to valleys or deep ocean bed.

The species involved and time required for reaching stability, all depend on the physical factors is called biological process.

The physical environment determines succession process. The stages of succession are called serial stages. The final steady state is called Climax (which remains stable).

Successions are of two types-Autotrophic and Hetrotrophic succession.

Autotrophic succession is governed by plants (having chlorophyll). This type succession is abundant and starts in inorganic environment.

Heterotrophic succession is governed by early dominance of hetrotrophs which flourish in *Organic Environment*

Succession is divided as primary and secondary succession.

Primary succession begins in a sterile area where conditions are not favourable. The first stage is nudation (bare) or exposure to the new surface over which seeds germinate and seedlings establish. This process is called colonization. Arrival of new migrants that support population rise is called aggregation. The first colonized organisms are called pioneers. Hydrarch succession begins in water. Hydrotropism is result of the response of roots to water. Water bodies like ponds, lakes are colonized by phytoplankton's, which consists of microalgae which forms pioneer colonizer. Autotrophs and animals that die in water body adds organic matter. This organic matter is used by bacteria, fungi and release minerals. Thus aquatic body gradually becomes the nutrient rich and support the growth of rooted hydrophytes like vallisnaria, ceratophyllum. When hydrophytes die, they are decomposed by microorganisms that release nutrients. This results in the decreases of aquatic body and hence decreases the margin of the pond. In this margin Nelunbo, Trapa and such plants grow in mud. Deposition of organic matter and evaporation of water makes nutrient rich pond. There by swampy ecosystem forms. Subsequently it will be gradually taken over by land plants and this further leads to the formation of jungle/deciduous forest. Finally (in long years) it will be occupied by steady plant community resulting in climax formation.

Pond Ecosystem → Phytoplankton → Rooted aquatic plants→ Free-floating and rooted plants → Reeds and sedges → Terrestrial communities → Climax vegetation

1.9.1 TYPES OF ECOSYSTEMS

There are two types of ecosystems. 1. Natural ecosystem and 2. Manmade ecosystem (or man modified ecosystem).

Natural Ecosystems are: Terrestrial and Aquatic. Terrestrial ecosystems are identified by forests, woodlands, savanas and grass lands.

Aquatic Ecosystems are: Fresh water or marine

Man made ecosystems are: Villages, towns, cities Agr-economy system, Aquat-culture and Reservoirs etc.

All ecosystems have the basic structure, function of energy sources primary producers consumers, decomposers with flow of energy and cycling of materials.

Xerosere: In exposed rocks or dry sand some plants grow, which are the habitats of xerach succession.   The   pioneer   plants   are   Lichens,   which disintegrate rocks through chemicals. Very small quantity of soil particles reach the rock or fall into crevices of rock. The crevices become the home for Mosses and Selaginella and aid to soil formation by increasing rock erosion. With the advance of time, grasses, animals and herbaceous plants grow on the soil formed on rocks. Subsequently mixed woody plants and then climax vegetation establishes.

1.10   TYPES OF CLIMAX

According to Braun- Blanquet (1937), there may be climate climax, edaphic climax and biotic climax. Succession may not always be progressive but at times may be regressive. Forest may get degraded into grass land patches. This is regressive succession. A climax community may be established through primary succession on sand dunes or barren area in lava flown area, which may take 1000 years. On an abandoned agricultural land or over degraded forest area secondary succession may take place in about 200 years for developing into a mature forest.

Importance of climax: climax community is stable with (i) greater biological diversity, (ii) larger biomass structure and (iii) balanced (or equatable) energy flow. These three factors improve the physical environment. As a result man gets food, fuel, fodder, medicine etc. with the stable community. This climax community controls climate and balance bio-geochemical cycles.

1.11 FOOD CHAIN

Green plants absorb solar energy, convert into chemical energy through photosynthesis process. This energy is stored in food material through a series of organisms and hence called food chain. Food chain represents a single energy pathway. Energy flow from autotrophs or green plants through consumer organisms in each trophic level is called food chain.

Food chain shows the energy flow process and feeding relation. Also shows interactions between living things in an ecosystem. All green plants of the biological community are known as producers

Herbivores: Animals which feed on producers (green plants) are called herbivores also called primary consumers eg. Deers, rabbits, goats, pigs.

Carnivores: Animals which feed on herbivore animals are called carnivores e.g., Lion, tiger, leopard. They are also called secondary consumers.

Note: Algae: Athallophyle plant containing chlorophyll.

Anabolism: Constructive process of metabolism

Catabolism: The destructive phase of metabolism

Lichen: A thallophyte composed of an alga and fungus living together for their mutual advantages.

Examples:

Food chain on land

Grass → goat → Lion

Grass → Deer → Lion

Green plants → grasshopper → frog → snake → peacock

Plant sap from leaves → plant lice → spider → sparrow → hawk

Food chain in a pond/lake

Microscopic Algae → protozoa → small aquatic insects → large aquatic insects → small fishes → large fishes

Food chain in the oceans

Phytoplankton → zooplankton → small fishes → large fishes → still larger fishes → crocodile

Type of food chains: Broadly can be differentiated as: (i) Grass land and (ii) pond types of ecosystems cattle and rodents graze grass land while zooplankton consume producers of a pond.

1.12   TROPHIC LEVELS

Through food, energy uses move one organism to another. All green plants (energy produces) categorised as trophic level one, while herbivores in trophic level two. Animals that feed on herbivores are in trophic level three. Animals that feed on the trophic level 3 are in trophic level 4 and that feed on the trophic level 4 are in trophic level 5.

Each food chain has at least three food levels (also called trophic levels) namely producer, herbivore and carnivore levels. It is worth nothing that if the herbivore is large in size then the food chain is shorter. If the herbivores are smaller in size then the food chains becomes longer.

Transfer of energy: In a food chain each organism dissipates a part of energy carrying out different processes of life implying transfer of energy one organism to the next is not 100%.

1.13   FOOD WEBS

The close knit relationship of living organisms and the threads is that constitute the web of spider. Because of this the biosphere is described as the web of life wherein each organism plays specific role:

Def: complex food chain in which one population feeds on a number of other populations is termed food web.

Food webs are of two types: (i) Grazing and (ii) Detritus (i.e., wornout matter).

In Grazing food web, energy and minerals move from green plants to herbivores and then to carnivores. Phytoplankton forms the primary Grazing food web in aquatic life. Zoo plankton (small floating animals) in turn are food for small fish and filter feeders.

Producers and consumers are part of the grazing food chain, while scavengers and decomposers are of detritus food chain. Producers are green plants. Primary consumers are herbivores, secondary and a tertiary consumers consists of carnivores (like lions, human beings, snakes). Scavengers like vultures, crabs eat the remains of other organisms. Micro decomposers are (mainly) earthworms, insect larva and micro decomposers are bacteria and fungi.

Food webs

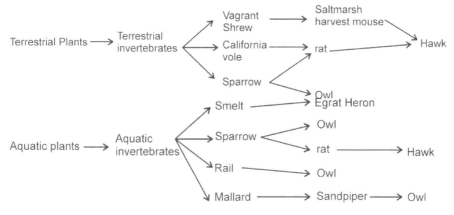

The food chains and food webs from a critical (delicate) balance in nature.

1.14 ECOLOGICAL PYRAMIDS

In nature, animals at the base of food chain are found relatively plenty and at the end relatively few in number. There is a gradual decrease between two extremes. This is called a pyramid of number and is observed in all types of ecosystems.

Ecological pyramids are constructed and quantified based on feeding relationships and energy transfer through biotic compoment of ecosystem. This type representation enables us to compare the ecosystem and their changes (variation).

Pyramid number: The pyramid that deals with the relationship between the number of primary producers and consumers of different order is pyramid number. In pyramids base is always shown by primary producers. Subsequent structures stand on this base with the number of consumers at successive levels. On the top lie the number of carnivores in the ecosystem.

Consumers are of three types-primary, secondary and tertiary. Primary consumers are herbivores while secondary consumers are carnivore and tertiary consumers are also carnivores.

Depending on the ecosystem the shape of the pyramid changes (one ecosystem to another ecosystem). In aquatic and herbaceous communities the size of autotrophs are small but their density per unit area is large. In forest ecosystem the size of producers is large but their population density (need not be large) is comparatively small. Both in grass land and aquatic ecosystem autotraphs are large (in number) but their size is small. The pyramid structure is upright. In parasite food web inverted pyramid (base small & stop large) is observed.

Pyramid Biomass: In pyramid biomass the base is the weight of primary producers and this pyramid is upright. The biomass of one tree is very large and the mass of birds feeding on it is very very small. Similarly the biomass of a very large number of parasites resting on or in the body of birds is far less compared to the bird. Hence the pyramid of biomass is upright, but in case of number it is converted. For example, a tree may require 5-10 years to produce first seed, where as a algae type bacteria (diatom) may take few hours to produce a large number. A diatom may reproduce billions and billions in 5-10 years. In all this biomass were to survive which would be heavier than a tree.

Pyramid of energy: This represents the total quantity of energy utilised by different trophic level organisms of an ecosystem per unit area over a period of time. Taking time factor into consideration the pyramid of energy is always upright. The quantity of solar energy trapped by green plants over a period of one year is highest as compared to other organisms of other trophic levels. Because of this the base of this pyramid is broad. In aquatic ecosystem the populations of phytoplankton quickly complete their life cycle and begin new generation of crops.

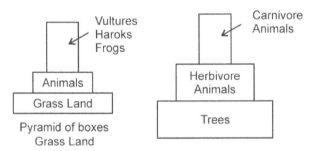

Terrestrial Ecosystem: The effect of climate on rock material and the available flora fauna results in the development of various types of Ecosystems. The major terrestrial ecosystems are forests grasslands, deserts etc.

Note: Biomes: Large recognisable communities in different parts of the world are termed biomes. Biomes are biological expressions of the interactions of organisms with physical factors in different regions of the world. Similar environmental conditions create similar biomes in different parts of the world.

Biomass: Any organic matter which is renewable or weight of living organism

Biota: The flora and fauna of a locality (or region)

Productivity: Formation of biomass by the use of solar energy is called productivity.

Biodegradable: Substances that breakdown naturally are called biodegradable.

CHAPTER - 2

Biological Environment of Sea

2.1 MARINE ENVIRONMENT

Water is essential to all life for its maintenance. Water constitutes about 80 % (by weight) of active protoplasm. Water is the most efficient of all solvents water carries in solution the necessary gases, oxygen, carbon dioxide along with mineral substances which are necessary for the growth of plants and animals and water is itself one of the essential raw material in the manufacture of foods by plants.

Terrestrial (environment) organisms have devised ways like impervious layers to conserve water and plants have roots and special vascular systems for transport of water to all growing parts. In contrast to this marine organisms have freedom from desiccation, except during high tide levels. Hence, require no highly specialized ways for conservation of water or for its transport in plants. Since water has high heat capacity and high latent heat of evaporation which the marine organisms from the dangers protect of rapid change of temperature in the environment. Further due to high degree of transparences of water, plant life is sustained in a deep layer and animals developed organs of vision and direction finding.

Sea water is a buffered solution (that is, changes are resisted from acid to alkaline or alkaline acid). This property has two important applications. (1) Sufficient supply of CO_2 to plants for photosynthesis, and (2) slight alkaline habitat sufficient for formation of $CaCO_3$ (Calcium carbonate) better than neutral solution.

For living cells the sea water is a most hospitable environment, because it contains all of the chemical elements essential for growth and maintenance of plant and animal protoplasm. Sea water contains a large number of dissolved salts and is important as an external fluid medium to the internal medium-namely the body fluids (blood, Coelomic fluid and so on) of the organism. Sea water contains salinity 35%, freezes at $-1.91^{\circ}C$. Ocean environment is suitable for existence of wide variety of living conditions, salinity up to 37%, temperature $30^{\circ}C$ to freezing point, light intensities from brilliant sunlight at the ocean surface to absolute and perpetual darkness in the deeper layers and pressure varying one atmosphere at the surface to more than 1000 atmospheres in the greatest ocean depth. Because of a great variety of environments in sea we find a very large species of faunal life in the ocean environments. A wide range of conditions, particularly observed along and off the coast regions and depth to bottom. The variations are because of physiographic character of coastline, topographic nature, inflow of land drainage, meteorological conditions etc.

2.2 MARINE BIOSPHERE

The part of the earth which is suitable for life, both plant and animal is called biosphere. The biosphere is subdivided into three main sub-divisions or habitats, known as bio-cycles. They are the terrestrial, the marine and the fresh-water bio-cycles. Each cycle has ecological features and associated plants & animals. A few animal species may migrate freely one type to another (e.g. Crocodile).

It is now known the oceans cover about 71% of the earth's surface. The marine environment provides about 300 times the inhabitable specie as compared to terrestrial and fresh-water bio-cycles together. Terrestrial environment provides habitable bio-space in a shallow zone of about a few feet, while marine habitat provides habitat, at least to some form life to the abyssal depth of several kilometers (3 to 10 km). The freshwater bio-cycle has a small fraction of the other two bio-cycles. The aerial portion of the globe (atmosphere) not treated as separate bio-cycles, because portion is a temporary journey of birds and insects.

2.3 MARINE ENVIRONMENT CLASSIFICATIONS

Broadly marine environment divided as primary biotic and secondary biotic division, on the basis of physical-chemical nature of biota. On physical basis the sub-divisions are recognized both biotically and abiotic ally without clear boundaries (see in the fig 2.1).

The two primary division of the sea are named the Benthic and the Pelagic. The benthic includes all life on the ocean floor, while the pelagic includes the life in whole mass of water.

The Benthic Biotic Environment

All of the ocean bottom terrain from the wave washed shoreline (at flood-tide level) to the greatest depth included in the benthic division.

According to Ekman a zoogeographic stand point, the Benthic division is subdivided as follows.

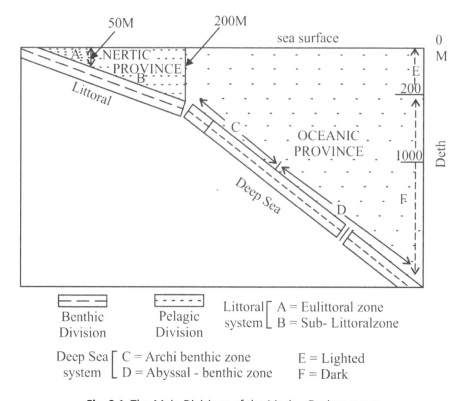

Fig. 2.1 The Main Divisions of the Marine Environment

Littoral system and the Deep sea system. Littoral system is subdivided into Eulittoral and Sub-littoral zone.

The Deep sea system as Archibenthic and Abyssal benthic zones.

The pelagic environment

The pelagic division includes all of the ocean waters covering benthic division. Horizontally, it is sub-divided in to two- Neritic province (an inshore an open area), oceanic province.

Neritic province horizontally extends from coast to littoral (slope depth). Water depth shallower than about 200 m in full or edge of continental shelf.

Ocean province (open sea) has an upper lighted zone and a lower dark zone (with no defined boundary) but 200 m depth is considered. The variable pelagic environment is important for population of the sea.

Potential features of open sea

1. Broad spatial width with great ranges of depth
2. Waters are very transparent with devoid of detritus of terrestrial origin. Predominantly blue water which supports blue surface faune
3. Only carnivores and detritus feeders seen in deep layers

4. Offshore water us relatively stable with uniform high salinity

5. Plant nutrients are relatively low in the upper layer

6. Biologically/chemically broader between Neritic and oceanic provinces not definable

7. Plant nutrients, nitrates, phosphorus are readily available on Neritic province, hence productive for marine life specifically economic fish greatly available.

Other Biotic units

The marine classification of benthic biotic and pelagic environment is based on geographical, physical chemical and biological characteristics, the zones are separate. Within each of these extensive zone, there are many and varied sets of ecological conditions.

The primary topographic unit used in ecological classification of the environment is the biotope or niche which is defined as an area of which the principle habitat condition and the living forms which are adapted to them are uniform. Thus the smaller habitat anomalies within the biotope are called facies. The community of forms in a biotope is called a biocoenosis. Biotopes near to the coast or estuarine locality is known as bio-chores.

Characteristics of population of the primary biotic divisions

Classification of marine environment: Population of the sea divided into three groups 1. Benthos, 2. Nekton and 3. Plankton. Benthos belong to Benthic region while the other two to Pelagic region.

In Greek benthos meaning deep/deep-sea. This include the sessile, creeping and burrowing organisms found on the bottom burrowing organisms found on the bottom of the sea. Benthos comprised 1. Sessile animals (like spurges, barnacles mussels oysters, corals, hydroids, bryozoa some worms, all of the sea weeds and eelgrasses diatoms), 2. creeping forms (like crabs, lobsters, certain copepods, amphipods and many other crustacea, many protozoa, snails and some bivalves and fishes, 3. burrowing forms including clams and worms, some crustacea and echinoderms.

The nekton (in Greek-swimming) is composed of swimming animals found in the pelagic division, which includes squids, fishes and whales (all of marine animals which migrate freely over long distances). These are no plants in this group.

In the plankton (Greek-wanderer) include all of the floating or drifting life of the pelagic division of the sea. The organisms, both plant and animal of this division are generally microscopic or very small, they float with currents. The plankton is divided into two divisions-the phytoplankton and the zooplankton. The phytoplankton comprises of all floating plants (like diatoms, dinoflagellates, sargassum weeds). Zooplankton include 1. Myriads of animals in a floating state 2. countless larvae and eggs of the animal benthos and nekton. Plankton and nekton belong to same biotic realm does not signify divergence of ecological relationship.

2.4   SEA LIFE

Evidence indicate that the sea life is the original environment of animal life. The whole of animal kingdom is divided into a number of primary divisions, each is known as a phylum. Each phylum is composed of animals having certain fundamental morphological similarities not having any animals of other phyla. Each phylum is divided into natural groups called classes and these in turn followed by other yet lower divisions in the following manner.

Phylum → class → order → family → genus → species

Species are formed of individual and the morphological features which are less fundamental (but of the genera). A generic restructure is less fundamental than those of families and so on to the highest divisions, which is based on structures of great antiquity.

According to taxonomic ranking, there are 17 phyla in the sea, of which the following five are exclusively marine-ctenophora, Echinodermata, phoronida, brachiopoda, chaetognatha.

Marine animals were abundant and became fossilized in the Cambrian period (500 million years ago). Several invertebrate phyla developed.

In summary, great part played by marine environment in the development and maintenance of a wide diversity of lower forms, while terrestrial environment produced less diversity form but a higher type of complexity.

The intertidal zone, the area where marine and terrestrial environments meet is a location of migration to land might have taken place.

2.5   POPULATIONS OF THE SEA

Populations of the sea is described as-

1. Plant groups of the sea and 2.The animal population of the sea.

1. **Plant groups of the sea:** In the sea the plants are the real producers. The existence is impossible without marine plants as synthesizers of primary food (for marine animals life).

   Marine vegetation is very small (in variety) as compared to the terrestrial vegetation. The production of primary food in the sea is different as compared to the land. It is said the animal kingdom mainly belongs to the sea. While forests (the plants) belong to terrestrial environment although the primitive plant groups, the algae are developed in the sea.

We know solar radiation (light) is essential for all photosynthetic plants and its attachment substratum is of secondary importance. From this point the eulittoral zone (about 2% of sea floor) is suitable for development of algae (plants), which do not posses true roots. For various other reasons the attached marine plants (a small percent) restricted to benthic are primary producers. These are floating plants (like diatoms, dinoflagellates) are microscopic and innumerable.

The entire plant kingdom is divided into four primary divisions-the Thallophyta, bryophyta, pteridophyta and spermatophyta. Of the four, thallophyta & spermatophyta are presented in sea.

Thallophyta: Almost all of the marine plants belong to this (thallophyta) botanical division (of primitive plants). This division do not show vegetative organs, (no root, stem or leaf). Thallus plants are the marine algae and the marine fungi, particularly the bacteria (which will be discussed separately). Most algae colored and some are iridescent. The pigment of the chromatophores absorb solar energy and synthesize organic compound. The pigment in algae present as colour, which is used for the classes.

Blue-green algae (Myxophyceae)

Green algae (Chlorophyceae)

Brown algae (Phaeophyceae)

Red algae (Rhodophyceae)

Yellow-green algae (a heterogeneous group)

The first four, except some blue-greens, are attached plants, while the yellow-greens are floating or planktonic forms.

2.6 ANIMAL POPULATION OF THE SEA

The animals of sea (fauna) is very large, where the number of species is also large. The important groups of marine animals-

(a) Invertebrates, (b) Vertebrates

Invertebrates are animals lacking backbone, vertebrates are animals having a spinal column (backbone) and endoskeleton.

Phylum: one of the large divisions in the classification of plants.

  (a) Invertebrates

      1. ***Phylum Protozoa***: Protozoa are single-celled organisms, microscopic. The pelagic forms inhabiting the plankton.

      2. ***Phylum porifera***: The sponges are multi cellular animals with spicules of silica or $CaCO_3$ with fibrous skeletons made of the

horny substance spongin. Sponges are benthic and marine. There are about 2500 species, mostly marine.

3. **Phylum coelenterata:** These are primitive forms and show a remarkable degree of polymorphism i.e., a single species present a variety of forms.

4. **Phylum ctenophora:** Ctenophora are small globular jelly forms. The abundant community is called comb jellies or sea walnuts. There are 80 species, all marine.

5. **Phylum Platyhelminthes:** They are flat worms found in the sea. It has class turbellaria and class Nemertinea. The planktonic forms are modified and found in about 550 species of nemerteans, all are marine.

6. **Phylum nemathelminthes:** The roundworms occur largely as parasites, but some found in the plankton. There are about 1500 species, many of which are nonmarine.

7. **Phylum trochelminthes:** Class rotatoria. These are tiny benthic or planktonic organisms having rings of cilia for swimming and for gathering food. They are about 1200 species of rotifers, most are freshwater habitats.

8. **Phylum bryozoa:** These are colonial animals known as "Sea mats" or 'moss animals'. They are more than 3000 species, about 35 of which are non-marine.

9. **Phylum brachiopoda:** These are ancient sessile animals, resembling bivalve molluscs. They grow permanently attached to rocks and shells in the littoral zone below low tide. About 120 living and 3500 fossil species are known.

10. **Phylum phoronidea:** Phoronidea are wormlike animals. These are about 12 marine species.

11. **Phylum chaetognatha:** These are numerous but small holopanktonic wormlike animals known as "arrow worms" or "glass worms". They oocur from the surface to great depths and are distributed far to sea in all latitudes. All 30 known species are Marine. Sagitta is the most abundant genus.

12. **Phylum annelida:** Annelida are true worms with elongated bodies composed of a series of similar segments

13. **Phylum arthropoda:** Arthropoda include animals with a segmented, exoskeleton and with joined appendages, variously modified for locomotion, feeding and other activities

14. **Phylum mollusca:** The molluscs are noted particularly for their construction of an infinite variety of calcareous shells encasing the body.

15. *Phylum Echinodermata*: Echinodermata are animals with calcareous plates forming a more or less rigid skeleton or with scattered plates and spicules embedded in the body wall. Many are provided with spines. All are marine, but a few sea cucumbers are benthic.

16. *Phylum chordata*: Chordate are animals which in some stage of their life have gill slits and a skeletal axis known as a notochord.

2.7 B – VERTEBRATES

Subphylum vertebrata: This group includes animals wave vertebrata. All but the classes Aves and Mammalia are cold blooded

1. *Class Cyclostomata*: The hagfishes, have circular sucking mouth. They live both in the sea and in fresh water.

2. *Class Elasmobranchii*: Primitive fishes, have paired fins and a lower jaw. There are many large forms like the giant manta and the whale shark (about 16m long). Nearly all are marine.

3. *Class Pisces*: True fishes with a bony endoskeleton, paired fins & gills, mostly carnivorous and marine, some are benthic but majority are pelagic.

4. *Class Reptilia*: Sea snakes and turtles. Breathe air, inhabitants of surface waters. Turtles deposit their eggs on sandy beaches. Snakes bring forth living young and found in Indo-West-Pacific & tropical waters of America. Length 1 to 2 m, more or very poisonous, largest of turtle class may weigh about 500 kg.

5. *Class Aves*: Most of birds are dependent upon the sea for food. Some frequent the land only for nesting and rearing of young. Eg. Albastrosses, petrels & auks.

6. *Class Mammalia*: Warm-blooded, air-breathing animals with hair and mammary glands.

Order carnivora: The sea otters & the polar bears (confined to the Arctic region)

Order Pinnipedia: Seals, walruses, nearly all marine

Order sirenia: Heavybodied mammals, live new shores in warm water and browse upon vegetation. eg, sea cows, manatees and dugongs.

Order cetacea: Whales dolphins

Suborder mysticeti: Baleen or whalebone, whales & blue whale. Blue whale is the largest of all animals with length about 34m, weight 150000 kg.

Life cycles in marine animals

Life cycles of marine animals in general three

1. The preponderance of animals which sessile, creeping or burrowing in the adult stage, possess swimming period during early stage of life
2. The very large number of number of young are produced by both pelagic and benthic animals, and
3. The similarity of larvae of deferent invertebrates groups

Reproduction: In reproduction, marine animals are either oviparous or viviparous. The oviparous forms deposit eggs that develop outside the mother's body, while in viviparous forms the young are nourished by the mother and are born alive in a postembryonic state. In an intermediate condition in ovoviviparous forms, where the eggs are incubated and hatched within the body (as in certain sharks, perch and blennies). Most number of animals of the sea are oviparous. In the marine population as a whole very little parental protection is given to the offsprings in the larval stages, & even the eggs.

Developments are mainly two types.

The direct development in oviparous species is associated with eggs, like the fishes, cephalopods. Direct development is common among the deep-sea benthic animals.

The indirect development the egg with little yolk develop quickly or the organism dies. The life cycle of many species of sea is not investigated.

The mammals of the sea bring forth living young, which are nursed by the mother for a long period.

Ecological groups of marine animals

The world distribution marine animals is a function of marine zoogeography. The present day fauna ecological groups are:

(a) Benthos, the animals of the sea floor

(b) Nekton, the swimming animals

(c) Zooplankton, the floating animals

2.8 PHYTOPLANKTON IN RELATION TO MARINE ENVIRONMENT

In Greek plankton mean wandered. Plankton includes all of the floating or drifting life of the pelagic division of the sea. The organisms, both plant and animal of this pelagic division are usually microscopic or relatively small. They float more or less passively with the ocean current.

The plankton is divided into two main divisions-the phytoplankton and the zooplankton. The phytoplankton comprises all of the floating plants (like diatoms, dinoflagellates, coccolithophores and sargassum weeds). The zooplankton includes myriads of animals that live permanently in a floating state and countless number of larvae and eggs of the animal benthos and nekton.

Phytoplankton are firstly, the diatoms and secondly the dinoflagellates. Oceanic phytoplankton are abundant in open sea and densities of about 220000 diatoms per liter. In general, coastal waters as a whole 50 times more productive than open sea waters. It may be noted that phytoplankton organisms are producers of the primary food supply at the sea.

Diatoms: By virtue of structural adaptations diatoms are of four types. 1. Bladder type, 2. The needle or hair type, 3. The ribbon type and 4. the branched type.

1. In bladder type the cell is relatively large and the cell wall and protoplasm form a thin layer inside the shell or test and the remainder of the cell is filled with a light cell fluid or sap. The shape may take the form of a disc.
2. The needle type is long and slender, which sink slowly.
3. The ribbon type cell are broad and flat. They are not numerous in the sea.
4. The branched type have many spines, grown as projections which resist sinking. Cells often form chain.

All of the pelagic species are thin shelled as compared to bottom or littoral forms. Some may secrete a coating of light mucus to remain floated.

Many diatom species conforming to the above types also unite to form chains of various types.

Dinoflagellates: They are passive have freeble locomotion. The longest structure found in the warmer waters.

2.9 FACTORS OF PHYTOPLANKTON REPRODUCTION

In the study of plankton the production is estimated as the amount of organic matter produced by the population under a unit area of sea surface or in a unit volume of water during a given period of time. The production is dependent upon radiant energy for the process of photosynthesis and the presence of inorganic nutrient substances, nitrates, phosphates, Iron etc. in the sea water and biological factors that work upon the availability of these dissolved substances. The principal direct factors (a) direct primary factors that operate directly on the growth and production of the individual alga and (b) direct secondary factor that directly affect the population density.

Direct Factors

1. *The energy factor*: CO_2 and water, sun's radiant energy

$$6CO_2 + 6H_2O \xrightarrow{\text{Sun's radiation}} C_6H_{12}O_6 + 6O_2$$

This process of carbon assimilation is endothermic reaction, where in solar energy absorbed. This energy stored in the complex organic substances of the plant.

2. **Factors of food supply**: CO_2 is sufficiently abundant at all times in the sea is never a limiting factor in phytoplankton production. According to Moore (1924), in the Irish sea 20000 to 30000 tons of CO_2 per cubic mile of sea water are passed through the biological cycle/year.

3. **Dissolved nutrient salts**: Nitrogen, phosphate, Iron and other trace elements available in sea water. Nitrate-nitrogen to phosphate-phosphorus found in the organic substances of mixed plankton.

4. **Accessory growth factors**: The necessary growth substances added with the sea water by the marine diatom, Ditylum brightwelli.

5. **Temperature:** The rate of metabolism, hence the rate of growth and reproduction is regulated by temperature. Temperature determines the composition of species that develop in any one region at various periods of the year.

The degree of viscosity or internal friction of water is vital to passively floating organisms changes in salinity have small influence on viscosity, but a change in temperature from 25° C (at surface) to O° C (at abyss) doubles the viscosity. Similarly summer to winter effects. In summer warm water forms have thinner shells.

The world distribution of marine animals is the functions of marine zoogeography. Complicated patterns depend on the nature of the organisms and the factors in the environment. Marine ecology is concerned with the organisms in relation to their environmental conditions. Hence marine ecology a vital link in the study of zoogeography and many other biological problems of the sea. The ecological groups: (a) Benthos, the animals of the sea floor, (b) Nekton, the swimming animals and (c) Zooplankton, the floating animals of the sea.

(i)   The ocean floors have been found to be inhabited by benthic animals form, the Arctic to the Antarctic and from shore to the greatest depth. The number of animals and the kinds of animals make up population differ, particularly the species genera and families.

The difference are mainly due to topography (as biotopes) and due to environmental divisions and they are the biological criteria for the vertical zones, littoral, archibentic and abyssal-benthic, beside the horizontal faunal areas.

(ii) In littoral zone, the upper or eulittoral zone up to a depth of 40-60 m exist a great diversity and variability physical-chemical conditions of marine life habitats. The sub-stratum changes from clear to soft muds with marked salinity gradient. In shallow areas, wave and tide actions are important. Morphologically animals are variously modified. Most of the sessile forms are flattened. Mussels are securely attached by strong and flexible byssus is threads. Littoral zone contains abundant supply of food for animals from plants, both attached and floating. As a result littoral zone produces abundance of benthic animals. Most animals of this zone have a great concentration of the species. Most littoral area are provided with good circulation due to irregular bottom configuration. However in some bays water may be sluggish, resulting shortage of free oxygen and produces H_2 S (Hydrogen sulphide), which is lethal, causes precarious living conditions. These conditions present an excellent example of the effect of physical-chemical environments on animal life. Much of the organic material produced from the planktons.

(iii) Littoral fauna may be divided into-Arctic, Tropical and Antarctic. Between these there are divisions like boreal temperate fauna. Some of the faunal divisions may further be sub-divided into east-west divisions. Example the tropical fauna, though homogeneous in many respects, the formation of coral reef consists of four parts (1) Indo-west Pacific, (ii) Pacific Tropical American, (iii) Atlantic Tropical American and (iv) the Tropical west African.

Deep-sea Benthos

All of the animals world below the littoral zone may be treated as deep sea fauna. There is no clear border line between littoral deep sea fauna. Similarly we cannot draw border between. Archibenthetic and the abyssal. According to Challenger Expedition [(1872-1876) systematic study of the biology, chemistry and physics of the ocean of the world] discovery, more than 1500 animal species below 1000 m at 6250 m, 20 specimens belonging to 10 species. The benthic region population of the deep-sea relatively small and decreases with increasing depth and also increasing distance from the shore to deep-sea.

2.10 MARINE ORGANISMS

All forms of life on earth are made of protoplasm. They carry identical processes and have similar problems of existence but live in a totally different environments. The life processes use inorganic compounds from soil and air, foods are organized and ultimately plants and animals assimilate their own substances. At the end are the destructive process which convert the complex organic matter (dead plants & animals) into simple compounds. Thus life cycle continues on the earth in natural way.

In marine organism food pyramid, phytoplankton and zooplankton stand as two huge volumes of organic substance at the base of food pyramid. The plankton are generally short-lived. Large phytoplankton and zooplankton population cannot exist simultaneously in the same area for a long time.

Bacteria plays specific role in sea (marine) nutritional cycle. Bacteria are unicellular organisms but some form chains or groups of cells. Morphologically they are three forms-cocci (which are spherical), Bacilli (elongated rods) and Spirilla (helical), structurally mostly mobile rods and of camma-shaped. There are fever spore-formers in the sea than on land. About 70% of marine bacteria are coloured whole only 15% on the terrestrial. Bacteria are the smallest of all organisms measuring only 0.0005 mm in diameter. Bacteria reproduced are very rapid by way of vegetative cell division. In nearly all instances in swarms organisms (bacteria, diatoms, dinoflagellates) occur in large numbers where water is discolored by accumulation of their bodies. The rate of division of bacteria may be frequent as every 20 to 30 minutes.

Associated with bacterial reproduction in the faculty of spore formation; whereby the individual bacterium of certain species withstand adverse conditions while remaining in a state of quiescence for long periods of life. Viable bacteria have been taken from strata of marine mud where they must have remained buried for thousand of years.

Autotrophic Bacteria: These bacteria prepare their food (carbohydrates and proteins) like green plants from CO_2 and in organic salts. Some of these, known as photosynthetic, possess colouring material (of bacteriachlorine) and use radiant it energy in building up protoplasm, while others known as chemosynthetic, derive their energy from the oxidation of various inorganic compounds like H_2S, S or N H_4.

Autotrophic purple sulphur-bacteria found in isolated inshore area impact a red tint to water to the surface algae or to the bottom.

Heterotrophic Bacteria: These bacteria obtain their energy by the oxidation of organic compounds. Hence they live as saprophytes or parasites. Most bacteria of the sea are of this type. Heterotrophic be bacteria makes natural balance cycle through mineralization of organic matter (dead organisms are converted to minerals).

In nature the solid surfaces are provided by all types of particulate matters animate or inanimate, on the bottom, in the plankton and on the nekton.

Oxygen Relations: There are obligate aerobes which use free oxygen and obligate anaerobic which function in the absence of free O_2, while facultative forms live in either type of environment. Most marine bacteria which require a reduced free O_2 are microaerophiles.

Bacterial chemical transformations: For convenience study of the cycles of substance in the sea, it is further classified as nitrifying, denitrifying, nitrogen fixing sulphur and iron bacteria, it is based upon their metabolic activities in transformation of these substances.

2.11 THE NITROGEN CYCLE

Nitrogen cycle is the role of bacteria in the chemical and biological cycles of the sea.

Nitrogen cycle is very important because of its importance to all forms of life. Nitrogen exists in the sea in combination with other element. Like NH_3, ions of NO_2 and NO_3.

Nitrogen exists in composition of all living things. It is one of the building blocks (nutrients) used by plants in forming the complex protein molecules of their bodies, from which animals derive their nitrogen. The complex nitrogenous compounds found in both plants and animals. Upon death of the organisms are decomposed along with their products of excretion into simpler compounds that could be used by plants. This process is achieved by the activity of proteolytic bacteria. [in the process of decomposition different bacteria acts in transformation, amino acids and nitrogen compound form]. The nitrogen compounds NH_3, nitrates and nitrites take part.

Nitrogen occurs in sea water both in compounds form and dissolved nitrogen gas. The chemical circulation of nitrogen is activated by organisms (plants, animals & bacteria) living within the sea large and complex protein molecules of plants and animal tissue is broken down into simple products containing nitrogen. The cycle is composed of six transformation known as:
1. Anmmonification 2. Nitrification, 3. Nitrogen assimilation, 4. Denitrifications 5. Nitrate reduction and 6. Nitrogen fixation.

Ammonia (NH_3) splits into amino acids and bacteria. $NH_3 \rightarrow NO_2 \rightarrow NO_3$

Nitrogen fixation: In terrestrial environment, the nitrogen fixation of free nitrogen is by bacteria called symbiotic bacteria. Fixation nitrogen mean conversion of atmospheric nitrogen into nitrogenous compound by natural or artificial methods. Nitrogen fixation in oceans not clearly known .

Marine phosphorus cycle: Phosphorus is an important plant nutrients which has a biologically activated cycle. Phosphate is quickly regenerated by bacteria and other agencies following the death of plants and animals upon death and autoanalysis of the bacteria the phosphate are returned in a few days in mineral form. Cooper (1935) found rapid regeneration of phosphates from animal plankton than from diatoms.

Marine carbon cycle: The complex carbon compound built up by marine plants and animals are decomposed through bacterial action with the formation of CO_2. This CO_2 together with respiration by all other organisms supply the photosynthetic processes of the diatoms and other algae. Ocean sensing satellite data used for measuring ocean colour as a proxy for plankton distribution.

The ice-core and deep sea sediment records show that the global carbon cycle undergone significant changes with the ice age fluctuations in climate.

The hydrological cycle is essential to the planets ability to sustain life. Models are used to reconstruct physical, biological process to test hypothesis and predict future changes. The General Circulation Model developed to understand the global physical climate system, global change phenomena include changes in climate, atmospheric composition and interaction with the changing land use.

Marine plankton in the sunlit upper portion of the ocean extract in organic carbon from water through photosynthesis process. About 90% of this carbon is respired during its transfer through the food web in the ocean layer. The rest 10% with the inorganic components of the skeleton of the organism settled to great depths. This carbon content in the upper ocean is replenished by the CO_2 in the atmosphere.

Over the time, the reservoir of organic and inorganic carbon in deep sea sediments is more than 100 times those presently circulating in the atmosphere, terrestrial biosphere and ocean. It is inferred about 50% of all fossil fuels burnt since industrial revolution has been absorbed in the ocean.

Marine Sulphate Cycle: Sulphur is one of the essential constituents of living matter and its compounds are acted upon by bacteria. Plants utilize a small quantity sulphur in their metabolism, sulphate is produced by chemical or biological oxidation. In decomposition (death) of organic compounds containing sulphur, H_2S (Hydrogen sulphide) is produced as disintegration product, which combines with O_2 to give sulpher.

$$2H_2S + O_2 \rightarrow 2H_2O + S$$

Bacteria and Bottom Deposits: The activities of bacteria in the processes of sedimentation on the ocean bottom. The activities of various types of bacteria occurs in abundance in the bottom sediments which play important role. The character of the bottom deposits and digenesis'of rock strata. Some of the results are-

(i)   Formation of humus a stable organic end product of decomposition

(ii) Calcium precipitation in the presence of calcium salts and high pH

(iii) Iron and manganese be precipitated by bacteria that form sheaths of compounds of these metals.

Distribution of Bacteria in the Sea

A very large number of marine bacteria found in the coastal waters where the largest plant and animal life is also produced. In vertical distribution, two main centers.

Water Sample	Depth in Meters	No. of Bacteria
Water	1	147. to 344
Water	10	238 to 400
Water	20	292 to 528
Water	50	86 to 620
Water	100	14 to 53
Water	500	2 to 0
Sediment	Bottom	768×10^3 to 162×10^5

2.12 ECOLOGICAL GROUPS OF MARINE ANIMALS

The world distribution of marine animals is a function of marine zoogeography. The main ecological groups of marine animals are:

(a)   Benthos, the animals of the sea floor

(b)   Nekton, the swimming animals

(c)   Zooplankton, the floating animals

2.13 ORGANIC PRODUCTION IN THE SEA

In the studies of marine sciences the production in sea is important. Particularly chemical, biological or geological, because of their relation on the extent, time and spatial distribution or organic and a few inorganic constituents of the water and of the bottom.

In a limited sense the term production is used to denote the product, that is, the amount produced of any give group of organisms in the sea. Thus we have plant production, phytoplankton production, zooplankton production and commercial production. In the sea, plants are the only organisms which synthesize organic matter from inorganic substances in significant quantities.

Phytoplankton comprises all plants and the phytoplankton production equals the gross sea production – the organic matter that is oxidized by the phytoplankton itself or secreted. The phytoplankton production can be expressed as grams of carbon/m^3/day. Zooplankton production is the amount of digested material that is converted into animal protoplasm.

Zooplankton production = amounted digested food (both plant & animal) + the amount used in katabolic processes.

Commercial production signifies to marine products of commercial value. It must be noted that production and population are different.

2.14 PLANT-NUTRIENT CONSUMPTION AS AN INDEX OF ORGANIC PRODUCTION

Changes in the concentrations of the essential plant nutrients coincide with plant growth. This is useful as index of production. It was demonstrated the existence of a constant ratio between C,N,P in the organic content mixed plankton given C : N : P = 41 : 7.2 : 1 by weight are in close agreement with the proportions of these elements in sea water. A measurement of decrease (fall/drop) in any of these elements in the mineralized state in the sea water is reasonably equal to incorporation into organic material during synthesis of carbohydrates or proteins. Similarly, when CO_2 utilization is used as an index, the respiration of animals and the tendency for CO_2 of the water to be equilibrium with that in the air must be taken into consideration.

The plant utilization of phosphate varies from year to year. For example the average maximum phosphate supply available for the vernal plant-production varies from an average of 0.67 μg atom /L to only 0.47 μg atom/L (L = liter). This has great effect on both plant and animal production. Years of low phosphate are correlated with scarcity of plankton animals and poor production of young fish. The difference between the winter maximum and summer minimum phosphate can give roughly minimal measure.

The use of nitrate by the growing phytoplankton gives an index of seasonal production. The different steps in regeneration are assumed to retard the process. Nitrate accumulation during winter is active in northern latitudes well-defined growing seasons. Nitrogen required by the plants acquired from ammonia and amino acids. Ammonia is a putrification product of organic material.

2.15 OXYGEN PRODUCTION AND CONSUMPTION AS AN INDEX OF ORGANIC PRODUCTION

The plants take carbon for carbohydrate synthesis from the (water) CO_2 and O_2 is liberated. $6CO_2 + 6H_2O \rightarrow C_6H_{12}O_6 + 6O_2$

A measure of this O_2 production gives the amount of carbon that is bound in organic compounds. For each 1ml of O_2 set free, 0.536 mg of carbon has been assimilated.

In nature, the O_2 accumulation in the layers of organic production ad thus its variations in time and space gives a measure of the intensity of phytoplankton outbursts. This gives minimal values of O_2 produced but obscured by respiration activities of animal & bacteria.

In conclusion the various indices of organic production in the sea require more experiments to authenticate.

2.16 ZOOPLANKTON PRODUCTION

The animals of zooplankton group in general mainly grazers. They are very large and regularly Distributed. The production of zooplankton is very very large. The average monthly-catch all sections of the Gulf was about 40 ml of water free plankton per m^2 of surface. The highest monthly value of 90 ml in September and the lowest about 10 ml in May. On the basis of this water-free values the standing crop for the whole Gulf about four million tons. Zooplankton production is assumed to be a measure of the phytoplankton production, however some factors must be taken into account.

Commercial productions

The amount of organic material shown in tons/year of fish over certain fishing grounds the poundage yield of oysters. On cultivated bed's or the season's yield in barrels of oil from whales. These are estimates of commercial important yields. They have roughly direct relation to the total organic production.

2.17 MARINE BIOSPHERE ATMOSPHERE INTERACTION

Marine plankton in the sunlit upper layer of the ocean extract inorganic carbon from water through the process of photosynthesis. About 90% of this carbon is respired during its transfer through the food web in the upper ocean layer. The carbon content in the ocean is replenished by the CO_2 in the atmosphere. The reservoir of organic and inorganic carbon in deep sea sediments is more than 100 times circulating in the atmosphere, terrestrial biosphere and ocean.

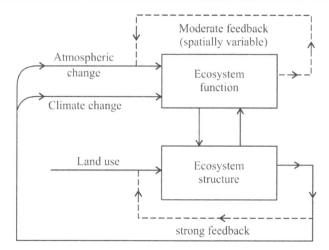

Fig. 2.2 Global change effects on ecosystems in terms of structure and functions

The ocean carbon cycle: Data from ocean sensing satellite used for measuring ocean colour as a proxy for plankton distribution. The ice core and deep sea sedimentation records reveal that the global carbon cycle underwent significant changes with the ice age fluctuations in climate. The marine biosphere involved in changes of nitrogen, phosphorus and to some extent in sulphur cycles.

CHAPTER - 3

Natural Resources

3.1 NATURAL RESOURCES – (A) LAND RESOURCES

Scientists are of the opinion, "life will cease on the earth when the energy resources of the planet earth are exhausted."

The basic needs of man are Roti (food), Kapda (clothes) and Makan (habitat). Man is a social being and hence developed villages, towns and cities. Ever since man learnt to eat cooked food, he is dependent on fire and now on various types fuels. For all this requirements, man is dependent on natural resources mainly fertile soil (for agriculture), sufficient water, green forests and rich minerals & biological diversity.

Natural resources may be categorized as inorganic (like air, water, mineral fuels etc) and organic (like plants, animals, microbes & their products). Besides the above two categories some are renewable and others are non-renewable. Thus environment weighs for man's safety and wellbeing. It is the duty of every citizen to safeguard the environment. Natural resources/environmental management must be catering (i) food security (ii) water security (iii) energy security and (iv) sustainable development.

Food security depends on land resources, which include 1. Land use, 2. Waste land mapping and updating 3. Corridor survey and 4. Land cover dynamics.

Land use Statistics (1970-71)

1. Total geographical area of India = 328 million hectares
2. The land use statistics available = 306 million hectares (93% of total land)
3. The arable land (the net area sown + fallow land) = 161.3 million hectares = 52.7% of 306 million hectares
4. Forest cover = 65.9 million hectares (or 21.6% of the total area)
5. Land put to non-agriculture use = 16.1 m h (or 5.2% of the total area)
6. The barren and unculturable land = 30.2 m ha (or 9.9% of the reporting area)
7. Permanent pastures and other grazing land = 13 m ha (or 4.2%)
8. Miscellaneous tree crops & groves = 4.3 m ha (1.4%)
9. Cultivable waste land = 15.2 m ha (5%)

Total 306 m ha (100%) = 52.7 + 21.6 + 5.2 + 9.9 + 4.2 + 1.4 + 5

Based on population figures as on March 1971 & FAO production year book 1972

Country	Hectares per head Geographical area	Arable land	Forest area
Austria	59.00	3.45	2.98
Canada	45.07	1.99	20.00
Russia	9.06	0.92	3.68
Brazil	8.65	0.30	5.26
US. America	4.48	0.91	1.46
China	1.20	0.13	0.12
India	0.56	0.29	0.12
World	3.56	0.39	1.07

(land utilization statistics and changes in the concept varies)

Note: Land use means, the land is used by humans. Forest, agriculture land, industry and urban, sub-urban land railways etc. are all land use.

Availability of land per capita in some states over the world.

State/country	Per capita area in ha	
	Agricultural land	Forest land
Australia	3.2	14.2
US America	0.9	1.3
Canada	0.4	0.5
India	0.2	0.09
UK	0.04	0.2
Nepal	0.3	0.3
World		1.08 or 33%

World forest area 108.49 million ha (= 33 % of land area)

According to FSI (Forest Survey of India) statistics 2011, India's geographical area is about 329×10^4 km^2, forest area is about 69×10^4 Km^2, per capita forest area is 0.0572 ha.

State of forest report 2013: Forest cover 21% of geographical area = 69.8×10^4 Km^2.

3.1.1 SOIL

Definition of soil

The top layer of land surface (earth's lithosphere) of the earth is called Soil. Soils are formed by mechanical and chemical weathering of rocks and decomposition of organic matter under the actions of weather. Soil serves a natural medium for the growth of plants.

The organic matter of soil which is formed by the slow decay of plants roots, stems, leaves and other vegetable matter and animal remains is called humus.

Soils contain mineral matter which is influenced by environmental factors like weather, topography organisms and acquire physical, chemical and biological properties. Soil changes due to parental material and morphology. Soils differ in color red, black due to the presence of iron (oxidation, hydration, diffusion), titamferous magnetite, drainage coxdition (black due to dissolved CO_2), some fine in texture which serves as a reservoir of nutrient and water for crops.

Components of soils are: The main components of soils are Mineral material, Organic matter and Inorganic matter.

All these required for plant growth.

Soil formation: The parent material of soils are rocks namely Igneous, sedimentary and metamorphic rocks.

Weathering: The physical, chemical disintegration and decomposition of rocks is called weathering. The factors that affect the physical weathering are–temperature, water, wind and plants & animals.

Chemical weathering is caused by solution, hydration, hydrolysis, carbonation, oxidation and reduction. Chemical weathering taking place in the lower layer of soil is called geochemical weathering. Weathering that is taking place at the surface and under the surface is termed pedochemical weathering. Mineral weathering is caused by oxidation and reduction.

Soil profile: The vertical section of the soil extending from horizon to parent material is called Soil profile. This shows soil formation, development and crops husbandry. Top layer contains rich organic matter mixed with mineral matter. The second horizontal layer consists of clay, iron, Aluminum of humus and third lower layer consolidated rock. All profiles do not show the three layers. Soil profile is significant in crop husbandry and in soil moisture.

Components of soils: The main components of soil – Inorganic or mineral particles, Organic materials, Water and Air. These components change from season to season and place to place.

Physical Properties of Soils

1. Water holding capacity 2. permeability to water 3. aeration 4. plasticity and 5. nutrient ability.

Particle sizes – USDA and international classification is given below.

	International system	USDS system
	Diameter in mm	
Gravel	2 mm & more	2 mm & more
Very coarse sand	1 to 2	-
Coarse sand	0.5 to 1	0.2 to 2.0
Medium sand	0.25 to 0.5	-
Fine sand	0.1 to 0.25	0. 02 to 0.2
Very fine sand	0.05 to 0.1	-
Silt	0.002 to 0.05	0.002 to 0.02
Clay	Less than 0.002	Less than 0.002

International system used to designate texture and the US system for classifying soil taxonomy.

Textural classes (particles size groups)

Textural classes are: 1. Clay, 2. Sandy clay, 3. Salty clay, 4. Clay loam, 5. Sandy clay loam, 6. Silty clay loam, 7. Loam, 8. Sandy loam, 9. Silt loam, 10. Sand, 11. Loam sand and 12. Silt

Colour of soil: Soil colour gives a ready clue to soil conditions. Colour of soil is due to mineral matter or organic matter and mostly both.

Red, yellow or brown soil colour related to oxidation, hydration and diffusion of iron oxides in the soils.

Dark–colours combination of one as more/several factors.

Density of soil: The true density of soil (T) is based on the individual densities of soil constituents. Apparent density or the bulk density (A) is weight per unit volume of dry soil as a whole.

Pore space   (P) % = $\left(\dfrac{T - A}{T} \right) \times 100$

Pore space (P): The pore space of soil is the portion occupied by air and water, and it is determined by structural conditions.

Sand pore space = 30%

Loam pore space = 40 to 50%

Plasticity and cohesion: Plasticity is the property that enables a moist soil to change shape on the application of force and retain the shape even when the force is withdrawn.

Sandy soils considered non-plastic

Clayey soils to be plastic

Cohesion is tendency of particles to stick to one another.

Soil temperature: Soil temperature is influenced by its colour, composition, slope aspect and water content. Dark soils absorb more heat during day and lose it during night more quickly than fine-texture soils.

Soil air: Soil air depends on pore space (30-60%). The pore space not filled by water is occupied by air.

Capillary pore space occupied by water. Pore space occupied by air is one-third and two-thirds pore space occupied by water. Soil air is composed of largely nitrogen and oxygen but differ. Soil air contains more moisture, more CO_2 and less O_2.

Table 3.1 Percentages composition of surface-soil air of different soil

Type of soil	Percentage composition		
	CO_2	O_2	N
Forest soil	0.87	19.61	79.52
Clay soil	0.66	19.61	79.35
Asparagus bed freshly manured	1.54	18.80	79.66
Vegetable mould compost	3.64	16.45	79.91
Swamp rice land	4.23-4.69	0.31-0.99	84.35-86.60
Vicinity of sunn hemp roots	12.12 to 16.99	2.23 to 4.67	79.77 to 81.63
Vicinity of maize roots	3.34 to 12.30	7.25 to 13.82	74.47 to 81.59

Soil air also contains soil solution, which contains about 90% CO_2, 10% N and trace O_2.

Soil water: Water has maximum influence on the growth and yield of a crop. It is required in large quantity compared to another substance that contribute to growth and yield of crop. This various functions of water in plant life are as follows:

1. Water is an essential part of plant food. It constitutes about 90% of plant tissues.
2. Water serves as a solvent and carrier of plant nutrients
3. Water maintains cell turgidity and regulates temperature

Water is held in the soil is the follow of forms

(a) **Hydroscopic water:** It occurs as a thin film (4 to 5 million microns) and is held tenaciously with a tension of about 31 atmospheres. This water is not available to plant.

(b) **Capillary water:** It forms continuous film around soil particles (outside the film of hydro-scopic water) and in the micro-pore spaces. It is held by surface tension. Capillary water is held at a tension 1/3 to 31 atmospheres

(c) **Gravitational water:** It is free water held at a tension of about 1/3 atmosphere. Gravity water saturates the soil and percolates downwards under the influence gravity.

Field capacity: The maximum amount of capillary water remaining in the soil after the removal of gravitational water (is called its field capacity). In general it represents the water held at a tension of 1/3 atmosphere.

Wilting point: Capillary water held at tension greater than 15 atmosphere is not available to plants. At this point of soil moisture the plant wilts permanently and hence the percentage moisture at 15 atmosphere is called its wilting point or wilting percentage. The moisture held at tensions between 1/3 atmosphere (field capacity) and 15 atmosphere (wilting-point) is the water available to plants to the maximum extent. Thus capillary water is the major source of water used by plants.

Capillary water is capable of movement upwards, downwards or laterally, the movement taking place from the thicker part of the film to the thinner part.

In general, a water-table lying more than 2 to 3 meters below the root zone is not of much use to crops.

3.2 MINERALS

The metals found in the earth either in native form or in combined state (the natural materials found in the earth in the combined state) are called minerals. The minerals from which metals are extracted easily and economically are called ores. The process of taking out the ores from the mines is called mining.

The process of extracting the metals from their ores is called metallurgy. The metallurgical process are:

1. Grinding or crushing the ore. Crushed ore fed to stamp mills or
2. Pulveriscrs to powder it, then taken out by steam water. It is then
3. Concentrated either by hand picking, froth floatation, electro-magnetic separation or leaching. The concentrated ore converted to
4. Metalic oxide (by calcination) or roasted. It is then subjected to
5. Extraction of metal from furnace.

Chemical Properties

Mineral matter: The principal minerals occurring in the earth's crust are

Mineral	Percentage	Composition
Feldspar	48	Silicates of K, Na, Ca
Quartz	36	Silica (Si O_2) & sand
Mica	10	Alumino silicates of K, Fe, Mg, Na
Limestone & dolomite	2	$CaCO_3$ with Mg CO_3
Hornblende and augite	1	Ferro-magnesium minerals of silicates of Ca, Mg, Fe, Na
Olivine & serpentine	1	Olivine green Fe magnesium serpentine hydrated silicate of Mg
Clay	1	Hydrated alumino silicates
Other minerals	1	

Inorganic components

Components of silicon AL, Ca, Mg, Fe, K, Na are principal constituents of the mineral matter in the soil and small quantities of P, Bo, Mn, Cu, S, Zn & Co.

Plants obtain C, H & O_2 from CO_2 and water. Soil is also the source of N for plants but ultimate source of N is atmosphere where fixation of N takes place physio-chemically and biologically.

Organic matter

Organic matter, a small part of mineral soils, plays vital role in the productivity and conditioning of soils. It serves as source of food for soil bacteria and fungi which are responsible for converting complex organic materials into simple substances readily absorbed by the plants. The inter mediate products of decomposition of fresh organic matter helps to improve the physical condition of the soil. Addition of organic matter improves the working quality or friability of soil. In association with clay and Ca forms aggregates of soil particles to produce the "crumb structure". Applied in sufficient quantity, organic matter looks as a mulch.

Humus: Black organic matter in the soil formed by decomposition of plant and animal remains.

Loam or topsoil: A mixing of mineral matter from subsoil and organic matter combine to form topsoil. Topsoil forms very slowly at a rate of about three continents (3 cm) in 500 years.

The soil micro-organisms: (i) Microflora and (ii) Microfauna.

Microflora: Bacteria, actinomycetes, fungi and algae

Microfauna: Protozoa and nematodes.

In addition to these, the soil has a large number of worms insets of different types and sizes.

Azotobactor is found to fix considerable amounts of nitrogen (N).

Actinomycetes Synthesize chemical substances which produce familiar, odour, when the sun-baked soil of a cultivated field receives the first showers of rain.

Humus: The organic matter in the soil contains largely of plant remains. Chemically humans is a mixture of decomposed or altered products of carbohydrates, protein, fats, resins, wax and other similar substances.

Humus is soluble in water in small extent, it can also absorb gases and water.

Fungi: Fungi disintegrate organic matter of the soil or plant roots in the upper strata below the surface. Many fungi are harmless saprophytes, living on dead organic matter, others are parasites which produce highly destructive epidemic diseases.

Algae: They are microscopic or larger plants containing chlorophyll. They are found in large numbers in the top layers of constantly moist soil, (like paddy fields). Some of them are capable of fixing atmospheric nitrogen.

Protozoa: soil protozoa are unicellular animal, feeding on soil organic matter or on bacteria. Larger forms are seen in very wet conditions or in swamps.

Worms and insets: Most of the soils inhabited by worms, insects and other animals of different sizes. They live (feed) on plant remains, which are ingested along with large quantities of earth, insets and some other larger animals promote aeration and water percolation.

3.3   SOIL SURVEY AND MAPPING

There are mainly two types of surveys.

1. Reconnaissance survey-scale 1 = mile or 1:63360
2. Detailed, survey of India scale 1:50000

Mapping is done after preliminary interpretation of the base maps and initial legend.

The purpose of survey maps is to identify bench mark soils. The survey maps provide. 1. The information of the soil, 2. Land problems, 3. Water use management, 4. Drainage and 5. Maintenance of soil fertility.

The land fit for ploughing or tillage is called arable land.

Arable lands are productive and suitable for intensive cropping: Non-arable lands with no erosion hazards are not suitable for common cultivable crops but they are useful for raising pasture and adaptable trees. They are used for cattle grazing, recreation and wild life.

According to soil taxonomy the soils classified with traditional nomenclature is given below.

Sl. No	Soil major group	Soil taxonomy orders
1.	Alluvial soil	Entisol Incepti sol Alfisol
2.	Black soil	Vertisol Inceptisol Entisol
3.	Red soil	Alfisol Inceptisol Uitisol
4.	Late rite soil	Alfisol Uitisol Oxisol
5.	Desert soil	Entisol Aridisol

Alluvial soils: The alluvial soils include the deltaic alluvium. Calcareous alluvial soils, coastal alluvium and coastal sands. Alluvial soils are the largest and most important soil group of India and has the largest share in respect of agriculture wealth.

Black soils: These soils vary with depth (from shallow to deep). In Deccan trap it is called regur or black cotton soil. Generally there is no change in colour up to a depth of two to three meters. Black soil areas are of high degree of fertility, contain high proportion of calcium and magnesium carbonates. These soils contain large iron, fairly high quantity of lime, magnesia and alumina and also potash, but poor in phosphorous, nitrogen and organic matter.

Red soils: The ancient and metamorphic rocks on weathering have given the red colour to the soils. The colour of the soil is due to the wide diffusion of iron (rather than to high proportion iron). Red soils are in general poor in nitrogen, phosphorus and humus content. The clay fraction of the red-soils is rich in kaolinitic type of mineral and found under forest vegetation. Morphologically red soils have two sun-groups-red loams and red earths. pH varies 6.6 to 8 and have deficient organic matter with poor plant nutrients.

Laterites and lateritic soils: laterite is a formation is peculian to India. It is a compact to Vesticular rock composed of a mixture of the hydrated oxides of aluminum and iron with small amounts of magnesium oxides. All lateritic soils are very poor in lime and magnesia and deficient in nitrogen. However there is occasionally contain high content humus. The soils are rich in nutrients, contain 10 to 20 percent organic matter with low pH 3.5-4.

Desert soils: The predominant component the desert sand and is quartz with small quantities of feldspar and with a fair proportion of calcareous grains.

Some of these soils have high percentage of soluble salts with varying percentage of caco₃ but poor in organic matter.

Large part of the arid region in western Rajasthan, Haryana, Punjab lying between Indus river and Aravalli range affected by desert conditions.

Acid soils: soils with pH less than 5.5 and which respond to liming are called acid soils. Acid soils found widely in the Himalayan region and eastern plains of extra Peninsular India, the peripheral Peninsula, coastal plains including gangetic delta. In there soils, rainfall is moderate to high. Soil acidity exceeding certain limit is injuries to plant growth.

pH measurement and degree of acidity given below.

pH	Reaction
6.6 to 7.5	Nearly neutral
6.1 to 6.5	Slightly acid
5.6 to 6.0	Medium acid
5.1 to 5.5	Strongly acid
4.6 to 5.0	Very strongly acid
4.5 & below	Extremely acid

Saltine and Alkali: Three types of saline and Alkaline soils

In arid and semiarid areas of India crop production is restricted due to salinity or alkalini city or both. It is estimated about 7 million hectares of India is out of cultivation or gives low yields of crop soils containing toxic concentration of chlorides, sulphates of sodium of soluble salts in the root-zone called saline-soils.

(a) **Saline-soils:** In saline soils have exchangeable sodium percentage is less than 15 and pH is less than 8.5. The saline soil is also called white alkali because of white encrustation due to salts.

(b) **Non-saline Alkali or sodic soils:** the decisive effect alkali soil on plants due to toxicity due exchangeable sodium and pH. Soils that do not contain neutral slats are called non-alkalini line or sodic soils. These soils have an exchangeable sodium percentage more than 15 and pH greater than 8.5 and low infiltration rate. The non-saline alkali soil also termed black alkali.

(c) **Saline-Alkali soils:** This group of soils is both saline and alkali. They have appreciable amount of soluble salts (electrical conductivity is more than 4 mm hos/cm). exchangeable sodium percentage is greater than 15, but pH is less than 8.5.

The soil salinity or alkalinity or both have many adverse effects which cause low yields of crop, poor quality of fodder and cause excessive run-off and flooding (due to low infiltration)

Main causes of salinity: In arid and semiarid areas, salts are formed during the weathering. During weathering of soils minerals are not fully leached. During heavy rains the soluble salts are leached from high to low lying areas. If the drainage system is restricted, salts accumulate on the surface as water evaporates. Similarly excessive irrigation of uplands (containing salts) contribute to the accumulation of salts in valleys. Irrigation of soil with saline water contributes to salinity of soil.

3.4 SOIL TESTING (PRACTICAL)

Soil testing is done to find mainly

1. pH-of soil: Is to find whether soil is acidic, alkaline or normal
2. Total soluble salts: This is determined by electrical conductivity, which gives the degree of salinity, alkalinity of the soil sample.
3. Organic carbon: it is a measure of available nitrogen in the soil
4. Available phosphorus (in soil sample)
5. Available potassium (in soil sample)

Soil testing in laboratory is made by standerdised rapid methods for the above five soil nutrients.

Rating chart for the soil sample test data

Sl. No	Nutrient	Low (%)	Medium	High (%)
1.	Organic carbon (as a measure of available nitrogen)	Below 0.5	0.5 to 0.75	Above 0.75
2.	Available nitrogen	Below 280 kg/ha	280-560 kg/ha	Above 560 kg/ha
3.	Available phosphorous	Below 10 kg/ha	10-25 kh/ha	Above 25 kg/ha
4.	Available potassium	Below 110 kg/ha	110-280 kg/ha	Above 280 kg/ha
pH				
5.	Acidic	pH below 6		
6.	Normal	pH 6 to 8.5		
7.	Tending to become alkaline	pH 8.6 to 9		
8.	Alkaline	pH above 9		
Total available salts (Electrical conductivity EC) in milliumhos/cm				
9.	Below	Normal		
10	1-2	Critical for germination		
11.	2-4	Critical for growth of the sensitive crops		
12.	Above 4	Injurious to most crops		

Contd...

Sl. No	Nutrient	Low (%)	Medium	High (%)
Soil-testing Laboratory (Report and Recommendation)				
13.	Sample No	Date		
14.	Texture			
15.	pH	Quantity of nutrient to be applied in kg/ha		
16.	EC	(m mhos/cm 25°C)		
17.	Organic carbon %, nitrogen	Nitrogen N		
18.	Available phosphorus (kg/ha)	Phosphorous (P_2O_5)		
19.	Available potassium (kg/ha)	Potassium (K_2O)		
20.	Compost or green manure	Tinnes/ha		
Opening of drains				
21.	Flooding	Times		
22.	Application of kg/lime/ha			
23.	Application of kg/gypsum/ha			
Signed: In charge (soil chemist) of the laboratory				

Note: In modern (state) time, survey and sampling is economically, quickly done with the help of remote sensing and GIS. This system provides end-to-end solutions for utilization data for natural resources and its management.

Land resources include-land use, land cover mapping, waste land mapping and updating.

In respect of agriculture remote sensing provides reliable, timely information and monitoring. It provides inventory of major crops, crop production estimation, agricultural drought assessment, horticultural crop inventory etc.

3.5 LAND UTILIZATION

Land use in India evolved by various factors like physical, economical and the structure of other resources (viz capital, labor availability) at the location particularly related to transport, industry and trade. Land use statistics gives a clear insight.

Total geographical area of India is about 300 million hectares of this 93% land use (306 m ha) is available.

According to 1970-71 statistics:

  (i)   Arable land 161-3 m ha (52.7%)
 (ii)   Under forest cover 65.9 m ha (21.6%)
(iii)   Non-agriculturable use 16.1 m ha (5.2%)
 (iv)   Barren and non-culturable land 30.2 m ha (9.9%)
  (v)   Permanent pasture and grazing land 13.0 m ha (4.2%)
 (vi)   Miscellaneous-tree crops and groves 4.3 m ha (1.4%)
(vii)   Culturable waste land 15.2 m ha (5%)

        Total 306 mha (100%)

According to state Forest-Report 2011-survey of India, geographical area of India is 3287263 Km^2, forest cover actual 692027 Km^2 (21% of geographical) per capita forest area 0.0572 ha (per per son).

According to Technical committee set by Ministry of Food and Agriculture (1948), the land use classified into 9 groups, which are given below.

Sl. No		During 1970-71
1.	Forest	21.6
2.	Non-agriculturable use	05.2
3.	Barren and unculturable land	09.9
4.	Permanent posture and other grazing land	4.2
5.	Tree crops and groves	1.4
6.	Culturable waste	5.0
7.	Fallow land (other than present fallows)	3.0
8.	Current fallows	3.6
9.	Net area sown	46.1
	Total	100%

The following table gives availability of geographical area (Ref: FAO production year book -1972), arable and forest area in some countries per head.

Hectares per head							
Country	Geographical area	Arable land	Forest land	Country	Geographical area	Arable land	Forest land
Australia	59.00	3.45	2.68	USA	4.48	0.91	1.46
Canada	45.07	1.98	20.0	Chaina	1.20	0.13	0.12
Russia	9.06	0.92	3.68	India	0.56	0.29	0.12
Brazil	8.65	0.30	5.26	World average	3.56	0.39	1.07

Source: FAO production year book 1972

The cropping pattern in India in 1970-71 & 1950-51 (percentage)		
	1950-51	1970-71
Rice	23.6	22.3
Wheat	7.6	10.8
Jowar	11.8	10.2
Cereals (total)	61.1	60.7
Pulses	15.6	13.7
Food grains (total)	76.7	74.4
Sugarcane	1.5	1.6
Fruits (total)	0.5	0.9
Vegetables (total)	1.1	1.3
Oil seeds	8.3	8.8
Cotton	4.3	4.7
Jute	0.4	0.4
Tea	0.2	0.2
Coffee	0.07	0.1
Irrigated area under food grains m ha (%) 1970-71 m ha % 1950-51 m ha%		
Food grains	30.6 (79.3)	18.3 (81.3)
Non-food grains	8.0 (20.7)	4.3 (18.7)
All crops	38.6 (100)	22.6 (100)

3.6   AGRICULTURE

India has three crops seasons.

(a) **Kharif:** Monsoon season crop. Sowing –June/July. Harvest-Sep/Oct. Main crops-Rice, Jowar, Bajra, Ragi, Maize, Cotton, sugarcane, soybeans, Groundnut, Jute.

(b) **Rabi:** Post monsoon season crop. Sowing Oct/Dec; Harvest-April/May. Main crops-Wheat, Barley, Peas, Rapeseed, Mustard, gram, Jowar.

(c) **Zayad crops:** Sowing-March/June. Main crops-watermelon, vegetables, Moong etc.

(d) **Commercial crops:** Oil seeds, sugar crops, fiber crops, Narcotic crops, Beverage crops. These crops raised for trade purpose and not for self-consumption of farmers.

3.7   DEFINITION OF TERMS USED IN LAND UTILIZATION STATISTICS

1. **Forest:** This includes all classified as forest under any legal enactment. The area of crops raised in the forest and grazing lands or the area open for grazing within the forests should be treated as forest area.

2. **Land under non-agricultural use:** This includes all lands occupied by buildings, roads & railways or under water [eg. Rivers, canals and other lands put to uses other than agriculture].

3. **Barren and uncultivable land:** This category covers all barren and uncultivable lands. It includes mountains, deserts etc. which cannot be brought under cultivation [which may be isolated blocks or within cultivated holdings].

4. **Permanent pastures and other grazing land:** This category covers all grazing lands whether they are permanent pastures or meadows or not. Village commons and grazing lands are included under this category.

5. **Miscellaneous tree crops and groves, not included in the net area sown:** This class includes all cultivable land which is not included under the net area sown, but is put to some agricultural use. Lands under casuarina trees, thatching grass, bamboo bushes and other groves used for fuel etc, which are not included under orchards are classified under this group.

6. **Culturable wasteland:** This class includes all lands available for cultivation, whether taken up for cultivation or not taken up for cultivation once, but not cultivated during the current years and the last five years or more in succession. Such lands may be either fallow or covered with shrubs and jungles which are not put to any use. Land once cultivated but not cultivated for five years in succession shall also be included in this class after five years.

7. **Current fallow:** This class comprises cropped areas which are kept fallow during the current year only. For example, any seeding area is not cropped again in the same year is treated as current fallow.

8. **Other fallow land:** This category includes all lands which were taken up for cultivation but are temporarily out of cultivation for a period not less than one year and not more than five years. The season may be (a) poverty of the cultivator, (b) in adequate water supply, (c) drought, (d) silting of canals & rivers or, (e) un-remunerative nature of farming.

9. **Net area sown:** This means the net area sown under crops and orchards, counting areas sown more than once in the same year only once.

3.8 SOIL EROSION

The wearing away of land surface by the action of natural agencies such as water and wind is called soil erosion. In this process soil fertility is lost.

Types of soil erosion in India is given below

1. **Geologic erosion** (also called normal erosion): Geological erosion takes place very slowly (at ages) and steadily. This may change major features of the earth's surface. The changes are always in equilibrium between the removal and formation of soil. The mature soil preserves character indefinitely at a constant depth.

2. **Accelerated soil erosion:** The removal of the surface soil from bare areas takes place by human and animal interference. This process takes place at a much faster than the soil-forming process. This is called accelerated erosion. It rapidly ravages the land.

   Natural process takes about 100 to 1000 years to build 2.5 cm of top soil, but the anthropogenic process may take only few years to destroy.

3. **Wind erosion:** During strong surface wind top surface fertile soil is blown off as dust storm or dust devil. The sub-soil over that place is exposed which is of less productive. In arid and semi arid areas (like Rajasthan), where vegetation cover is comparatively very less but aids winds to be stronger. As a result surface soil (humus) is blown off as dust-storm when the dust bearing wind slows down, coarser soil particles are deposited in the form of dunes. Thus results in loss of fertile soil. It becomes unproductive or desertification, that is unfit for cultivation.

4. **Water erosion:** Water erosion (of soil) caused in three ways: (a) Sheet erosion, (b) rill erosion and (c) gully erosion.

   (a) *Sheet erosion*: During rainy season surface water run-off is common. Surface water runoff removes top sheet of soil cover from large areas, often from entire fields uniformly. This type of erosion is insidious, slowly deteriorates soil. Its existence can be detected by the muddy colour of the run-off from fields.

   (b) *Rill erosion*: If sheet erosion is continuous & unchecked, the silt laden run-off forms finger-shaped groves over the field. This creates thin channeling which is called rill erosion.

   (c) *Gully erosion*: When rill erosion is prolonged, the tiny groves transform into wider and deeper channels. This is called gully erosion. The gullies deepen and widen with every heavy rainfall event. In course of time large fields convert into small fragments which make them unfit for cultivation.

Factor influencing erosion: (i) Precipitation (ii) slope steepness (iii) soil type and (iv) nature of ground cover and land use.

(d) *Landslides or slip erosion:* A land slide is an outward and downward movement of materials along the slope. The materials may be natural rocks, soil, artificial fills. Land slide depends on topography of the region, geological structure, the kinds of rocks. The cause of slide may be an earthquake or a heavy rainfall /snowfall over ground structure/road. Himalayan region, is susceptible to such slides.

(e) *Stream-bank erosion:* In the event of very heavy rain/snowfall, sudden and violent flow may take place. These flows carry enormous quantities detritus of boulders, shingle, sand and silt (of region/terrain). These debris get deposited in the torrent bed and widening channel. As channel emerges from the hills it uproots trees, moves boulders, it overflow the channel banks and cause erosion.

3.9 SOIL FERTILITY

Plant nutrients are essential elements. These elements are carbon (C), hydrogen (H), oxygen (O), nitrogen (N), phosphorus (P), sulphur (S), potassium (K), calcium (Ca), magnesium (Mg), iron (Fe), manganese (Mn), zinc (Zn), copper (CU), molybdenum (Mb), boron (B) and chlorine (Cl). Green plants obtain these elements from soil and air (viz C, N & H) N, P and K are primary nutrients; Ca, Mg & S are secondary nutrients; Fe, Mn, Cu, Zn, B, Mb and Cl are trace elements or micronutrients. The primary and secondary nutrient elements are called major elements.

It is known that no two soils are alike either in respect of their nature or in respect quantities of plant nutrients they contain. The productivity of soils crop yields depend on soil (nutrients availability). The various crops remove different plant nutrients from soil. The materials which are commonly used to maintain and improve soil fertility by use of manures and fertilizers.

The fertility of soil is measured by its ability to support plant life. The productivity of soil is a function of availability of water, air and nutrients. The nutrient of soil is reduced by soil erosion, over cropping and by leaching of mineral nutrients. To supplement the soil deficiency for plant growth we use fertilizers. Essential features of fertilizers-nitrogen present in the fertilizer must be readily available to the plant, soluble in water, converted into a form (in rain water) which the plant easily assimilates.

Types of fertilizers: Mainly 5 types of fertilizers depending on the presence of N, P & K.

(i) *Nitrogenous fertilizers*: These fertilizers mainly supply nitrogen to plants/soil. Ammonium sulphate [$(NH_4)_3$ SO_4), calcium ammonium nitrate, calcium cyanamide, urea etc, are examples of nitrogenous fertilizers.

(ii) *Phosphate fertilizers*: These provide phosphorus to the soil. Ex: superphosphate of lime, phosphate slag, triple superphosphate etc.

(iii) *Potash fertilizers*: These supply potassium to plants/soil, which are useful to plants particularly grass, tobacco, cotton, coffee, potatoes and corn. The common fertilizers that occur in nature are KCl, KNO_3, K_2SO_4.

(iv) *N P fertilizers*: These contain manly nitrogen and phosphorus. Ex: dihydrogen ammonium phosphate (NH_4 H_2 PO_4), calcium superphosphate nitrate [(Ca (H_2 PO_4), 2Ca (NO_3)] and ammoniated phosphate sulphate [(NH_4) H_2 PO_4, $(NH_4)_2$ SO_4]

(v) *NPK or complete fertilizers*: These fertilizers supply all the three essential elements N, P & K to the soil. These fertilizers produce much better crops. NPK fertilizers are produced by Zuari Agrochemicals Ltd, Goa.

Some common fertilizers

1. Ammonium sulphate or Sindri fertilizer (NH_4) SO_4. Manufactured at Sindri Fertilizer factory, Sindri, Bihar fertilizer factories at Sindri, Durgapur, Berhampore, Jamshedpur, Bhilai, Alwaye Bansjora, Digboi and Hanumangarh, are manufacturing (NH_4) SO_4.

   These fertilizers contain 24-25% NH_3 which is converted by nitrifying bacteria present in the basic soil into nitrates.

2. Calcium ammonium nitrate [CAN] or Nitrolimestone or Nangal fertilizer $Ca(NO_3)$. NH_4NO_3, Manufactured at Nangal in Punjab and Rourkela. It contains about 20% nitrogen. It is directly assimilated by plants. It is highly soluble in water.

3. Basic calcium nitrate or norwegion saltpeter or nitrate of lime, $Ca(NO_3)$, CaO.

   Basic calcium nitrate is hygroscopic and is very soluble in water. It is basic in nature. It is directly assimilated by the plants.

4. Calcium Cyanamide or nitrolin $CaCN_2$.

   It is derivative of cyanamide (H_2CN_2). $CaCN_2$ is moderately soluble in water. It added to the plants before sowing below the surface of soil and not when they are actually growing. In the soil it undergoes a series changes and ultimately produces NH_3. Ammonia thus produced is converted into nitrates by the nitrifying bacteria, which are assimilated by plant.

5. Urea or carbamide NH_3CONH_2. Urea is the diamide of carbonic acid. Urea is an important industrial chemical. It is used as a solw acting nitrogenous fertilizer and in the manufacture of several well-known drugs e.g. burbiturates. Urea is also end product in the metabolism of proteins in human body. It is produced in the liver indirectly from ammonia and CO_2 and excreted is the urine. A normal adult excretes about 30 grams urea per day. The name urea is the result of the occurrence of this compound in urine.

Uses: Urea is used:
   1. In agriculture as a fertilizer, has nitrogen content 46.6%
   2. As a protein supplement in cattle feed
   3. In the manufacture of urea-from aldehyde resins,
   4. Urea is not subject to fire or explosive hazards (as is ammonium nitrate)
   5. As a stabilizer for nitrocellulose explosives,
   6. In the manufacture of barbiturates-a group of drugs which induce sleep,
   7. In the manufacture of non-ionic detergents

6. Superphosphate of lime or calcium superphosphate [Ca $(H_2PO_4)_2$ + 2 $(CaSO_4.2H_2O)$]

Superphosphate lime contains about 16-20% of P_2O_5. It provides phosphate in a water-soluble form, which is readily assimilated by plants for their growth. There are several public sector private sector factories in India which manufacture superphosphate of lime.

3.10   NATURAL RESOURCES – (B) WATER RESOURCES

Water is a renewable natural resource. Water resources can be divided into two categories. (i) Surface water and (ii) Ground water. Surface water resources are: Rivers, Lakes, Swamps, Glaciers and sea water. Underground water resources are: Wells and Spring water.

Rain water is the purest form of natural waters. However it contains traces of dissolved material & gases compounds. Rivers, lakes, swamps waters are all accumulated rain water. These have some dissolved minerals and suspended matter. Sea water contains dissolved salts about 35 gm/Kg or 3.5% by weight. Six ions of sea salts together will be more than 90%. They are ions of chloride (55%) Sodium (30%), Sulphate (8%), Magnesium (4%), Calcium (1)% and Potassium (1%).

Water, in the form of vapor, is exchanged between sea and air. More than 80% of water vapor in the atmosphere comes from the evaporation of sea surface water. Along with water vapor, energy also transferred from sea to atmosphere in the form of latent heat of water vapor. On condensation it releases latent heat. Condensed water returns to the sea in hydrological cycle, but energy remains in the atmosphere. Part of the heat energy is converted to mechanical energy (wind). Two-thirds of the precipitation on land returns to the atmosphere by soil evaporation and plants. The remaining one-third precipitation either percolates into soil or runs off on the land surface. The bulk of the fresh water on land is stored in the form of ice in glaciers (volume 24×10^6 km^3) which is about 20000 times worlds river waters. Ice melted water flows into rivers and lakes. Water stored in lakes (750×10^3 km^3) is about 625 times the water in river. This water is of great importance in many parts of the world. They are the fresh water source for cities and agriculture. Because of this importance artificial lakes are created by damming rivers.

Percolation of water through large pores into the soil is called gravity water. This gravity water accumulates at some depth over rocky layer and fills all cracks and pore spaces. The level below which soil and rocks are saturated with water is called **water table**. Water below water table is called **ground water**. The depth of water table varies from place to place and with seasons. In dry season the depth will be more than in wet season. During good rains water tables slowly increases and this results in ground-water flow into nearby steams. When there are persistent heavy rains, stream channels will be overflowing into adjacent areas. This causes floods. Surface run-off of water depends on precipitation and varies with intensity of precipitation. Ground water flow on the other hand is more steady. The ground water (volume = 60×10^6 km^3) is about 50,000 times the volume of river water.

3.10.1   HYDROLOGY

Hydrology may be defined as the science that deals with the processes governing the depletion and replenishment of surface water and ground water resources of the earth. Consequently it may be viewed as part of physical geography and hydrometeorology. Hydrology includes the study (of water) of rivers, lakes reservoirs, marshes and accumulations of moisture in the form of snow cover, glaciers and ground water. Hydrometry is the technology of water measurement spanning all aspects of water movement within hydrological cycle.

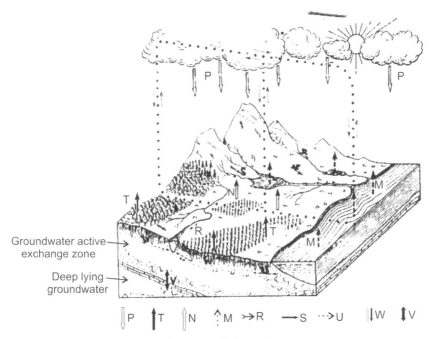

Fig. 3.1 Hydrological cycle

Importance of water: All life on earth is sustained by water. It is the second most vital matter for life after air and then follows food, clothing, shelter etc. Man can survive without food for a few weeks, but without water he cannot survive even a few days/week and without air even a few minutes. It is estimated nearly 70% of man (by body weight) is water. Water is inorganic matter and does not provide energy to tissues. Much of water in body is in the form of body-protoplasm and in spaces between cells. Plasma (fluid part of the blood) contains 91 to 92 per cent water. Water is essential for food digestion, blood circulation and removal of waste products from body. Man's average daily loss of water is 2500 CC. Of which 1500 CC as urine, 500 CC in perspiration, 400 CC in exhaled air and 100 CC in faeces. Any loss of water more than 250 CC (about 10%) is fetal. On an average man drinks about 1500 CC of water and the remaining he gets it from food. Fruits and vegetables contain about 80% water. Kidneys remove excess water along with cellular waste.

Plants absorb minerals, salts dissolved in water from soil through roots. Plants retain only 1% of water absorbed by their roots and the remaining 99% is lost in the form of vapor/aerial parts.

Water acts as a reagent in photosynthesis as a universal solvent. It is let off as a product in respiration. In plants water absorptions mainly takes place through root hairs.

Water has maximum density at 4 °C, and its density at other temperatures given below.

Temperature °C	4	0	10	20	30
Water density	1.0000	0.999878	0.99739	0.99827	0.99577

On a global scale the fresh water is only 2.5%. Of this, ground water is depleting at a fast rate and other sources of fresh water are threatened by pollution. In recent times, man has been mining ground water for agriculture, domestic and industrial use. This is depleting ground water resource which was reliably dependent through the ages. Unknowingly man has made many blunders in water use and its conversation. The worst part is pollution of lakes and rivers. Pollution is the main factor that is threatening the exhaustion of fresh water resources. One cubic meter of swage dump contaminates more than 12 cubic meters of clean water in rivers and lakes. Use of polluted water is a severe health hazard and is the most concern of mankind. Experts are of the opinion that water famine may arise only through inefficient use of natural water resources rather than the lack of water in the world. Further they concluded that the water resources existing on the earth are fully sufficient to meet all the growing needs of man for an indefinite time period provided man must practice to avoid water pollution, extend recycling of suitable water that is used in agriculture, muncipalities and industries. Consumption of water is a sign of country's economy and civilization, more use means more developed. A country economic development largely depends on the water resources and its geographic distribution.

3.11   WORLD'S WATER

[Ref. WMO No.3643 Compendium of Meteorology General Hydrology]

1. World's surface area = 510 million sq km (WSA)
2. World's oceans and seas area = 361 million sq km. (70.8% of WSA)
3. Land part of earth's surface area = 149 million sq km (29.2% of WSA)
4. Area of continental water bodies area = 4.47 million sq km (3% of land area)
5. Area occupied glaciers = 16.4 million sq km (11 % of land area) NH 39% + SH 19% = 58% surface area NH = Northern hemisphere SH = Southern hemisphere
6. Amount of water in world's oceans and seas = 1370 million cubic km
7. (i) Water in earth's lakes = 750000 km^3 (ii)   Water in earth's rivers = 1200 k^{m3}

Total fresh water on earth = 751200 km^3

The distribution of water over the earth's surface given in the following table

Total amount of water in the world

Nature of water	Volume (in km^3)	Percentage of total amount of water
1. Total amount of water in the world	1455×10^6 km^3	100
2. Water in the oceans and seas	1370×10^6 km^3	94.2
3. Fresh water in the world	85×10^6 km^3	5.8
of fresh water in rivers		
(i) Rivers	1.2×10^3 Km^3	0.0001
(ii) Lakes	750×10^3 km^3	0.05
(iii) Glaciers (11% of land area cover)	24×10^6 km^3	1.65
(iv) Groundwater in	60.065×10^6 Km^3	4.11
(a) zones of in groundwater in active water exchange	4.065×10^6 Km^3	0.27
(b) Other groundwater in the earths crest	56×10^6 Km^3	3.84
4. Water in the atmosphere	14×10^3 Km^3	0.001

An estimate of annual water balance of the world by MI LVOVICH is given below

	Volume in Km^3	Average depth in mm
(a) Over periphery of land (Area = 116.8×10^6 Km^2)		
(i) Precipitation	106000	910
(ii) River discharge	41000	350
(iii) Evaporation	65000	560
(b) Inland Area = 32.1×10^6 Km^2		
Precipitation	7500	238
Evaporation	7500	238
(c) World ocean (area = 361.1×10^6 Km^2		
Precipitation	411600	1140
Inflow of river water	41000	111
Evaporation	452600	1251
(d) World total (Area = 510×10^6 Km^2)		
Precipitation	525100	1030
Evaporation	525100	1030

According to another estimate, the volume of earth's water is about 1386×10^6 Km^3. Of this only about 2.5 % (34.65×10^6 Km^3) is fit for consumption (that is fresh water). Of this fresh water about 68.7% (23.80×10^6 Km^3) is present in glaciers and ice caps, and about 30.1 %

$(10.43 \times 10^6 \text{ Km}^3)$ is ground water and the remaining 1.2 % $(0.42 \times 10^6 \text{ km}^3)$ is surface and other fresh water.

In addition to the above estimate, another global water distribution is as below.

(i)   Earth's water volume $1386 \times 10^6 \text{ Km}^3$

(ii) Oceans contain 96.5% $(1337.5 \times 10^6 \text{ Km}^3)$

(iii) Other saline water 0.9% $(12.5 \times 10^6 \text{ Km}^3)$

(iv) Fresh water 2.5% $(34.65 \times 10^6 \text{ Km}^3)$

Of the global fresh water $(34.65 \times 10^6 \text{ km}^3)$

(i)   69% is ground ice and perma frost

(ii) 20.9% in lakes

(iii) 3.8 % in soil moisture

(iv) 2.6 % in swamps and marshes

(v)   0.49 % in rivers

(vi) 0.26 % in living things and 3 % atmosphere

Based on water use, water resource are categorized into three groups.

Group I: Man's use of water in streams and Lakes, like navigation, recreation, hydroelectric power, preservation of fish, wild life and pollution abatement

Group II: Diversion of water for consumption use, like-irrigation, municipal, rural supplies and wide variety of industrial uses.

Group III: Control of water to avoid damage from its excesses (rather than its use). This purpose include-agricultural damage, flood control, urban storm drainage and erosion and sedimentation.

In project planning regionally includes average annual (rainfall & evaporation), evaporation from reservoirs availability of water downstream.

Water uses & care: Increase in population and increase in per capita use, demand more fresh water.

Navigation on inland waterways requires operation of reservoirs and controlled release to maintain minimum flows.

Recreation: Outdoor recreation involves mostly water. It includes hunting, fishing and wildlife preservation. Water quality and pollution abatement is necessary for preserving fish, wildlife and other uses of water.

Water carries waste and dilutes and stabilizes them chemically & biologically.

Solar radiation plays an important role in purification of water.

Hydroelectric power generation depends on availability of water. Similarly, irrigation, heating & air-conditioning and lightning also depends on water availability.

Water supply for municipal, industrial and rural use purposes is drawn from ground (bore wells). Water, from streams or from impounded (in reservoirs, dams etc). All these ultimately depend on precipitation.

India is rich in water resources. In order to feed the increasing population, the available water resources have to be developed. The water demands are mostly for irrigation, human and livestock consumption, expanding industries, hydroelectric power generation, recreation, navigation etc.

The main water resources are: (i) Surface water, and (ii) Groundwater. They are interdependent.

For a given basin, the surface water source depend on: (i) stream inflow (i.e. rainfall), (ii) stream outflow from basin, (iii) loss of evaporation and (iv) the influent recharge to the ground water.

Similarly the groundwater resources of a basin depend on (i) evapotranspiration from the groundwater table, (ii) outflow to the neighboring groundwater basin, (iii) the effluent discharge to the stream and (iv) the addition to the groundwater storage.

The surface water resources that contribute to the groundwater recharge are: (i) the influent recharge from the stream, (ii) the seepage from natural lakes & ponds, (iii) seepage from artificial storage reservoirs, cannal systems etc. and (iv) the returnflow from irrigation.

The above factor contribute to about 25% of the country's total groundwater resources.

The main three factor that affect the water resources are:

1. Climatic factors (i) Rainfall (intensity & duration) and distribution (ii) snowfall & melt (iii) evaporation

2. Physiographic factors (i) geometric factors, like drainage area, shape, slope and steam density (ii) physical factor-land use, surface infiltration and soil types, (iii) channel-carrying capacity and storage capacity.

3. Geological factors (i) Litho-logic (ii) structural and (iii) hydrologic

   (a) Lithologic (including composition), texture, sequence of rock types and the thickness of rock formation.
   (b) Structural (including chief faults & folds)-rock types & bed, joints, fissures, cracks etc.
   (c) Hydrologic-The aquifer (characteristics), permeability, porosity, transmissivity, storage ability.

In short, surface water resource are rivers, lakes, swamps, glaciers & sea water-underground water resources are wells and spring water.

Rain water is the purest form of natural waters but contain traces of dissolved materials & gases compounds. Rivers, lakes, swamps water are all accumulated rain water. These have some dissolved minerals and suspended matter. Sea water contains dissolved salts about 35 mg/kg by weight.

Water, in the form of vapour, is exchanged between sea/ocean and air (atmosphere). More than 80% of water vapour in the atmosphere comes from the evaporation of sea surface water. Along with water vapour energy is also transferred from sea to atmosphere in the form of latent heat of water vapour. On condensation water vapour releases lantent heat. Condensed water returns to the sea in hydrological cycle, but energy is converstied to mechanical energy (wind). About two-thirds of precipitation on land returns to the atmosphere by soil evaporation & plants. The remaining one-third precipitation either percolates into soil or runs-off on the land surface. The bulk of the fresh water on land is stored in the form of ice in glaciers (volume 24×10^6 Km^3) which is about 20000 times the worlds river water. Ice metted water flows into rivers & lakes. Water stored in lakes (about 750×10^3 Km^3) is about 625 times the water in rivers.

Gravity water: percolation of water through large pores into the soil is called gravity water. The gravity water is accumulated at some depth over rocky layer and fills all cracks and pore spaces. The level below which soil and rocks are saturated with water is water-table. Water below water table is called **ground water**. The depth of water table varies from place and with season. During good rains water table slowly increases and this results in groundwater flow into nearly streams.

When there is persistent heavy rains stream channels will be over flowing into adjacent area. This causes floods. Surface run off of water depends on precipitation and varies with intensity. Groundwater flow, on the other hand is more steady. The ground water volume is 60×10^6 Km^3, which is 50000 times the volume of river water.

3.12 SURFACE WATER RESOURCES ESTIMATES FOR INDIA

Actual gauge discharge records and river run-off are incomplete. Keeping this in view Dr. A.N. Khosla, using IMD records of rainfall, temperature etc. (this recorded data is available for more than 100 years) estimated run-off using the following empirical formula.

$$R_m = P_m - I_m, I_m = 0.481 \ T_m$$

Where R_m = monthly run-off (cm)

P_m = monthly precipitation (cm)

I_m = monthly evaporation losses (cm)

T_m = mean monthly temperature ($^\circ$C)

Dr. Khosla's mean annual run-off of six water resource (viz Indus, Ganga, Brahmaputra, West flowing rivers, east flowing rivers and Luni, Ghaggar basins) is 167.23 million ha m (ha = hectare, m = meters).

According to Central Water and Power Commission, the Irrigation commission (1972) the water resources of India is 178 m ham, which includes surface run-off and ground run-off. Of this, utilizable flow is 66.60 m ha m.

Region-wise surface water resources of India

S.NO	Name basins	Catchment area (m ha)	Annual average precipitation (cm)	Total pptn (m ham)	Mean annual temperature ($^\circ$C)	Average annual run-off (m ha m)
1	Indus	35.40	56	19.82	12.6	7.94
2	Ganga	97.60	111	108.34	16.8	48.96
3	Brahmaputra	50.62	122	61.76	8.2	38.08
4	West flowing rivers	49.16	122	59.98	25.5	31.06
5	East Flowing rivers	121.03	109	131.92	26.1	41.19
6	Luni & Ghaggar	16.80	29	4.87	26.2	-
	Total	**370.61**	**549**	**386.69**	-	**167.23**

In India, the irrigation projects are designed for 75% dependability, so that the designed quantity water availability for at least 75 % of the year. Irrigation commission (1972), estimated that the total utilizable water flow 66.6 m ha m, which is given in the following table.

Region wise the total available and the utilizable surface water resources

S. No	Name of basin	Average annual run-off m ham	Utilisable flow m ha m
1	Indus	7.70	4.93
2	Gange	51.00	18.50
3	Brahmaputra	54.00	1.23
4	West-flowing rivers	28.80	6.92
5	East flowing rivers	34.80	33.80
6	Luni, Ghaggar	1.70	1.22
	Total	178.00	66.60

Utilization of water resources in India

According to Central Water and Power Commission:

1. Geographical area of India is about 328 mha = $328 \times 10^6 \times 10^4 \text{ m}^2$
2. Average annual rainfall of India is about 112 cm = 1.12 meters
3. Total annual precipitation estimated = $328 \times 10^{10} \times 1.12 \text{ m}^3$
   (mi = million), = 3700000 mi m^3
4. Rainfall during summer monsoon 80% of annual precipitation = $\left(\frac{80}{100} \times 3700000 \right) = 2960000 \text{ mi m}^3$
5. Seepage of water into ground 800 000 mi m^3
6. Surface flow (river flow) of water is about 1, 700, 000 mi m^3
7. Surface (soil) and plant evaporation is about 1200, 000 mi m^3
   i.e. sum of (5) + (6) + (7) = 3 (S.NO)

Scientific utilization of water resource depends suitability of land and water for irrigation. It is then the planning of crops and water management practices, which include irrigation and drainage.

Efficiency of field irrigation, a guide for different soil classes given below.

Type of soil	Irrigation efficiency (%)
Sandy	60
Sandy loam	65
Loam	70
Clay loam	75
Heavy clay	80

The quantity of water required for agricultural operation, as a guideline given below

Operation	Details	Quantity of water (mm)
Land preparation	Light soil	50 to 60
	Heavy soil	70 to 80
Puddling and transplanting of rice	Light soil	300
	Heavy soil	200
	Other crops	75 to 100
Leaching down of salts	Low salinity	10 % of the comparative use
	Medium salinity	20% of the comparative use
	High salinity	25 to 40% of the comparative use

Depth of Irrigation: The quantity of water required for net irrigation on different soil types per meter depth of soil profile at 50% soil moisture availability given below

Soil type	Irrigation depth (mm)
Sandy	25 to 50
Sandy loam	45 to 80
Loam	70 to 110
Clay loam	80 to 120
Heavy clay	100 to 140

3.13   GROUNDWATER RESOURCES

The assessment of groundwater resource is much more difficult. The estimate of average contribution to groundwater recharge in the country as a whole is 10% of the total rainfall but 12.5% considered as more realistic.

Total annual volume of rainfall is about 400 m ham, assenting a mean rainfall 120 cm over the geographical area 328 m ham. With this assumption ground-water recharge is about 50 m ham. Considering other aspects like ground water run-off evapotranspiration, the ultimate usable groundwater resource assessed at 35 m ham.

3.14   UTILIZATION OF WATER RESOURCES

Water required to live life on earth. The scientific utilization of water resources depend on crop, which in turn depend on land & water. Water management include irrigation and drainage. Irrigation comprises-when to irrigate, how much to irrigate and how best to irrigate. Similarly, drainage comprises-how much to drain, how best to drain and how rapidly to drain, under a given situation of soil, water and crops.

According to the National commission on Agriculture, by 2050, the requirement of water will be 25% more than the availability from water resources. To meet this need, there is urgent emphasis on conservation by efficient utilization of available water resources. Further, the degradation of environment through pollution, there is urgent need for artificial recharging, recycling of used water and even desalination of sea-water.

According to IPCC 2007, the per capita availability water in India in 2001 record was 1820 m^3 per year. It is now projected, it will decline to 1140 m^3/year by 2050. It is further warned that India could reach state of water stress before 2025 when the availability of water per capita could be 1000 m^3/year, this fall would be the result of both climate change and population growth.

Indian rivers altogether carry 1880×10^9 m^3 of water every year. Of this total rainfall 53% is lost as evapotranspiration and retention in soil. Remaining 47% flows in rivers. Of this only 28% is available for use and the rest lies in inaccessible places or turned brackish.

3.15   CRYOSPHERE OR ICE SPHERE

Forms of Ice: 1. Seasonal snow, 2. Continental ice sheets 3. Sea ice 4. Mountain glaciers:

On the time scales of: seasons to million years. Cryosphere influences the properties of atmosphere-earth interface significantly.

Distribution of global water: 1. Sea water 97.4% 2. Frozenice/snow 2.0% 3. Ground water (at great depths) 0.5% 4. Fresh water (non-saline) 2.6% (= 2 + 0.5 + 0.1), 5. River and lake waters (surface water) 0.1% 80% of fresh water is found in ice/snow.

Perennial ice covers 10% of the earth's surface.

Earth's ocean surface ice cover 5%

Seasonal snow estimated based on satellite data

(Both land and sea)	Area unit 10^6 k^{2}m		
	NH	SH	Globe
January	58.4	18	76
April	41.2	18	59
July	14.3	25	39
October	22.8	34	57

Total area of earth's surface 510×10^6 Km^2.

Area covered by ice-snow cover annually 59×10^6 Km^2 (or 12%). Of this 12% ice-snow cover, year to year variation 10%. A discontinuous snow cover along with vegetation cover, lakes & micro land relief.

3.16   WATER USES

On a global scale, about 2.5% of water on earth is fit for use (i.e. fresh water). Of this, groundwater is fast depleting and other fresh water resources are also threatened.

In India, fresh water resource = 4000 b m^3

[b = billion = 10^9, m^3 = cubic meters, 1m^3 = 1000 liters

Of this fresh water (4000 bm^3):

1047 b m lost in evaporation

1084 b m^3 non-available water

1869 b m^3 available water.

Of the available water 1123 bm^3 usable water.

Of the usable water, 395 b m^3 groundwater, 728 b m^3 surface water.

India's water demand: In 2006, 829 bm^3, and estimated demand in 2025 is 1093 bm^3 and in 2050 it would be 1047 bm^3. Demand decreases with population decline and water use efficiency increase.

Water use shares (%) in	India	World	Europe
Agriculture	83	69	33
Industry	12	23	54
Domestic	5	8	13

India consumes most of its usable water for agriculture (83%) purposes, while household only 5%.

Global water: Earth's water 1386 million km^3. Only 2.5% of earth's water (34.65 mi km^3) is fit for consumption. Of this fresh water (34.65 mi km^3), 68.7% is present in glaciers & ice 30.1% ground water and 1.2% surface & other fresh water.

Distribution of global water: Earth's water 1386 mi Km^3.

Oceans contain 96.5%,

Fresh water 2.5% [2.5/100 × 1386] = 34.65 mi km^3 and other saline water 0.9%.

Of the global fresh water: 69% ground ice & perma forest, 20.9% lakes; 3.8% soil moisture; 2.6% swamps, marshes; 0.49% rivers; 3% atmosphere and 0.26% living things

CHAPTER - 4

Environmental Degradation and Protection

From the dawn of civilization man has been trying to use the natural resources for making life comfortable. Science has helped him to understand the nature of animate and inanimate objects around him. This and scientific knowledge helped him to understand the physico-chemical laws that govern the behaviour of living and non-living objects and their inter-relationship. The beneficial applications of science to humanity is called Technology. The application of technology in industry largely modified the life of man in twentieth century. Along with benefits, technology has also brought decline in the environmental quality and threatening to affect available natural resources for the present and future generations.

Besides causing great damage to the health of living beings, it has threatened the extinction of some species of life. Thus technology proved both boon and bane to mankind and life on earth.

Industrialization contamination lead to the pollution of water, air and soil and environmental degradation and effecting the human health and biological diversity. In order to prevent further decline, world community resolved to protect the environment. In this direction the United Nations Conference on human environment was held in Stockholm in June 1972. The conference recommended to enact a comprehensive law to take action for environmental protection. As a result Govt. of India introduced Environmental Protection Bill and was passed in both houses of Parliament, received the assent of the President on 23rd May, 1986. It came on the statute books as The Environmental (Protection) Act, 1986.

The Act (1986) aims at the protection and improvement of environment and prevention of hazards to human beings and other living creatures, plants and property.

Def: *Hazard:* Any event that is likely to cause death to life and destruction to property is called hazard.

Ex: A thunderstorm is a hazard (in aviation) but becomes a disaster when, it effects life and property causing death and destruction.

Disaster: Disaster means a catastrophe, mishap, calamity or grave occurrence in any area, arising from natural or man-made causes, or by accident, or negligence which results in substantial loss of life or human suffering or damage to, and destruction of property, or damage to or degradations of, environment and is of such a nature of magnitude as to be beyond the coping capacity of the community of the affected area.

Natural hazards have no political boundaries and no one can stop them. The hazards are associated with geological, atmospheric, hydrological, ecological and biological events. Geological hazards are earthquakes, Tsunamis, Volcanoes, Atmospheric (including hydrological) cyclones, hurricanes, typhoons, floods droughts, forest fires, deluge etc.

Note: According to WHO report, about one-fourth of global deaths are caused by environmental risks, like polluted air, dirty water, hazardous work places and dangerous roads. The WHO estimates 12.6 million deaths in 2012 or about 23% of total world deaths were attributed to such environmental factors. The burden is greatest on the poor and the youngest. Majority from environmental risk is highest in sub-Saharan Africa and Low and middle income countries in Asia.

4.1   DISASTER MANAGEMENT

According to WHO, "A disaster is an event that causes damage, economic disruption, loss of human life and deterioration of health and the health of societies on a scale sufficient to warrant an extra-ordinary response from outside the effected community or area".

Most of the severe floods are caused by the rivers that originate in the Himalayas. To combat floods the Rashtriya Barh Ayog was established in 1978, Task Force on flood management and Erosion control in 2004. The primary responsibility of flood control rests with State Governments. The responsibility of NDMA (2005) is (i) prevention, (ii) preparedness, (iii) mitigation, (iv) rehabilitation (v) reconstruction (vi) recovery and (vii) formulation of appropriate policies and guidelines for disaster management.

Flood management includes structural measures like flood embankments, dams, reservoirs, channel improvement, drainage improvement, diversion of flood waters, afforestation, catchment area treatment and anti-erosion works. Flood management courses being introduced in professional institutions.

On an average annually 5 to 6 cyclones form in the Indian seas of which 2 to 3 become severe. East coast is more vulnerable to cyclones than west coast of India. 13 coastal States and Union Territories with 84 coastal districts are directly affected by cyclones. India Meteorological Department provides four stage cyclone warnings. Under NDMA, National Cyclone Risk Mitigation Project (NCRMP) drawn up to mitigate cyclone risk in all coastal districts. The project is being financed by World Bank assistance. NDMA shall arrange: (i) Mass awareness, (ii) mock drills for community capacity building to combat cyclone disaster. Government of India undertook a comprehensive Disaster Risk Mitigation (DRM), a programme with UNDP in 169 hazard prone districts to train the people at grass root level.

According to Atlas prepared by the Building Materials Technology Promotion Council (BMTPC), 229 districts from 21 States and Union Territories, comprising 58.6% of geographical area of India is vulnerable to moderate to severe earthquakes. The NDMA proposes to launch the National Earthquake Risk Mitigation Project (NERMP) during the 11th five years plan, to identify critical gaps and formulate guidelines for management of earthquakes. The NDMA and the Government of India seek the support and cooperation of various stakeholder groups, NGOs and common people, in ensuring for the institutionalization of earthquake risk reduction initiatives in the country.

According to a UN report the incidence of natural catastrophic have increased from about 100 per year in 1975 to 395 in 2006. It shows the urgency of an institutionalized mechanism for disaster management with a well organized elaborate medical preparedness in every country. In view of recent chemical tragedies and terrorism explosions, the medical preparedness for CBRN (Chemical, Biological Radiological and Nuclear) management requires Standard Operating Procedures (SOPs) for CBRN management at the incident site for triage (Sorting), personal protection, decontamination, resuscitation, causality evacuation followed by treatment of exposed victims at the hospital level. The critical infrastructure for medical management includes CBRN causality treatment centers/wards and training facilities for specialized response to deal with covert CBRN attacks. NDMA, Government of India with all stakeholders and experts in the country taking a proactive initiative for institutionalization of medical management by formulating national guidelines for all stakeholders and in the public domain. The main object of medical preparedness is to convert over capabilities in terms of human resource and supportive infrastructure into capacities for quick and efficient medical response.

The estimate annual loss due to natural disasters is about 2 to 3 percent of National income, 85% of the country is prone to natural disasters. It is highest in Southeast Asia. Drought prone area is about 68% of the cultivable area. Flood prone area is about 12 to 13 percent of the geographical area. About 2000 sq. km area in India is exposed to avalanches and landslides. According to US State Departments annual report, global terrorism fatalities increased from about 15000 in 2005 to 20500 in 2006 (about 40% rise).

During 20th century global industrial activity increased 20 to 25 folds, global population grown about 2.5 billion to about 6 billions, fossil fuel consumption increased 30 to 35 folds, water demand increased enormously. All this lead to the destruction of green lungs of the earth, increased and expeditions disposal of cases arising out of any Accident occurring while handling any hazardous substances to give relief and compensation for the loss incurred to the persons, property and environment. An industry or enterprise engaged in activities which are hazardous to the health and safety of persons residing in and around the areas of the factory owes the responsibility to the community that it does not harm in any way on account of hazardous nature of activities. In case any accident the Tribunal decides or sets standards of compensation for damages to human health, property and environment, particularly contamination of air, water, land in the neighbourhood of the industry/factory.

The 1992, Bill was passed by the both the houses of parliament in 1995 received and the assent of the President on 17th June 1995. This is called the National Environmental Tribunal Act 1995.

Def: Accident: An accident involving fortuitous or sudden unintended occurrence while handling any hazardous substance resulting in continuous or intermittent or repeated exposure to death of, or injury to, any person or damage to any property or environment but does not include an accident by reason only of war or radioactivity.

4.1.1 MITIGATION

Means measures aimed at reducing the risk, impact or effect of disaster or threatening disaster situation.

Preparedness: Means the state of readiness to deal with a threatening disaster situation or disaster and the effects thereof.

1. *Environment Includes*:
   (i) Atmosphere
   (ii) Hydrosphere
   (iii) Lithosphere and
   (iv) Biosphere

2. *Environment is (the sum of)* = Animate world + Inanimate world = Biotic environment + Abiotic environment

3. *Environmental Pollution*:
   (i) Natural
   (ii) Anthropogenic or man-made

4. *Types of Environment Hazards*:
   (i) Natural
       Planetary hazards (associated with Earth events)
       Extra planetary hazard (associated with outside extras terrestrial events)
   (ii) Man-made or induced (wars, nuclear hazards)
   (iii) Planetary hazards

Endogenous [volcanoes earthquakes]

Exogenous [cumulative atmospheric] cyclones, land sides floods, droughts etc.

A thunderstorm is a hazard but a heavy thunderstorm together with lighting, heavy rain fall causes loss of life and property (the thunderstorm) becomes a disaster.

A flood is a hazard, but a great flood which causes heavy loss to life or property or both it is a disaster. Similarly an earthquake is a hazard, when it is of slight or moderate intensity (which does not cause extensive damage to life and property) but a severe intensity earthquake which is causes extensive damage to life and property it is a disaster. Thus any extreme hazard event becomes a disaster.

Science is both a boon and a bane to mankind

From the down of civilization man has been trying to use the natural resources for making life comfortable. Science has helped him to understand the nature of animate and inanimate objects around him.

This scientific knowledge helped man to understand the physic-chemical that govern the behavior of living and living (objects) things, and their inter-relationship. The useful/beneficial application of science to humanity is called Technology. The application of technology in industry greatly modified the life main in twentieth century. Along with the benefits Technology has also brought decline in the environmental quality and threatening to affect the available natural resources for the present generation and future eve ration. It also causing great damage to the health of living beings and threatening extinction of some species of life earth.

Biosphere, Ecology and Ecosystem

It is now known that all forms of life is made up of protoplasm and carry identical process but they live in a totally different surroundings (environments). There exists a critical relationship between living things and its physical environment. Biosphere is the area/space near the earth's surface which encompasses all living organisms. This region includes parts of Atmosphere, Hydrosphere and Lithosphere. Environment of an organism is its surrounding (habitat) media. Ecology is the relationship of an organism to its environment. Ecosystem is the functioning of living and non-living components of the environment. The main physical factors of an environment are: temperature, soil condition, inclination of the sun (or light), water, atmospheric conditions and topography.

All living organisms are composed of carbon, Oxygen, Nitrogen, Hydrogen which are also the basic elements of water and air shells of the earth A large part of the living matter contained in green plants, which entrap solar energy and make complex compounds by photosynthesis process.

The main plants source of plant feeding are carbon dioxide and water. Plants use about 2% of insolation (incoming solar radiation) or photosynthesis process, about 7 w/m^2 is consumed by plants.

The total mass of CO_2 in the atmosphere is about 0.023×10^{17} kg, while the mass of CO_2 in the ocean water is about 1.4×10^{17} kg which is more than 60 times the CO_2 in the atmosphere. Sea water plays a pouters role in respect of dissolved natural gases N, O_2, CO_2 and H_2S.

These gases are closely related to living matter on the land and in the sea. CO_2 enters the atmosphere by human and animal breathing, decay and burning of materials containing carbon and volcanic activity. More than 90% of the earth's CO_2 is dissolved in sea water. The stability of CO_2 in changes with temperature.

4.1.2 CLIMATE DISASTERS

The word climatology is derived from the Greek word kilmo, meaning inclination or slope and refers to the angle of inclination/incidence of sun's rays falling on a particular place. The parameters of climate are pressure, temperature, wind (direction and speed) humidity and precipitation, solar radiation. Traditionally long term averages of these parameters together with their extreme values are considered in climate.

The hazardous events like droughts, floods, cyclones tornadoes etc. cause of hardship. Extreme cold witnessed in parts of north America in 1977 and unusually hot summers in the same region in 1980 were also hazardous. The environment abnormal conditions cause great human stress and suffering particularly in old people, children and infirm, which intern leads to significant increase in the fatality rate. Large departures from seasonal and annual of climatic parameters come under the heading of *Climatic fluctuations*. On longer time scales from decades to centuries the earth is subjected to climatic changes. When the average temperature of a whole hemisphere or even the whole globe may increase or decrease by significant amounts. During 1880-1940 (60 years) northern hemisphere warmed up by about 0.5 °C and by 2°c rise in some parts of the far north.

The past civilizations show large climatic changes caused major impacts on human society. Only a few thousand years ago, the present subtropical deserts, such as the Sahara and Rajasthan were much weather than today and were able to support much larger populations. Some 18000 years ago, at the climax of the most recent ice age, much of north America, northwest Europe were covered by sheets of ice. If such large changes that occurred in the part they can surely be repeated in future and even the polar regions could be habituated. On such climatic changes some places may be covered by deserts and thick forests.

If we look at the sun's evolution in the process of nuclear burning of hydrogen indicate that about 4 billion years ago, the radius of the sun was 93% of the present radius and temperature of its surface was 3 to 4% less than present one this means sun's luminosity was 25-30% lower in those days than it is today.

PETM (Paleocene-Eocene Thermal Maxima)-high latitude air and ocean temperature soared by more than 7°c over of few thousand years. According to one study, during the PETM temperature rose by about 23°C (73°F).

During Mesozoic and rose Cenozoic eras (180 million years ago) earth passed through most favorable climatic periods which is called Cretaceous paleogene period.

At that time climate of sub-polar latitudes resembled modern sub-tropical climate. Arctic islands and Antarctic were covered with forests. During late cretaceous period the surface water temperature of Arctic Basin in Alaska area and Siberia was 14°C. The present or very recent glacial conditions about 53% of the surface water of the world oceans temperature varied between 20 to 28°C and 13% area have temperature of 4°C or less (or mean temperature 17.4°C). during the Cretaceous, Pathogen period there were no deserts. Deserts came into existence at the end of Neocene (between Miocene Pliocene epochs). The most arid regions smaller than present day deserts were covered with tropical Savannah, with oases in river valleys. The same plants are found the geological deposits of Greenland, the Bear islands, North America, western Europe, Russia, Australia, Antarctica and Africa.

During the glacial epochs there was considerable increases planetary cooling reduction in condimental humidity. On the contrary interglacial epochs were associated with disappearance of ice sheets from continents (except Antarctica and Greenland) and from the surface of Arctic Basin. During interglacial epochs the thermal conditions of the earth's surface become more favorable than they are today, the aridity and continental character decreased. The Pleistocene was the peak of aridity observed in Russia, Central Asia.

4.2   THE NATIONAL ENVIRONMENT APPELLATE AUTHORITY ACT 1997

In Public Interest Litigation cases involving Environment issues the pronouncements by the Supreme Court had made it necessary to set up an independent body for quick redressal of public grievances. Consequently, the National Environment Appellate Authority Ordinance, 1997 was promulgated by the President. The Authority to hear appeals with respect to restriction of areas which any industries, operations or processes or class of industries, operations or processes shall not be carried out subject to certain safeguards under the Environment Protection Act 1986.

We shall now briefly study about Air pollution and control, water pollution and control, forest degradation and conservation and wild life protection (biodiversity).

4.3 WORLD ENVIRONMENTAL DAY 5TH JUNE

World Environmental Day is observed every year on 5th June, all over the world. It stimulates worldwide awareness about Environment and increases political attention and action. WED aims to give human face to environmental issues, empower people as active agents of sustainable and equitable development. Encourages communications to change attitude towards environmental issues. It advocates partnership for all nations to enjoy secure future. The day is observed by street processions, competitions in schools and colleges, tree planting, recycling and cleaning up campaigns etc.

The earth charter seeks to inspire all people a sense of global interdependence and shared responsibility for the well being of humanity

In connection with World Environmental Day 2017, United Nations Environmental Programme selected a theme, "Connecting people to Nature". This theme implores all people of the world to get outdoors, to appreciate its beauty and also its importance and take forward the call to protect earth that we all share.

4.4 ENVIRONMENTAL ISSUES

During last 150 years, all species of life on earth are affected by environmental issues, which are invading and threatening the air, water and soil pollutions. The main issues are given below.

1. Global warming and climate change
2. Deforestation, weakening the lungs of the earth
3. Energy shortage crisis
4. Environmental pollution
5. Harmful toxic wastes/Radioactive waste
6. Oil spills which harms/extinct some marine life. Plastic dumping in lakes and seas will have adverse effect on marine life
7. Depletion of natural resources
8. World population explosion, 74 million people per year. 2010 world population estimated to be 6909×10^6 and it would be 8012×10^6 by 2025
9. *Nuclear Issues:* Nuclear weapons-dangerous to environmental issues

10. Strengthening world greenery or lungs of the earth-Go Green

11. Globally, phytoplankton absorbs as much as CO_2 as tropical rainforest. It is therefore, very important to understand their response to global warming.

4.5 ENVIRONMENTAL DEGRADATION

Environmental pollution is the product of direct or indirect (i) changes in physical, chemical constituents, (ii) changes in energy use patterns, (iii) radiation levels and (iv) widespread organisms. These changes may affect man directly or through water, agricultural and other biological products. A few important environmental pollutants are given below.

1. Emission of gases like sulphur dioxide and oxides of nitrogen

2. Particulate matter like smoke particles, lead, aerosols and asbestos etc let into the atmosphere.

3. Pesticides and radioisotopes let into the atmosphere and water bodies (rivers, lakes, ponds etc)

4. Letting of sewage, toxic organic chemicals and phosphates into water

5. Adding of solid wastes on land and water ways. Some of the above pollutants are introduced into atmosphere, soils and water bodies naturally and others are introduced by human activities. Major sources of pollutants are:

    (i)   Industrial emissions from chimneys

    (ii) Automobile exhaust.

The other details will be discussed in chapter Environmental pollution:

The effects of particulate matter on human beings are: chronic bronchitis, bronchial asthma, emphysema, lung cancer.

According to WHO report, nearly one fourth of all deaths world over are caused by environmental risks like polluted air, dirty water, hazardous work places and dangerous roads. The WHO estimates 12.6 million death in 2012 or about 23% of total world deaths were attributed to such environmental factors, or environmental beating. The burden is greatest on the poor and the youngest. Majority from environmental risks is highest in sub-Saharan Africa and low and middle income countries in Asia.

According to recent WHO report (2015) at least 600 million people or one in ten worldwide fall ill from contaminated food each year and 420,000 die, many of them children.

Children are especially vulnerable to diarrhoeal diseases, often caused by eating raw or under cooked meat, or eggs. Fresh produce and diary products that are contaminated in same form.

Food borne diseases-caused by bacteria such as salmonella viruses, parasites, toxins and chemicals-mostly cause temporary symptoms such as nausea, diarrhoea and vomiting.

Figures suggest Africa that has more than 91 million people are estimated to fall ill and 1, 37,000 die each year. South-east Asia has 150 million cases, 175000 deaths per year.

4.5.1   AIR POLLUTION CONTROL BOARDS

Central Pollution Control Board (CPCB) constituted, under the 1974 Act, with powers to perform the functions.

Similarly State Pollution control Board (SPCB) constituted under the 1974 Act with powers and functions. SPCB consists of : a chairman, having special knowledge and experience, nominated by state govt. and not exceeding fire members of State Govt. employees, three members at most of non-official members (nominated to represent agriculture, fishery, industry/trade or labour. Besides these members, two nominated members to represent companies corporations (controlled by govt.) and a fulltime member Secretary.

Board must meet at least once in three months.

Powers and functions of boards

1. *CPCB:* The main function of CPCB is to improve the quality of air and to prevent, control or abate air pollution in the country. In particular, to advise Central Govt. to plan and execute nationwide improvement of quality of air, coordinate the State PCBs to resolve disputes, provide technical assistance, guidance, sponsor /investigation and research, organizes training and mass media programme. The CPCB may establish or recognize laboratory/laboratories to perform its functions efficiently.

2. *SPCB:* The principal functions of SPCB is to plan and execute programmes for the prevention, control/abatement of pollution, advise state govt. ways to minimise pollutions, collect and disseminates with information about pollutants, collaborate CPCB and organize training persons, organise mass education programme, inspect industrial plants and give then correct directions assess pollution, laydown pollution standards for industrial emission, automobiles, advise state govt. where to establish industries and other functions entrusted to it (by CPCB). In addition to above establish or recognise laboratories for efficient functioning.

Automobile pollution control: It is now well known automobiles contribute a major part of air pollution particularly in metropolis cities. Govt. may put a cap for vehicles exhaust (like emission of smoke, visible vapour, sparks, ashes, grit or oil) reduce noise by vehicles, plying of vehicles restricted on odd dates with odd vehicle plate numbers and on even dates with even number vehicle plates etc.

Mitigation: Creation of awareness and publicity helps in mitigation or balancing the weather from extremes to moderation. This requires education in masses (general public), policy makers, intellectuals and decision makers. For this purpose all means of communication, books, journals, mass media, films, TV etc have to be utilised.

Environmental changes will have grave impacts on human beings, all living creatures and vegetation.

4.6 ENVIRONMENTAL HAZARDS

In 2002, at world summit on sustainable development, in Johannesburg, world members of UN reaffirmed their commitment to sustainable development. Sustainable development, a collective efforts necessary at local, national, regional and global level which ensures economic, social development and environmental protection.

Biological environment includes parts of atmosphere, hydrosphere and lithosphere (which are briefly presented in different chapters). Environment thus includes air, water and land and human beings, other living creatures, plants, microorganisms and property. Environmental pollution means presence of solid. Liquids or gaseous substance in such concentrations as may be injurious or tend to injurious to environment.

Concept of Hazard: Hazardous substance means any substance or preparation which by reasons of its chemical or physico-chemical properties or handling is liable to cause harm to human beings other living creatures, plants, microorganisms, property or environmental. An introduction to atmosphere, oceans and lithosphere, cryosphere had been given. This introduction creates awareness and understanding of the environment, components and processes and the hazards associated with them. Hazards signifies a danger or put at risk, while disaster is sudden event causing great damage or loss of life or sudden misfortune.

A thunderstorm is a hazards but a heavy thunderstorm associated with heavy lightning, rain, squall etc, that cause loss of life and property loss is a disaster. A flood is a hazards but a great flood which causes heavy loss to property, loss of life becomes a disaster. Similarly an earthquake is a hazard with slight/moderate intensity (which does not cause damage to life and property) but it becomes a disaster if it is of severe or very severe intensity which causes extensive damage to property/environment or loss of life or both is a disaster.

Past experience world over shows it is a futile exercise to try to stop natural hazard events like thunderstorms, cyclones, hurricanes, volcanoes, earthquake, tornadoes etc. The main purpose of studies about these hazard events to mitigate its ill effects by knowledge, technology by creating awareness among masses and timely appropriate retardant actions.

4.6.1 TYPES OF HAZARDS/DISASTERS: NATURAL OR MAN-MADE

Natural hazards may be planetary (terrestrial) or extra-planetary. Extra-planetary hazards/disasters include meteor falls, satellites/sky labs or cosmic body falls. There is no mitigative way/protection for extra planetary hazards/disasters. These are of two types– Endogenous (pertaining to inner layers of the earth) and Exogenous (pertaining to outside the earth). Endogenous hazards/disasters include volcanoes, earthquakes etc, while exogenous hazards/disasters include tropical cyclones, floods droughts, cloudbursts, a under storms, snowstorms, tornadoes, heat waves, cold waves etc.

Man made hazards/disaster include air, water, soil pollution chemical hazards, radiation hazards etc.

According to National Academy of science USA, humans form a minuscule parts of life on earth. Plants outweigh people by about 7500 to 1 (ratio) and makeup more than 80% of the worlds biomass. Bacteria are nearly 13% of the worlds biomass, fungi, yeast, mold and mush-rooms makeup about 2%.

Man is a highest form in natural species the earth/ecosystem. He has developed brain and evolved language for communication. He has developed science and technology, agriculture. He has constructed reservoirs for agricultures and cities, transports, medicines etc, for better/comfortable living. Thus man created new environments and ecosystems. However his inadvertent application of science and technology becoming a bane than a boon to mankind and in general to all life and environment on earth. Man's creation of hazards/disasters include toxic chemical hazards, atomic weapons, nuclear bombs and as a consequence pollution/contamination of air, water, soils.

Of all hazards/disasters, meteorological and hydrological events are very important (which are more frequent). Man achieved proficiency in forecasting weather nearly accurate to a greater. As for other natural events/hazards like volcanoes, earthquakes, tsunamis etc, man is still out of the reach in forecasting. In general if we study the disaster management of cyclones/hurricanes in detail, the other disaster management emulate to a larger extent. Management of any hazard event without adequate knowledge will not yield expected favorable results.

Keeping this point in view and experience more stress is giving on cyclone disaster management which is of at most important world over.

4.6.2 THE CHEMICAL WEAPONS AND WEAPONS OF MASS DESTRUCTION

The organisation for prohibition of chemical weapons (OPCW) is a Hague-based international body that works for the elimination of chemical weapons. It was formed after the chemical weapons convention (an arms control treaty that bans the production, stock piling and use of chemical weapons) entered into force in 1997. The OPCW is the implementing body of the convention. All its 192 member states are required to destroy the existing stockpiles of chemical weapons and stop large scale production. These actions are subject to verification by the OPCW. According to the OPCW, over 96% of the worlds declared chemical weapons stockpiles have been destroyed. OPCW has powers to inspect production facilities. Israel is a member state but Egypt, South Sudan and North Korea are not members. OPCW noted chemical weapons were used in Syria's Ghouta (in 2013) and Khan Sheikhoun (in 2017). A chemical attack over north western province of Idlib (4 April 2017) in Syria killed about 100 people and wounded more than 500 people. The aerial attack caused many people to choke and some (people) to foam at the mouth. Sarin gas, killed hundreds of civilians in Ghouta (near Damascus) in August 2013. OPCW found particles of chlorine, sulphur, mustard gas and Sarin in the Syrian war. Use of chemical weapons may cause large scale death and disabilities like the Bhopal gas tragedy. The central nodal ministry that looks after (management) in such disasters in India is assigned to *Environment and Forest Health and Family Welfare.*

India is a signatory to the Chemical Weapons Convention held in Paris (1993). It was enacted by parliament of India in 2000. Act is called the Chemical Weapons Convention Act 2000. This Act extends to the whole of India and apply to the citizen of India both inside and outside India. This Act was amended on 11 September 2012 and came into force on 23rd November 2012 vide so 2776 (E) published in the gazette of India.

4.6.3 PROHIBITION OF CHEMICAL WEAPONS

According to the amended Act 2012: No person can: 1. develop, produce and acquire; stockpile retain or use chemical weapons or transfer, directly or indirectly to any person; 2. use riot control agents as method of warfare; 3. engage in any military operations to use chemical weapons; 4. assist, encourage or induce in any person to engage in (a) the use of any riot control agent as a method of welfare, (b) any other activity prohibited by a state party under the convention.

(A) **Chemical Weapons** mean include all the toxic chemicals and their precursors (except some stated), munitions and devices, specially designed for direct use of munition.

Toxic Chemical means, any chemical which causes death or temporary/permanent harm to humans or animals.

Precursor means, any chemical reactant which takes part as a toxic chemical.

The prohibition also includes to toxic chemical precursors or direct organic chemical (containing elements of phosphorus, sulphur or fluorine), chemical weapons, old chemical weapons, Abondoned chemical weapons, riot control agents, chemical weapons production facilities etc.

Note: Chemical used in industries, agriculture, research, medicine, pharmaceutics etc. are not prohibited. Chemicals used for military purposes are not included in prohibition.

(B) **Biological Weapons** These are (i) microbial or other biological agents or toxins, (ii) weapons equipment or delivery system specially designed to use such toxins for hostile purposes or in armed conflict.

(C) **Nuclear Weapons or Nuclear Explosive Devices** These are any nuclear weapons or nuclear explosive device (determined by Central Govt.)

(D) **Weapons of Mass Destruction** These are any biological, chemical or nuclear weapons.

4.6.4   PROHIBITION RELATING TO WEAPONS OF MASS DESTRUCTION

No person shall unlawfully (i) manufacture, acquire, process, develop or transport a nuclear weapon or other nuclear explosive device and their means of delivery (ii) transfer directly or in directly to any one a nuclear weapon or other nuclear explosive device or transfer control over such a weapon, knowing it to be a nuclear weapon or other nuclear explosive device (iii) manufacture, acquire, possess, develop or transport a biological or chemical weapon or their means of delivery, (iv) transfer directly or indirectly to any one biological, chemical weapons, or any one missiles specially designed for the delivery of weapons of mass destruction.

No person shall directly or indirectly transfer to a non-state actor or terrorist, any material, equipment and technology, except such transfer allowed lawfully by Govt. of India.

Empowerment of Authorities for Various W.M.D and their Delivery System

Authority	Powers conferred
1. Ministry of External Affairs	(a) To coordinate between concerned ministries and department for the purpose of the Act. (b) To implement provisions of section 11 of the Act
2. Department of Atomic Energy	To implement provisions relating to weapons of the Act
3. Department of Chemical and Petrochemicals, as the case may be, The National Authority under the chemical weapons convention Act 2000	To implement provisions relating to chemical weapons and prosecution of offences of the Act
4. Ministry of Environment Forest and Dept. Biotechnology	To implement provisions relating to Biological weapons and provisions of offences of the Act
5. Director General of Foreign Trade and the Central Board of Excise and Customs	To implement provisions relating export controls and prosecution of relevant offences of the Act
6. Ministry of Defense	To implement provisions relating to Delivery System of WMD and prosecution of relevant offences under the Act

4.6.5 IPCC (INTER GOVERNMENTAL PANEL ON CLIMATE CHANGE)

IPCC was established jointly by WMO and UNEP (United Nations Environmental Program) in 1988.

It is open to all member countries of WMO and UNEP. Its role is to asses in a comprehensive, objective, open and transparent basis the largest scientific, technical and socio-economic. Literature produced worldwide relevant to the understanding of the risk of human induced climate change its observed and projected impacts and option for adaptation and mitigation.

IPCC is a scientific body, it provides information is based on scientific evidence. Its reports are taken as authentic/standard works.

4.7 INDUSTRIAL CHEMICAL WASTES AND TOXIC CHEMICALS

The economic development of twentieth century fully rested on industrial and technological development, though chemicals played vital role in the development of industries and wealth of the world. However at the same time the chemical wastes caused adverse impact on human health and damage to environment and ecosystem. The industrial wastes contain a variety of toxic inorganic, organic and metallic compounds like Arsenic, Mercury, Chromium, Zinc, Cyanide, Chloroforms and various kinds of pesticides. Toxic industrial waste materials deposited in landfills. This creates health problems, because landfill deposits attract rats, flies which transmit diseases. However by burning solid combustible materials in incinerations reduce the volume and heat produced can be used. Similarly air and water discharge problems can be solved by the reuse of renewable/recycling industrial refuse products. The agricultural and synthetic organic chemicals form carbon-tetra-chloride, vinyl-chloride, chlorobenzene and PCBs (Polychlorinated Biphenyls) which are all toxic. The disposal of hazardous chemical waste is of great importance. The industrial emissions of oxides sulphur and nitrogen into the atmosphere in Europe and North America resulted in widespread environmental ill effects that include acid rains, acidification of soils, surface waters, ground waters, injury to vegetation, corrosion of building materials and reduction in atmospheric visibility by way of fog and smog formation. Similarly gases pollutants, suspended particulate matter in industrial areas proved to be hazardous to plants, animals and health of human beings. In view of recent chemical industrial tragedies and terrorism explosions, the medical preparedness for CBRN (Chemical, Biological Radiational and Nuclear) management WHO recommended standard operating procedure. The industrial accidents leachate contamination (pollution caused by liquid water that leaches from landfill) is a serious problem (stink). Stink causes respiratory disorders. Capping dump prevent rainwater sinking in, which will solve leachate problem and air pollution to some extent Mobile leachate treatment plant operated for a trial basis (20 days), which proved to be successful. The leachate treated water could be used for irrigation purposes.

4.7.1 SOME TOXIC CHEMICALS

1. **Phosgene:** In Greek phos means light and gennao means I produce. Phosgene gas is produced by direct combinations of CO with Cl_2 in the presence of bright sunlight. Phosgene is highly poisonous (toxic) when inhaled in high concentration. It is so toxic that lungs abruptly stop functioning and death occurs in a few minutes. Phosgene was extensively used in World War I as a used chemical warfare agent.

2. **Chloropicrin:** It is used as insecticide and also as a war gas.

3. **Carbon disulphide (CS_2):** It is a colorless mobile poisonous liquid. Its vapor is highly inflammable. It forms a explosive mixture with NO. It forms carbon tetra chloride when Cl_2 is passed into boiling CS_2.

4. **Carbon monoxide (CO):** High concentration of CO in air reduces oxygen supply to vital organs (hearts and brain). At very high levels it causes dizziness, confusion, unconsciousness and even death.

5. **Chlorine (Cl_2):** It is a greenish-yellow gas with a pungent irritating odour. If inhaled in small (quantities) it produces headache, in large quantities prove fetal. It is used for the manufacture of poison gases like phosgene and mustard gas.

6. **Some Banned Toxic Chemicals:** During September 2017, a few farmers lost their lives by way of dusting pesticides. The following seven are extremely hazardous pesticides (banned). The total number of pesticides registered in Indian 279, of which 66 hazardous.

   Class I pesticides 1. Monocrotophos, 2. Dichloro, 3. Triazophos, 4. Phosphamidon, 5. Carbofuran, 6. Phorate, 7. Methonyl .

   Farmers death attributed to the use of pesticides (i) Acephate (ii) Profenofos (iii) Monocrotophos and (iv) Oxydemeton-methyl. WHO considers the last two pesticides as class I pesticides

7. **Cyanogen (CN_2):** It is a poisonous and clourless gas.

8. **Hydrogen Cyanide:** Alkyl cyanides are moderately poisonous, while hydrogen cyanide (HCN acid) is deadly poisonous. It is colourless liquid has bitter almonds smell. It is one of the deadliest poisons, the fetal dose the fetal dose is 0.05 gm. its anti-dose is ammonia and chlorine water.

   The salts of this acid (HCN) are called cyanides.

9. **PETN (Penta Erythritol Tetra Nitrate):** It is a very high explosive and one of the most powerful plastic explosive. It is legally used by military and mining company's

   PETN explosive belong to the nitroglycerin and nitro cellulose chemical family.

   PETN is a colourless, odourless crystals and can be detonated through a mobile phone. PETN is often mixed with another explosive TNT. (trinitrotoluene) and RDX to obtain powerful explosives used in grenades and war-heads-with RDX it forms Semtex, another powerful explosive. It is generally does difficult to detect at x-ray machines at airport. It generally does not explode if dropped or lit up. It requires a wire detonator or shock cap to set-off.

10. **TNT** is a high explosive, used for filling bombshells, hand grenades and torpedoes alone or with other explosives. TNT is a reasonably safe explosive, requiring strong detonation to set it off. Mixed with ammonium nitrate it forms the blasting material *Amatol,* which is used for demolishing old building and rocks.

Some Terms

Export means-it related to foreign trade Act 1992.

Fissile material means-material equipment and technology notified in relation to Act, Atomic Energy Act 1962.

Non-state Actor means a person/entity not acting under the lawful authority of any country.

"Nuclear Weapon or other nuclear explosive device" means any nuclear weapon or other explosive device as determined by the Central Govt. of India.

4.7.2 NATIONAL AUTHORITY

Govt. of India established the National Authority for implementing the provision of Chemical Weapons Convention (Act 2012) and the Weapons of Mass Destruction and their delivery system Act 2005.

The National Authority is empowered to interact with the organisations and other State Parties for implementing the aims of the Chemical Weapons Convention Act 2000 and A mendement Act 2012.

The National Authority is empowered to monitor compliance of the provisions of the Act 2000 and to regulate and monitor the development, production, processing, consumption, transfer or use of Taxic Chemicals etc.

The Steering Committee oversees the function of National Authority and its performings. The Steering Committee consists of the following members.

1. Chairman-the Cabinet Secretary in the Govt. of India
2. Five members
   (i)   The Secretary Ministry of Defence
   (ii) The Secretary dept. of Chemicals and Petrochemicals (– is the Ministry of Chemicals and Fertilizers in Govt. of India)
   (iii) Foreign Secretary (Ministry of External Affairs)
   (iv) The Secretary DRDO (Ministry of Defence, Govt. of India)
   (v)   The Secretary, Dept. of Revenue (Ministry of Finance)
   (iv) The Secretary, dept of Committee (Ministry of Commerce and Industry Govt. of India)
   (vii) Member Secretary, the Chair person of the National Authority Chemical Weapons Convention.

CHAPTER - 5

The Atmosphere

The envelope of air surrounding the earth to great heights is called atmosphere. The atmospheric air is a mixture of gases and contains the particles of liquid, solid (which are suspended in air). These are bound to the earth by the gravitational attraction of the earth. Atmospheric gases obey the ideal gas laws and for practical purposes treated as fluid.

5.1 COMPOSITION OF THE ATMOSPHERE

The composition of the dry atmosphere has:

Constituent	by volume	by wt.	Total weight × 10^{17} kg
Nitrogen (N_2)	78%	75.5%	38.65
Oxygen (O_2)	21%	23.1%	11.84
Argon (A_r)	0.93%	1.29%	0.66

The other minor gases include Neon, Helium, Methane, Krypton, Hydrogen, Xenon, Carbon dioxide, Ozone and water vapour. The last three gases are variable both in time and space. This composition of atmospheric gases are practically in the same proportion up to an altitude of about 80-90 km. The non-gaseous constituents are dust, smoke, salt particles from sea spry and water particles, which are all variable.

More than 50% of the mass of atmosphere lies below an altitude of 5.5 km and 98% of mass below 30 km altitude. The lowest one kilometer of atmosphere contains about 10% of the mass of atmosphere. All biological and human activities except aircraft flights are confined to this lowest one kilometer, which is called Planetary Boundary Layer (PBL).

The mass of the atmosphere is about 5.6×10^{18} kg. The mass of the oceans water is about 1.4×10^{21} kg. The density of dry air at surface (msl) is 1.225 kg/m^3 which reduces to 50% (0.6125 kg/m^3) at 6 km altitude, 25% (0.30625 kg/m^3) at 12 km, at 18 km altitude to 10% (0.1225 kg/m^3) and at 30 km altitude to about 0.013 kg/m^3 (about 1%).

On an average at msl nitrogen exerts a pressure of about 760 hPa, oxygen 240 hPa and water vapour about 10 hPa.

Atmospheric mass (gases, liquid and solid particles together) is treated as fluid and it obeys gas laws. The main properties of atmospheric mass are:

(i) Molecular mobility; (ii) capacity for expansion and compression with adiabatic heating or cooling

5.2   ATMOSPHERIC HEAT PROCESS

Heat is a form of energy, which produces sensation of warmth in us.

The degree of hotness of a body is called temperature. Heat flows from one body (at higher temperature) to another body (at lower temperature) till they are at equal temperature. The measurement of temperature is made by thermometers. Heat is transferred in the atmosphere in five ways: (i) conduction, (ii) convection, (iii) radiation (iv) advection and (v) condensation.

Conduction: The transfer of heat energy from hot surface to the adjacent cooler surface without the movement of molecules is called conduction. This process is important in transfer of heat very close the ground surface.

Note: air is a poor constructor of heat.

Convection: The transfer of heat energy with the movement of molecules and applies to fluids (liquids and gases). This process is important in atmosphere.

Radiation: The transfer of energy in the form of electromagnetic waves is called radiation. In this case no medium is required. All objects in the universe radiate as long as its temperature is more than 0 °K.

Advection: The transfer of heat energy from one area to another, through horizontal wind motion is called advection (i.e., by the movement of air masses).

Condensation: In atmosphere precipitation in the form of rain or snow release latent heat, which warms the atmosphere.

The important methods of heat transfer from the earth to the\ atmosphere are: (i) convection (ii) Advection, (iii) latent heat of condensation of water vapour (which was transported upwards). Sun is the main source of heat energy for the earth and atmosphere, which is received in the form of radiation (insolation). The sun's rays do not heat the air in the atmosphere but heat the solid and liquid particles suspended in it. Because of this about 85% of the insolation hits the earth's surface, while only about 15% heats the atmospheric (aerosols) particles. The surfaces of the earth, which are heated by the suns' rays, pass heat energy to the air film in contact with the surface by way of conduction. By convection warm air moves up and it is replaced by cooler air. The transfer of heat from the surface of the earth to the atmospheric air takes place by conduction and convection, but it is called re-radiation or back radiation. Thus earth acts as the secondary source of heat for the lower atmosphere. In meteorology the incoming solar radiation is called insolation (also shortwave radiation) and the back radiation from the earth to space is called Terrestrial radiation (or long wave radiation or **IR** radiation). The absorption of solar energy at the surface of the earth depends on the angle of incidence of the sun's rays or inclination of the sun and the materials of the earth on which it falls. Different types of surfaces absorb differently on earth and create uneven heating (such as sea water, black soil, sandy soil, green trees etc.) which in turn results in horizontal motion of air. The unequal heating of the surface of the earth is the cause for Cell pattern (Hadley Cell, Ferrel Cell) of atmospheric circulation.

5.3 THE VERTICAL STRUCTURE OF THE ATMOSPHERE BASED ON TEMPERATURE

In 1962, world meteorological organization (WMO) decided to divide the atmosphere into four regions (strata) based on temperature change with altitude. They are:

(i) Troposphere, (ii) Stratosphere, (iii) Mesosphere, (iv) Thermosphere. The salient features of these layers are given below.

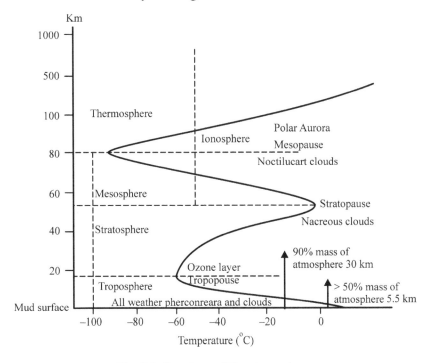

Fig. 5.1 Structure of the atmosphere

Troposphere: The lowest layer of the atmosphere adjacent to the earth's surface is called troposphere. The altitude of this layer is about 16-18 Km in the equatorial regions (temperature at the top – 75 °C), about 11 Km in middle latitude (lat 30°), (temperature at the top –60 °C) and about 8 km at the poles (with temperature at top – 50 °C).

All weather systems, clouds practically confined to this layer. In this layer (mostly) temperature decreases with increasing height. The troposphere contains about 75% of the mass of the atmosphere and virtually all water vapour of the atmosphere. The average lapse rate (rate of change of temperature with height) is about 6.5 °C/km. The upper boundary of the troposphere is called tropopause, pause means break.

Tropopause

Definition: The tropopause is the lowest level at which the lapse rate (of temperature) decreases to less than or equal to 2 °C/km at least for a layer of 2 km and above does not exceed this.

Tropopause is not a continuous surface. In middle latitudes two tropopauses are found. Lower one with tropical characteristics and the other with extra tropical characteristics. In between these two tropopauses lies the subtropical jet stream (around lat 30°). There is a sharp rise in temperature above tropical tropopause but in subtropical tropopause there is slight fall.

Troposphere is warmed up by the underlying earth surface and hence instability exists.

Stratosphere

The layer above tropopause is called stratosphere, which extends up to an altitude of 50 km, where its temperature is above 0 °C. There is an isothermal layer (lapse rate zero) above tropopause up to 20 km altitude and thereafter temperature rises (inversion) generally up to about 32 km altitude and thence rises rapidly and equals to the earth's surface temperature (positive) at its top.

This layer has large concentration of ozone between altitudes 18-35 km (60000-115000 ft) with a maximum density at about 25 km (80000 ft). The increase of temperature in stratosphere is attributed to the presence of ozone, which is a green house gas. Ozone absorbs harmful solar UV-radiation (ultra-violet radiation) in the wavelengths 220 nm to 290 nm (1 nm = 10^{-9} m). UV-radiation induces skin cancer, damages eye and suppresses immune system in human beings. It effects the productivity of aquatic and terrestrial ecosystem. 1% decline in ozone concentration in stratosphere may result in 3% increase in the incidence of skin cancer among humans. Ozone layer absorbs about 2% of insolation (incoming solar radiation).

Stratosphere is stable, because it has cold temperature at its base and warm temperature at the top.

In equatorial region of stratosphere biennial wind oscillations are observed.

The boundary surface which separates stratosphere from Mesosphere is called stratopause (where temperature is about $0°$ C).

Mesosphere

The layer above stratopause is called Mesosphere, which extends up to 80 km altitude. In this layer temperature falls above stratopause and attains the lowest temperature about -95 $°C$ at 80 km altitude. Noctilucent clouds belong to Mesosphere, which are observed in higher latitudes during summer. The top of Mesosphere layer is called Mesopause.

Troposphere, Stratosphere and Mesosphere together is called **Homosphere** because the ratio of the constituents of the air in this layer (region) is practically constant (except ozone, carbon-dioxide, and water vapour which are variable).

Thermosphere

The region above Mesopause is called thermosphere which extends to great height (700-1000 km altitude). Thermosphere is also called hetrosphere because the composition of the ratio of constituent gases is heterogeneous. The main gases stratify according to their molecular weights. The temperature rises above Mesopause to about 1000 to 1200 $°C$ at about 400 km altitude and the density of air falls to 3×10^{-12} kg/m^3 (very thin density) and pressure 10^{-8} mm of mercury.

The lowest layer contains oxygen and Nitrogen molecules (up to 500 km) and thereafter hydrogen (up to 1000 km). There is no boundary to the thermosphere. The upper part of the thermosphere merges with the interplanetary gas with temperature about 2000 $°K$ which is called Exosphere. The thermospheric gases are found to be mostly in atomic state due to photo-dissociation by the insolation.

Ionosphere

The lower thermosphere and upper Mesosphere contains gases mostly in ionized state and hence it is called Ionosphere. The characteristic property of these ionized gases is that it reflects radio waves, which helps in long wave radio communication. Ionosphere is divided into D-region (50-90 km altitude), E-region (also called Kennelly Heaviside layer) 90-140 km altitude and F-region 140-500 km attitude. F-region is further divided into F_1 region 140-250 km altitude (also called Appleton layer) and F_2 region 250-500 km altitude.

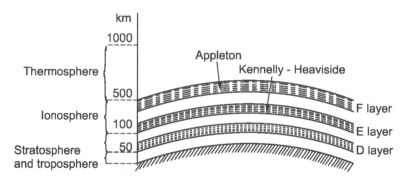

Fig. 5.2 Ionosphere layers.

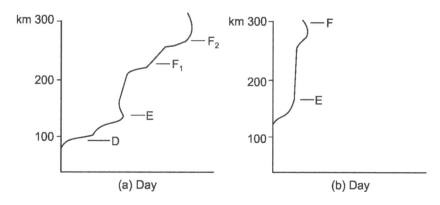

Fig. 5.2(a)&(b) Ionosphere diurnal effects

D-region (50-90 km altitude) reflects low frequency radio waves but absorbs medium and high frequency radio waves. This region disappears during nights (in the absence of solar radiation). In this region air density is more than electron density.

E-region (90-140 km altitude) strongly reflects medium and high frequency radio waves. This region weakens during might, but does not disappear. However during polar nights, E-region disappears. In this region electron density is $(10^5/cm^3)$ and air density is less.

F_1-region (140-250 km altitude) is important for the fact that it propagates medium and high frequency radio waves.

F_2-region (250-500 km altitude) is important for long distance radio communication. When the sun is low and during night, F_1-region merges with F_2-region.

In F-region electron density is high $(10^6/cm^3)$ and air density is low.

5.4 THE SUN

The mass of the sun is about 1.99×10^{30} kg, radius 6.97×10^8 m (or diameter $\simeq 14 \times 10^5$ km), while the mass of the earth is 6×10^{24} kg, radius 6.4×10^6m. The gravitational attraction of the sun is about 274 m/sec^2, which is about 25 times that of the earth (gravitation of the earth is 9.81m/s^2). About 99.87% of the mass of the solar system is contained in the sun.

Photosphere

The visible surface of the sun is called photosphere, its temperature is about 6000° K, thickness about 161000 km and consists hot gases under high pressure. Photosphere produces radiations of the visible part of the solar spectrum, but the energy is produced in the deeper central region of the sun's sphere. The surface of the photosphere changes constantly without any rhythm. It appears as mottled or granular due to boiling motion of the gases, however the granules are not visible to the naked eye. They are about a million in number with about 10 minutes life. Besides granules the photosphere shows bright (hot) regions called faculae and dark (cold) regions called sunspots. Sunspots are the regions of strong magnetic field.

Chromosphere

During total solar eclipse a reddish ring of light observed around the sun's photosphere, which is called chromosphere or colour sphere of the sun. The depth of the chromosphere varies 3000 to 5000 km above the photosphere. It consists of hydrogen and helium gases at low pressure but has temperature about 20000 K Fig. 1.1. Fibrilles and spicules are observed in chromosphere. Fibrilles are extended structures approaching granules, located around the centre of the solar disc, while spicules are located at the edge of the disc with life period of about 5 minutes, diameter 750 Km. In chromosphere temperature rises slowly from 4400 K to about 1.5 to 2.0 million K. Between photosphere and chromosphere there is a cooler gas layer with temperature about 4000 K, thickness about 500 km. This is called reversing layer and is responsible for Fraunhofer absorption lines. Above this layer temperature rises.

Corona

Above chromosphere lies corona, which spreads far out into space with temperature 1.5 to 2.0 million K. During total solar eclipse corona shows spectacular light of halo. Solar winds originate in Corona of speed exceeding 100 km/s and the plasma particle of corona exceed the escape velocity of the sun's atmosphere. This is called solar wind, depends on the solar activity. Associated with solar wind, solar mass of about 4×10^8 kg/s escapes from the sun.

The central core of the sun has temperature 15 – 20 million °K. Above this convective core there is a radiative equilibrium layer with temperature ranging 10 million °K to 6000 °K. The convective core and radiative equilibrium region is characterized by nuclear reactions, which is the source region of solar radiative energy.

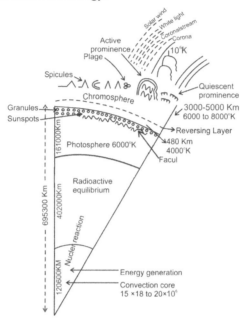

Fig. 5.3 Schematic representation of structure and activities of the sun with the exaggerated photosphere, chromospheres

Above this lies the visible surface of the sun called photosphere. Its temperature is about 6000 °K . It consists of hot gases under high pressure. Photosphere emits radiations of the visible part of the solar spectrum, but the energy is produced in the core region & neighbouring region. Photosphere shows bright regions called faculae and dark regions called sun spots. Sun spots are regions of strong magnetic field.

Solar Activity: when the frequency of observable features of the sun (sun spots, solar wind, solar flares) are more than normal, the sun is said to be Active or disturbed, but when they are absent the sun is said to be "Quite." The solar activity has about 11 year cycle. The magnetic field in sunspots observed to exceed thousand times as compared to the magnetic field outside the sunspots. The polarity of the magnetic field in sunspot region (area) changes in 22 years, which is twice the period of sunspot cycle. No sunspots observed during 1645 to 1715 (70 years), and there does not appear 11 year sunspot cycle before 1645. These facts were brought to the light by

E. Maunder (superintendent of Greenwich observatory) in 1890. The period 1645 – 1715 of sunspot minima or complete absence of it is called

Maunder minima. There appears a relation between sunspot cycles and climate change. The solar flares are the result of gigantic nuclear explosions in the central core. There may be hundred odd flares daily but a few large flares may occur in a year. The large flares emit electromagnetic radiation and high energy charged particles of plasma, which interact with atmosphere and earth's magnetic field, cause ionospheric storm (high frequency radio blackout) and magnetic disturbances. The impact of solar flares on climatic change is not clearly understood.

The sun's hot gases produce electrically charged particles, which produce electromagnetic waves. These waves constitute electromagnetic spectrum of the sun, which extends from gamma rays. ($\lambda = 10^{-13}$ m) to radio waves ($\lambda = 10^3$ m). The total range of λ (wave length) is 10^{16} m.

About 99% of the solar radiation lies between ($\lambda = 0.15$ μm to 4.0 μm). About 9% of this lies in UV-radiation ($\lambda = 0.15$ μm to 0.38 μm). This causes photochemical effects, bleaching and sunburn. About 44% of this lies in visible portion, $\lambda = 0.38$ μm to 0.7 μm (i.e., violet to red). About 46% of this lies in IR region, $\lambda = 0.7$ μm to 2.3 μm. This causes radiant heat with some chemical effects. The sun emits maximum radiation energy at wave length $\lambda_m = 0.5$ μm, while the earth emits maximum radiant energy at wavelength $\lambda = 10$ μm (in IR).

Solar Constant: The amount of solar radiation (energy) incident on unit area in unit time on a surface held at right angle to the solar beam at the outer boundary of the atmosphere is called solar constant and is given by S = 2.0 ly/min or 1359 W/m^2, where 1 ly (one langley) = 1 cal/cm^2.

Solar Energy

The fusion furnace of the sun converts about 4 million tons (4 × 109 kg) of hydrogen into helium and radiates an amount of energy 3.8 × 1023 Kw/s as a byproduct. The solar energy incident on 1m^2 area held normal to the sun's rays at the outer boundary of the atmosphere in one second is 1.38 × 10^3 W/m^2. It is virtually constant and called solar constant (accepted by IGY).

A house with peak load requirement of 2.5 KW would require a solar collector of 3.6 m^2 with 100% efficiency. The total solar energy received at the earth in one day is equivalent to 35 lakh nuclear explosions or 10000 hurricanes or 10^8 thunderstorms or 10^{11} Tornadoes. If this energy were stored it will satisfy all the world needs of domestic and industrial use for about 100 years.

According to one estimate, the solar radiation received at the outer boundary of the atmosphere is about 1.74 × 10^{14} **KW. Of this about (30%)** 5.2 × 10^{13} KW is reflected back to the space as shortwave radiation 8.2 × 10^{13} KW (about 47%) is directly absorbed by the land, oceans and atmospheric system and the remaining 4.0×10^{13} KW (about 23%) is utilised in driving the hydrologic cycle through evaporation, convection and precipitation. About 3.7×10^{11} KW ($\sim 0.21\%$) of the incident solar energy is utilised (as input into Biosphere) for driving winds, waves, convection and sea currents. 4×10^9 KW (0.023% of the incident solar energy) is used in photosynthesis process on earth. The green canopy of plant kingdom consumes solar energy about (0.01 ly/min) 7 watt m^2. The total solar energy intercepted by the earth $= 2.55 \times 10^{18}$ Cal/min or 3.67×10^{21} Cal/day.

Solar energy per unit area expressed in langley (ly) or Kilo-langley (kly).

$$1 \text{ ly } = 1 \text{ Cal/cm}^2,$$

It is estimated that sun radiates each minute about 56×10^{26} Cal of energy in all.

The energy received from the sun is not remained constant over the 4500 million years of the earth's existence. The sun is growing slowly hotter over the millennia and emitted more and more energy. It is now emits about 30% more heat than it did when the earth was first formed. With all these variations, during the present millennia the temperature of the earth's surface has remained remarkably stable with all interactions of plants, animals and by changes in the heat absorbing capacity of the earth's surface – the extent of its water, lands, vegetation and ice.

The fluctuations in composition of the atmosphere and changes in temperature over the past 200,000 years can be determined by analysis of tiny air-bubbles trapped in ice that has accreted year by year in the Antarctic and Arctic. The cores drilled in ice caps, a record of changes can be measured and deducted. The Vostok core in Antarctica (in Antarctic station Vostok) by USSR, the CO_2 concentration and temperature were estimated for the past 160000 years. The analysis shows CO_2 fluctuated from about 180 to 220 ppm in the atmosphere and temperature rose and fell over the range of 9 to 10 °C, in more or less perfect rhythm with the rise and fall of CO_2 concentrations (or GHG concentrations).

According to prof. George Voitkevich, the mobile gas components of the atmosphere and water experience the fastest cycle, while a very slow cycle in the continental matter. The time required for a full cycle of matter is 7 years for atmospheric carbon dioxide, 4000 years through photosynthesis for atmospheric oxygen; nearly one million years for ocean water by way of evaporation; and 80-100 million years for continental matter by way of weathering and removal from land surfaces.

According to WMO, the life time of GHGs in atmosphere vary from hours/weeks to more than 100 years; tropospheric ozone has a life period a few hours/ days; CFCs have about 75 to 110 years; N_2O (nitrous oxide) 150 years; CH_4 (methane) 7-10 years, CO_2 (carbon dioxide) $50 - 200$ years.

CHAPTER - 6

The Oceans

INTRODUCTION

Oceans cover about 71% of the earth's surface area, and it is one major part of the earth system. The total mass of the hydrosphere is about 1.37×10^{21} kg. The oceans help in the processes of the atmosphere by the transfer of mass, momentum and energy through its surface. Oceans receive water and dissolved substances from the land. This dissolved substances settle down as sediments, which ultimately become rocks. At a lower level physical oceanography and meteorology are merging. The ocean provides the feedback leading to slow changes in the atmosphere.

In a broad sense, oceans provide us food (fishes etc.), affects weather and help in ocean transport, marine navigation. Ocean beds are sources for oil and gas extraction and are used for boating, fishing, navigation, surfing, swimming, recreation and extracting energy from waves. Because of these activities we are interested in the study of sea waves, currents, winds and temperature. It is a known fact that of natural disasters tropical cyclones (which form over the oceans) cause highest damage. We study the oceans to predict the formation, intensification, movement and ferocity of tropical systems to mitigate its adverse effects. In addition to tropical cyclones, the studies contribute for transport of heat energy from ocean to atmosphere in tropical latitudes to extra-tropical latitudes. Marine voyages also require the help of ocean studies. We shall study briefly the composition of sea water and exchanges that occur across the sea-air interface and motion of the sea in deep and shallow waters and discuss the effects of the wind on ocean currents.

6.1 COMPOSITION OF SEA WATER

From a very long time men living near the sea obtained common salt for cooking by evaporating/boiling of sea water. It is now known that there are a number of salts dissolved in sea water. The dipolar nature of water molecule is responsible to dissolve salts by breaking and the ions of the salt are completely free from one another. More than 99% of sea salts contain six ions/atoms which are given below. The common salt (NaCl-Sodium chloride) dissolved material comprises about 85%.

Approximate composition percentage of sea water

Salt	Atoms	Ions
Chloride	55	55
Sodium	30	30
Sulphate	4	8
Magnesium	3	4
Calcium	1	1
Potassium	1	1

The above composition is more or less same throughout the global sea water.

Salinity: The number of grams of dissolved material in 1000 gm of sea water is called salinity. The average salinity of sea water is about 35 gm/kg or about 3.5% by weight.

Dissolved earth material enters ocean waters mainly through rivers on land entering into seas in the form of ions. Besides these ions, rivers carry eroded particles of soil and rock which are deposited as sediments on the sea floor. Dissolved salts in sea water differs from the average composition of the earth's crust, this is because certain elements are dissolved more readily than others and some chemicals are removed from ocean waters by living organisms. The shells and skeletons of living organisms in sea contain calcium (Ca) and silicate (SiO_2) and sea plants and animals too remove some elements from the sea/ocean water.

Sea plants, animals consume very little soluble sodium chloride (NaCl) hence they accumulate in sea water at a faster rate than other ions. Sea waters dissolve most types of ions except calcium. Calcium carbonate ($CaCO_3$) is deposited near sea shores, which form limestone rocks.

6.2   THE EXCHANGE OF EARTH MATERIALS BETWEEN SEA AND ATMOSPHERE

Earth's matter exists in three states viz. solid, liquid and gas. The structure of solids is crystalline or amorphous. Liquid state matter exists as molecules or group of molecules. The gaseous state matter exists in molecular form but they are widely separated. The three states of matter generally changes one form to another with temperature (or heat).

As regards to exchange of matter between the oceans/seas and the atmosphere, some salts move from sea to the atmosphere as breaking waves toss the water droplets into the air, which generally evaporate leaving tiny salt crystals into the air. These crystals act as or become condensation nuclei. Sea plants release oxygen near surface, some escape into the air and the remaining goes to the depth of sea by currents. Water is exchanged between the sea and air. More than 80% of water vapour in the atmosphere is pumped from the oceans by way of evaporation. This occurs when the sea is warmer than air. The salinity of sea surface water is affected by the loss or gain of water which takes place in evaporation and precipitation processes. Thus the salinity of sea surface is greater in the vicinity of sub-tropical high pressure belts in both hemispheres.

There is no sharp boundary between the hydrosphere (sea) and the atmosphere. The water of hydrosphere contains solid materials and gases in solution while the atmosphere contains solid and liquid particles. Salts from sea enters the atmosphere by breaking of sea waves. The evaporated tiny particles of salt act as condensation nuclei to form clouds. Oxygen and carbon dioxide are exchanged across air-sea interface. On land animals inhale oxygen and release (exhale) CO_2 from their bodies. Fish and other marine animals use O_2 that is dissolved in sea water which is replaced by O_2 (oxygen) of the atmosphere.

Plants absorb CO_2 and release O_2. Sea plants release O_2 near sea surface, some of which escape to the atmosphere. The remainder is carried to the depths of the sea by ocean currents. Water is exchanged between sea and the atmosphere as water vapour. About 80% of the water vapour in the atmosphere enters from the ocean by evaporation. This occurs readily when sea is warmer than the air. This water vapour returns to the sea when condensed in the atmosphere (in the form of precipitation) flows through river and falls into the sea. Some water is retained on land. The salinity of sea surface water is affected by evaporation and precipitation processes. By evaporation of water salinity increases, while with precipitation salinity decreases. Because of this salinity of ocean surface water is greater in sub-tropical anticyclone high pressure belts (where evaporation is more than precipitation water content).

Ocean processes are non-linear and turbulent. Ocean (like atmosphere) is a stratified fluid on the rotating earth. The air and water (both fluids) have many similarities in their fluid dynamics but there are some important differences. Water is practically incompressible. Atmospheric moisture plays very important role in water (in terms of latent heat). In case of ocean thermodynamics there is no counterpart. All oceans are bounded by countries (laterally). The ocean circulation is forced in a different way as compared to atmosphere. Atmospheric motion is transparent to incoming solar radiation (i.e., insolation) and heated from below (i.e., at surface of the earth). Ocean exchange heat and moisture with atmosphere at the ocean upper surface. Convection in the ocean is by buoyancy loss from above. Wind stress over the surface drives ocean circulation, particularly upper one kilometer of depth. The wind driven and buoyancy driven circulations are inter wind. Ocean circulations affect climate and paleoclimate. As noted earlier about 71% of the earth's surface is occupied by oceans, with an average depth of about 3.7 km. Ocean basins are very complex, bottom topography notched (jagged) much more than land surface. Abyssal ocean currents are comparatively weak and temperature changes are very little. Because of this submarine ocean relief erosion is very slow as compared to the mountains on land.

The ocean volume is about 3.2×10^{17} m^3, mass is 1.3×10^{21} kg and has huge (enormous) heat capacity, which is 1000 times the heat capacity of the atmosphere. Because of this (reason) it plays an important role in climate. The important features of ocean are given below.

Ocean surface area: 3.61×10^{14} m^2

Mean ocean depth: 3.7 km

Ocean Volume: 3.2×10^{17} m^3

Mean density of ocean water: 1.035×10^3 kg/m^3

Mass of the ocean: 1.3×10^{21} kg

The following table gives the albedo of different surfaces.

Surface	Albedo (%)
Ocean	2-10
Frost	6-18
Cities	14-18
Grass	7-25
Soil	10-20
Grass land	16-20
Desert land (sand)	35-45
Ice	20-70
Cloud (thin, thick status)	30, 60, 70
Snow (old)	40-60
Snow fresh	75-95

6.3   THE CRYOSPHERE

About 2% of the water on the surface of the earth is frozen and this is known as Cryosphere. In Greek Kryos - meaning "frost" or "cold". The Cryosphere includes: Ice-sheets, sea-ice, snow, glaciers and frozen ground (permafrost). Most of the ice is contained in the ice sheets over the land masses of Antarctica (89%) and Greenland (8%). These ice sheets store about 80% of the fresh water on the earth.

The Antarctica ice sheet average depth is about 2 km while the Greenland ice sheet is about 1.5 km thick. Climate is affected by the surface area covered by ice (not by the amount of ice). The albedo of ice varies 40-95%, about 70% in the mean, which reflects the incident radiation on it. The perennial (year-round) ice cover 11% of the land area and 7% of the ocean area.

Two forms of ice observed in Antarctica ocean.

(i) Sea ice, which is formed by freezing of sea water and

(ii) Ice bergs, which are broken off pieces of glaciers. Sea ice is important because it regulates the exchange heat, moisture and salinity in polar oceans and insulates relatively warm ocean water from the cold polar atmosphere.

6.3.1 SEA ICING

The physical properties of sea ice depends on the salt content. Salt content is a function of the rate of freezing, age, thermal history. The composition of salts in sea ice more or less same as in brine. For practical purposes the chlorinity and salinity of sea ice have the same meaning as for water (although the salts are not uniformly distributed in the ice).

The sea ice of salinity 10 ‰ at 3 °C is a mush (soft pulp) having 200 gm of brine per kilogram. Sea ice contains small bubbles of gas which changes the properties. The gases occur as small bubbles in the ice. The ice which has been frozen rapidly contain large gas quantity and in this case bubbles represent gases originally in solution in the water or in old ice (that has undergone partial thawing and has been refrozen in which case atmospheric air is trapped in the ice)

Pure water at 0 °C has density 0.9998674 kg/m^3 and pure ice at 0 °C has density 0.91676 kg/m^3. The specific heat of ice depends on temperature and changes in narrow limits. Whereas sea ice varies largely and depends on the salt content and temperature. The change in sea ice temperature depends on either melting or freezing and the amount of heat required depends on the salinity of the ice.

The specific heat of pure ice is less than that of pure water. The very high specific heat of ice of high salinity at the initial (near) freezing point is due to the formation of ice from the enclosed brine or its melting. The latent heat of fusion of pure ice at 0 °C and at atmosphere pressure is 79.67 cal/gm.

The vapour pressure of sea ice has not been determined but it could be very near to that of pure ice, which is given below.

Temp °C	0	−10	−20	−30
Vapour pressure h Pa	6.11	2.61	1.04	0.39

The Latent heat of evaporation of pure ice is variable.

6.4 SOME PHYSICAL PROPERTIES OF SEA WATER

The most abundant element of hydrosphere is oxygen. It comprises 88.9 % oxygen and 11.1 % nitrogen by weight. Waters of oceans, lakes and rivers contain dissolved elements of earth's crust in small amounts. Sea water has about 3.5% dissolved minerals, of which sodium and chlorine ions are the largest, whose combination is found sodium chloride (NaCl, common salt) and hence the salty taste to the sea water.

Liquid water made up of multiple groups of H_2O molecule having one, two or three elementary molecules called Monohydrate, dihydrate and trihydrol. These forms depend on temperature, immediate past history of water and other factors. The degree of polymerization decreases with increasing temperature. Neutral waters have variable amounts of heavy hydrogen (deutrium – isotope of hydrogen) and oxygen. This modifies the density and other properties of water. Fresh water or rain water have lower heavy isotopes as compared to sea water. The important properties of water are given below.

Density of pure water at 4 °C is 0.999×10^3 kg/m^3

While the average density of sea water is 1.035×10^3 kg/m^3

Specific heat (C_ω) 4.18×10^3 J/kg °K

Latent heat of fusion (L_f) 3.33×10^5 J/kg

Viscosity (μ) 10^{-3} kg/m, sec

Kinetic viscosity (υ) = $\dfrac{\text{viscosity of water}}{\text{density}} = 10^{-6} m^2$ / sec

Thermal diffusivity (K) 1.4×10^{-7} m^2/sec

Heat capacity of water is the highest as compared to all solids and liquids except liquid Ammonia. This does not allow extreme range of temperature but allows large quantity of heat transfer water to atmosphere.

Ocean temperature ranges from about –2 °C to + 30 °C, salinity varies from 33 ‰ to 37 ‰. For chlorinity 19.00 ‰ salinity is 34.325 ‰.

Temperature, salinity of deep ocean and bottom ocean water vary 4 °C to – 1 °C, salinity 34.6 ‰ to 35 ‰ and high pressure. (‰ Stands for per thousand or per mile).

Sea water diurnal temperature range is less than 2 °C and maintains uniform (water) body temperature.

6.5   ENERGY EXCHANGE PROCESSES BETWEEN SEA AND ATMOSPHERE INTERFACE

In the process of evaporation of sea water, energy is transferred from the sea to the atmosphere. 80% of the world's atmospheric water vapour goes from the sea (oceans). Energy that is required to evaporate sea water is derived from the insolation. This energy is transferred from sea to the atmosphere as latent heat of the water vapour. When water vapour condenses it releases latent heat to the atmosphere, which remains as a heat in the atmosphere.

A tropical cyclone resembles a great heat engine that derives its energy mainly from the transfer of sensible and latent heat from sea to air. The main input is water vapour, a form of latent heat.

The oceans cover about 71% of the earth's surface, and so a large part of insolation (incoming solar radiation) in shortwave radiation is absorbed by the oceans. The oceans then radiate back a great portion of terrestrial long wave radiation. This long-wave radiation by the oceans absorbed by the atmospheric greenhouse gases (like CO_2, O_3, water vapour, CH_4 etc). This energy heats the atmosphere.

In equatorial low latitudes more energy (insolation) is absorbed by the earth-atmosphere system than energy radiated back to the space as terrestrial radiation. Thus ocean surface warms the tropics. In contrast to this, at high latitudes where energy is deficit and sea surface temperatures are low in polar regions.

Most of the insolation (short wave radiation) is absorbed in the top few meters of the oceans. A part of this absorbed heat energy transmitted downwards by vertical mixing by the winds and waves. As a result there is a surface layer with a uniform temperature (in the sea). This layer may extend to about two to three hundred meters depth. Below this (surface) layer, temperature decreases rapidly for (another) a few hundred meters because warm surface waters do not reach this depth. In boundary (range of depth) where temperature changes rapidly with depth is called the thermocline. Thermocline tends to seal off vertical water movements in many parts of the ocean. This thermocline depth is also a zone of highest density gradient is called pycnocline. Below the thermocline temperature decreases gradually. Even in tropics the temperature of the ocean water at a depth of one kilometer is only a few degrees above freezing point.

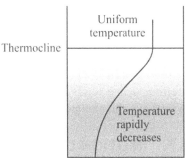

The exchanges of matter and energy are related to the events of atmospheric circulation and its waters. Because of this the study of transfer processes that occur across air-sea interface is important, which helps in understanding of weather and its forecasting.

6.6 OCEAN CURRENTS

Introduction

The sun drives the oceanic circulation through the atmospheric circulation (wind), and in turn ocean circulation greatly exerts influence on world's climate by way of winds, temperature, precipitation and humidity (i.e., climatic controls).

Drifts: Movement of ocean water top shallow layer with speed about 3-4 kmph (2-3 mph) is called drift.

Ocean currents: Movement of ocean water with deep effects, with speed exceeding 16 kmph (10 mph) is called ocean current.

Ocean water circulation mainly depends on wind-stress on ocean water and different densities within the ocean water itself. (Density depends on temperature and salinity of sea water).

Circulation in both the atmosphere and the oceans is driven by insolation and the earth's rotation (Coriolis force). Radiation of the earth-atmosphere system is positive at low latitudes and negative at higher latitudes. Heat is redistributed from low to higher latitudes through wind system in the atmosphere and ocean current systems. There are two principal components of ocean circulation – wind driven surface ocean currents and density-driven (thermohaline) deep circulation.

Changes in temperature are caused by fluxes of heat across the air-sea boundary. Changes in salinity is brought by the removal or addition of fresh water through evaporation and precipitation. In polar regions freezing and melting of ice lends support to variation of salinity. All those processes are linked to solar radiation directly or indirectly.

6.7   MOTION OF THE SEA AND THE EFFECT OF WIND ON OCEAN CURRENTS

Ocean waves are largely produced by the action of wind. The stronger wind and longer it blows it produces larger waves. Even after the wind has stopped, energy transferred to water causes ocean waves continue to travel for hundreds of kilometers. The waters of oceans are always on the move. Mariners have made use of ocean currents in their journey across the sea for many centuries and even now wind systems are the major driving forces of the ocean currents. It may be noted that, even in the absence of wind (speed), temperature differences and the force of gravity sets ocean waters to move.

The energy that drives the atmosphere and ocean is received from the sun's radiation. Solar heating in equatorial region and cooling near the poles create temperature differences coupled with gravity convection develop ocean currents (due to the tendency of cold water to sink and less dense warm water to rise).

The direction of ocean currents are affected by the rotation of earth (i.e., coriolis force = f). The Coriolis force deflects the ocean currents to the right in the northern hemisphere and in the left in the southern hemisphere. The combined effect of winds and gravity, coriolis force produce an inter connected ocean system develop clockwise and anti-clockwise currents, which are the main features of the world ocean currents pattern.

Ocean currents carry heat from equatorial region to polewards. In contrast dense deep currents flowing toward equator transfer cold water from high latitudes to equatorward.

The driving force of ocean currents in equatorial region (east to west) is the trade wind system in each hemisphere. These are north equatorial current and south equatorial currents respectively. Water piles up on the west side of the oceans, sloping upwards towards the west about 1 cm/200 km. The ocean waters flow up this slope as long as the trade winds are blowing.

Within the equatorial trough the winds are generally light or variable. In response to this ocean waters flow in opposite direction from west to east down the sea slope. This flow is called the equatorial counter current. This counter current is seen clearly in eastern parts of Pacific. North equatorial and South equatorial currents flow west wards, which are diverted by land barriers northward along the east coast in the northern hemisphere. The warm Gulf stream in the Atlantic ocean and Kuroshio current in the Pacific ocean are permanent. In the southern hemisphere the south equatorial currents are diverted to southwards along the east coast of the continents. Heat energy is also transported polewards in these ocean currents.

In middle latitudes strong westerlies (called roaring–fifties) force the ocean waters to travel west to east, particularly in southern hemisphere. There are land masses which divert the ocean waters. They drive the currents continuously around the southern hemisphere. This is called West wind Drift. Ocean currents are slow as compared to the wind speed. Narrow currents, like the Gulf stream, flow at about 8 kmph but in mid ocean, the ocean speed is less than 2 kmph.

In some areas of ocean, there is upwelling (upward flow of deep sea water). When winds blow equatorward along the west coasts of continents, the surface sea water is forced to move towards equator. This is coupled with coriolis force deflects the water away from the coast. This results in upwelling of deep cold water to replace surface water which was diverted. The best example is the upwelling off Peru coast, called the Peru current or Humbolt current. When the Peru current moves northward along west coast of south America, and when nears the equator it joins the south equatorial current. The upwelling of sea water, like near Peru, is also observed in Coastal regions of California, Western Australia, Vietnam and South Africa. The upwelling of cold water near northern California produces fog.

6.8 OCEAN CURRENTS CAUSED BY DENSITY DIFFERENCES

Density of sea water depends on temperature and salinity. Water becomes denser when cooled or when more salts are dissolved in it. At the air-sea interface transfer of energy or mass takes place which changes the density of sea water. It is hypothesized that density differences cause ocean deep water to move in slow currents.

Deep ocean currents can be gauged or assessed indirectly by measuring temperature and salinity of sea water at different depths. Measurements of oxygen content (in sea water) also gives some information because oxygen is transferred to deepest parts of ocean as it is required for the survival of marine life.

In all oceans including in the tropics, cold water lies at the bottom whose temperature would be a few degrees above zero degree Celsius. Very cold (freezing) water forms at surface near polar regions – Antarctica, Greenland, where the water is densest. (i) In Antarctica dense surface water sinks and spreads out at the ocean floor. Freezing (like in evaporation) leaves the salt in the unfrozen water. As ice forms at the surface, the salinity of the remaining water increases. High salinity and very low temperature causes water densest, which sinks the water to ocean floor and moves away north words (from Antarctica); (ii) In Arctic region, the cold water is relatively fresh and light (less dense) by river outflows. This water at surface moves south ward and when it meets warmer (less dense) waters of the North Atlantic Drift, it sinks and moves southwards at great depths. Thus in both (cases) hemispheres denser deep water moves towards equatorial region. The following figure shows some important features of deep ocean currents.

A_1, Antarctica bottom current moves slowly, crossing the equator. Finally meets North Atlantic deep current A_2, which overrides A_1 on its journey towards Antarctica. The Atlantic intermediate current A_3 which rises near Antarctica. This water returns to the North Atlantic Ocean with the surface currents. A_4 is another Antarctic intermediate current. This moves northwards from Antarctica and crosses the equator flowing at a depth less than 2 km at equator. This current in Atlantic Ocean sends tongues of cool low-saline water under the warm salty surface waters of the Sargasso sea.

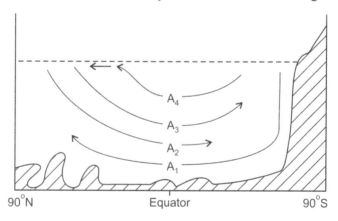

Fig. 6.1 Deep ocean current

Another current exchanges water between the Mediterranean sea and the North Atlantic ocean. Due to the high rate of evaporation Mediterranean water is the saltiest on the earth. Thus dense salty water sinks as it is not made lighter with local precipitation and river flows.

Water exchange takes place through into the Straight of Gibraltar. North Atlantic ocean surface water flows into the Mediterranean sea (about 2×10^9 kg/sec). Under this a heavy salty water flows out via Straight of Gibraltar as a compensating current and spreads out at an intermediate depth into (south-east portion of) the North Atlantic ocean.

The deep waters of other oceans is less known, but they have also layered water masses.

6.9 SOME OCEAN CURRENTS

The main ocean currents are:

I North Pacific
1. Cool-Oyashio
2. (Cool) Alaska Currents
3. Warm-California Current
4. Warm North Pacific Current
5. Warn-North Equatorial Pacific Current
6. Warm Equatorial Counter Current

South Pacific
7. Off china-Kuroshivo Current
8. South Equatorial Pacific Current
9. Cool Peru Current or Humboldt Current

II Indian Ocean
1. North Equatorial Current
2. Equatorial Counter-Current
3. South Equatorial Current
4. Off east Africa-Agulhas Current

III North Atlantic
1. Cool Labrador Current
2. North Atlantic Drift
3. Warm Florida Current
4. Canaries Current (North Africa Coast)
5. Guinea Current (West Africa Coast)

South Atlantic
6. Warm Brazil Current
7. Falkland Cold Current (East South America)
8. Cold Benguela Current (South Africa)

IV Antarctic
1. Antarctic Circumpolar Current Westward drift

6.10   INDIAN OCEAN CURRENTS

In certain polar regions, water is subjected to extreme cooling-sinks and flows equatorward in the thermohaline circulation.

In order to know the net pole ward heat transport in the oceans at any latitude, one would require to know the direction and speed of the flow of water and its temperature at all depths.

Relatively a thin layer close to the solid earth has frictional coupling between moving water and the earth and the same holds for air masses. Except these the frictional coupling is very weak. In case of a projectile moving above the earth and thermocline region the frictional coupling is practically zero.

6.11   SOME DEFINITIONS AND EXPLANATION OF TERMS

1. **Oceanography:** It is the study of ocean in relation to environment. It deals with description quantitatively, which enables to predict future state with some confidence.

2. **Geophysics:** It is the study of physics of the earth.

3. **Physical Oceanograhy:** The study of physical properties and dynamics of the ocean. The main aim is to understand the interaction of the ocean with the atmosphere, water masses formation, ocean currents, the ocean heat budget and coastal dynamics. It is viewed as a sub-discipline of geophysics.

4. **Geophysical Fluid Dynamics (GFD):** It is the study of dynamics of fluid in motion with respect to the scales influenced by the rotation of the earth. In meteorology and oceanography the GFD used to calculate the planetary flow fields.

5. **Hydrography:** The preparation of nautical charts, including charts of ocean depths, currents, internal density field of the ocean and tides.

6. **Earth System Science:** It is the study of the earth as a single system. Which includes many interacting subsystems like the ocean, atmosphere, cryosphere and biosphere and changes in these systems in response to human activity.

7. **Important Physical Properties of Water:**
   (i) Heat Capacity of water is the highest of all solids, liquids except liquid ammonia. This property prevents extreme ranges in temperature. Heat transfer by water movement is very large. Water tends to maintain uniform body temperatures.

   (ii) Latent heat of fusion is highest except Ammonia. Thermostatic effect at freezing point owing to absorption or release of latent heat.

(iii) **Latent heat and evaporation:** Highest of all substances. Largest latent heat of evaporation is very important in heat and water transfer to atmosphere.

(iv) **Thermal expansion:** Temperature of maximum density decreases with increasing salinity. For pure water it is at 4 °C. Fresh water and dilute sea water have their maximum density at temperature above the freezing point. This property plays an important role in controlling temperature distribution and vertical circulation in lakes.

(v) **Surface tension:** Highest in all liquids. This property is important in physiology of the cell. Controls certain surface phenomena and drop formation and behaviour.

(vi) **Dissolving power:** In general water dissolves more substances and in greater quantities than any other liquid. Obvious implications in both physical and biological phenomena.

(vii) **Dielectric constant:** Pure water has highest of all liquids. Of most important in behaviour of inorganic dissolved substances because of resulting high dissociation.

(viii) **Electrolytic dissolution:** Very small. A neutral substance, yet contains both H^+ and OH^- ions.

(ix) **Transparency:** Relatively great. Absorption of radiant energy is large in IR and UV. In visible portion of energy spectrum there is relatively little selective absorption, hence is colourless. Characteristic absorption is important in physical and biological phenomena.

(x) **Conduction of heat:** Highest of all liquids. Although on small scale as in living cells, the molecular processes are far outweighed by eddy conduction.

6.12 GEOPHYSICAL PROPERTIES OF OCEANS AND EARTH

Introduction

The earth is an (oblate spheroid) ellipsoid which rotates about its minor axis. The equatorial bulge of the earth is due to earth's rotation.

The equatorial radius of the earth 6378 km.

The polar radius of the earth 6357 km.

The measurement of the earth is made in units of degrees of latitude and longitudes, nautical miles or meters. Longitude measurements made from Greenwich meridian and latitude with respect to the earth's great circle.

1° latitude =111 km

$1°$ longitude $=111$ cos ϕ km

1nautical mile (nm) = length of the arc of 1 minute of a great circle of the earth.

Equator is a great circle midway between the poles from which latitudes are measured.

$$1 \text{ meter} = \frac{P_{axis} \text{ meridian distance between equator to pole}}{10^7 \text{ (ten million)}}$$

1 nm = 1852 m or 1.852 km

Earth's volume = 1083×10^{12} km^3 or 1.083×10^{21} m^3

Average density of the earth = 5.41 g/m^3

Density of continental crust = 2.7 g/m^3

Density of ocean crust = 3.0 g/m^3

Density of upper mantle = 3.4 g/m^3

Density of outer core =10 g/m^3

Density at the centre of the earth $\sim$ 13.6 g/m^3

Mass of the earth = 5974×10^{18} tonnes $\left.\right\}$

$\qquad\qquad \sim 6 \times 10^{24}$ kg

Equatorial circumference of the earth = 40 077 km

Surface area of the earth = 510 100 000 km^2

Earth's dry land area = 149 400 000 km^2 (29.3% of the earth)

Surface area of oceans = 360 700 000 km^2 (70.8% of the earth)

Average oceanic crust = 6 km in thickness

Maximum thickness of the continental crust $\sim$ 70 km

Age of the earth $\sim$ 4600 million years $\left.\right\}$
or $\qquad\qquad 4.6 \times 10^9$ years

Volume of the oceans and seas $\sim$ 1 285 600 000 km^3 $\left.\right\}$
or $\qquad\qquad$ 97.2% of the total worlds water

About 2.15% is frozen in bodies of ice, the remaining lies on or under the land.

0.001% of global water lies as water vapour in atmosphere.

Total Area of oceans $\sim$ 360 700 000 km^2

Area of Pacific ocean $\sim$ 179 700 000 km^2 (more than 1.2 times the area of dry earth's land area)

Area of Atlantic ocean $\sim$ 106 100 000 km^2

Area of Indian ocean $\sim$ 74 900 000 km^2

Average global ocean depth $\sim$ 3550 m

The greatest depth of ocean $\simeq$ 11033 m, in the Marianas trench in the Pacific.

The lowest ocean surface temperature = –2 °C in the white sea.

The warmest ocean surface temperature = 35.6 °C in shallow parts of the Persian gulf.

Largest Bay is Hudson Bay, Area $\simeq$ 822 300 km^2

Largest gulf is, gulf of Mexico, Area $\simeq$ 1500 000 km^2

Largest sea area, south china sea, Area $\simeq$ 2 974 600 km^2

Salinity and main constituents of sea water already given.

6.13   MAJOR OCEANS

The Pacific, the Atlantic and the Indian ocean. The oceans are interconnected and not separate bodies. Around the north pole a large body of sea water called the Arctic ocean and around the south pole a large body of sea called Antarctic ocean. These two are the polar extensions of the main oceans.

SEAS

Parts of oceans are called seas. The major seas of the world are given below.

South china sea, Area $\simeq$ 29 746 00 km^2

Caribbean sea, Area $\simeq$ 2 589 800 km^2

Mediterranean sea, Area $\simeq$ 2 512 150 km^2

Bering sea, Area $\simeq$ 2 273 900 km^2

Sea of Okhotsk, Area $\simeq$ 1 507 300 km^2

The ocean floor: The echo-sounders mapping presents the following features.

The continental shelves: The gently sloping areas around land masses are called continental shelves. The higher parts of continental shelves are islands. The shelves extend outward to a water depth of about 200 m. Geologically shelves are submerged of the coast lines and are the real boundary of the continents. The continental shelves end where the slope (gradient) suddenly changes at the continental slopes. The slopes fall sharply to the Abyss.

The Abyss: It is largely covered by oozes consists of volcanic dust or the remains of marine organisms and large plains and some other interesting features.

Ocean Trenches: The deepest of the abyss are the ocean trenches. The maximum depth of trench is 11033 m in the Pacific ocean, called the Marianas Trench. These trenches are zones of crustal instability being associated with much earthquake activity.

Sea Mounts: Many volcanic sea mounts rise from the abyss and some of them surface as islands (example in Hawaii). The underwater high mountains are the long ocean ridges, extending from about 4000 m deep to about 1000 m with isolated peaks (as in Iceland) emerging above the surface.

Main Ocean Trenches

In Atlantic Ocean: Puerto Rico trench, depth 9212 m, Sandwich island trench, Meteor chasan, depth 8250 m.

In Pacific Ocean

North Pacific: (i) In Japan trench, Rampo chasm, depth 10230 m, (ii) Cyrillian trench,10550 m (in Iturup island) (iii) Aleutian trench 7435 m (iv) Marianas trench 11033 m   (Ifalik island) (v) Philippines trench 10497 m (Cape Johnson chasm).

South Pacific: Tonga trench 10888 m (Ata island).

Kermadec trench 10000 m (L'Esperance Rock island).

Atacama trench 8050 m (Clule, Lagarators Promontory).

In Atlantic Ocean

Puerto Rico trench (Dominican Republic) Milwaukee abyss 8385 m.

Sandwich island trench, Visokoi island, Meteor chasm 8250 m

In Indian Ocean: Sound trench, Island of Java 7450 m

TERMINOLOGY OF CRUSTAL DEFORMATION FEATURES

6.14 SEA FLOOR FEATURES

Ridge: A long and narrow elevation with sides steeper than those of rise.

Rise: A long and broad elevation. The elevation rise is gentle from the ocean bottom.

Isolated rises from the ocean bottom which appear like mountains are called sea mountains. Ridges are curved particularly if parts of them rise above sea level (which are also called arcs). The broad top of a rise is called plateau.

Sill: A submerged elevation dividing or separating two basins is called a sill. The sill depth is greatest depth at which there is free horizontal link or communication between the basins.

Depression: Trough, Trench and Basin are the large scale depressions on the ocean bottom.

Trough: A long, broad depression with gentle sloping sides is called a trough.

Trench: A long and narrow depression with steep sides is called a trench.

(i) ***Basin***: A large depression, roughly circular or oval shape is called basin. A depression having depth more than 6000 m is called deep depression (3000 fathoms or 5480 m).

(ii) ***Basin shelf***: The zone extending from the low water line (permanent immersion) to the depth about 120 m, where there is a marked or deep descent toward the great depths.

Continental shelf: The feature bordering the continents is called continental shelf.

Insular shelf: It is used for the feature surrounding island.

Slope: The downward slope (declivity) from the outer edge of the shelf into deeper water. Continental slope and insular slope are used to describe the slopes bordering continents and islands respectively.

Bank, Shoal, Reef: The upper parts of elevations which reveal effects of erosion or depositions (used for Bank, Shoal, Reef).

Bank: Approximately flat-topped elevation over which the depth of water is relatively small but at the same time it is sufficient for surface navigation.

Shoal: A detached elevation with deep depth which is hazardous for surface navigation but that is not composed of rocks or corals.

Reef: An elongated rock or coral elevation which is hazardous to surface navigation.

Canyon: The steep-walled fissures that penetrate the slope and cut across the shelf are called canyon and valley, also called gully, gorge, mock-valley.

6.14.1 BASIN

A basin is a depression that is filled with sea water and separated by land or submarine barriers from the open ocean (with which horizontal link or communication is restricted to less depth to great depths. in the basin.)

The maximum depth of an entrance from the ocean to a basin is called threshold depth or the sill depth of that entrance.

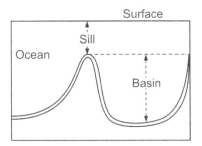

Fig. 6.2 Basin

The water in the basin is restricted horizontal communication with adjacent sea at all levels above that of the lowest sill depth, but below the sill depth renewal of water in the basin can take place by vertical motion only. Below the sill depth water is nearly uniform (homogeneous) and has the character as the water at sill depth.

The Beach: The zone extending from the upper and landward limit of effective wave action to low tide level is called beach.

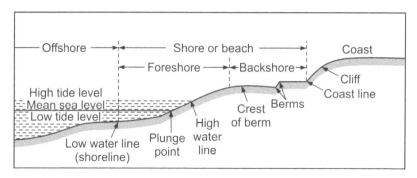

Fig. 6.3 Beach profile (various parts)

The upper part of the beach is covered only during periods of high waves (when storm coincides with spring tides). The various parts of the beach are shown in the figure.

Berms are small impermanent terraces formed by deposition during clam weather and by erosion during storms.

Plunge point is the variable zone where the waves break.

CHAPTER - 7

Soils

7.1 SOIL EROSION TYPES

The wearing away of land surface (soil) by the action of natural agencies like wind and water is called soil erosion

There are about 9 types of soil erosions in India which are briefly described below.

1. **Geological or Normal Erosion:** Geological erosions take place slowly and steadily over a longer period (ages) that makes marked changes in the major features of the earth's surface. There is always an equilibrium between the removal and formation of soil.

2. **Accelerated Soil Erosion:** The removal of the surface (top) soil from denuded (removed) of their natural protection cover by human and animal actions at a much faster rate than at which it was built by soil-forming processes. We are seriously concerned by this type of soil erosion. It may be noted here that nature builds up about 2.5 cm of top soil (on average) in about 1000 years. In wrong farming this will erode in a few years from its average slope.

3. **Wind Erosion:** In arid and semi-arid areas, where vegetation is very little wind erosion takes place during strong winds. The soil particles on the land surface are lifted and blown-off as dust storm or sandstorm. When the dust/sand storm wind slows down (weakens) the coarser soil particles are deposited in the form of dunes and thus fertile lands are rendered unfit for cultivation. In places where fertile soil is blown off the such soil is exposed which loses considerably its productive capacity.

4. **Water Erosion:** It is of three types. Sheet erosion, rill erosion and gully erosion.

    (a) **Sheet Erosion:** During (monsoon) rain run-off a thin uniform field soil covering is washed off. This type of erosion is very insidious. It is detected by muddy colour of run-off from the fields. Soil erosion takes place slowly and imperceptibly

    (b) **Rill Erosion:** In the wake of continuous sheet erosion, the silt laden run-off forms minute finger shaped groves over the entire field. This thin channeling groves is called rill erosion.

(c) **Gully Erosion:** When rill erosion is neglected, the groves develop into wider and deeper channels. This is called gully erosion. The gullies tend to deepen and widen with every heavy rainfall and cause destruction of soil.

5. **Landslide or Slip Erosion:** A landslide is defined as an outward and downward movement of the slope forming material, composed of natural rocks, soil, artificial fills etc. The basic causes of landslides are topography (steep slopes) of the region formed by geological structure of rocks and soil. The slides are caused by earthquakes, heavy rainfall which unduly saturates the ground or a part of road. These are accidents rather than fundamental causes.

6. **Stream Bank Erosion:** Torrents are defined as hill streams characterized wide spreading beds on emergence from the hills with ill-defined banks, flashy flows and swift currents. Generally they are dry water-courses except in rainy season. During very heavy rain in their catchment the streams would be swollen and subside to normal size just after rain stops.

The sudden and violent flows move large quantity of detritus, comprising boulders, shingle (small stones), sand and silt depending upon the geology of the terrain. These debris gets deposited in the torrent bed and creates islands. The bed level of the torrent raised by these deposits. The deposits in turn create overflowing and the meandering of the course and in the erosion of the banks.

7.1.1 MECHANISM OF EROSION

Water Erosion: Soil erosion caused by rainfall is the result of the application of energy from two distinct sources. (i) falling rain drops (ii) the surface flow. Falling rain drops detach soil particles while surface flow transports the soil. Falling drops also helps the movement of soil on sloping levels by splashing.

Wind Erosion: wind helps in three types of soil movement. (i) saltation, (ii) suspension and (iii) surface creep.

(i) **Saltation:** the soil carried by the wind is moved in a series of short bounces called "saltations". In the process of saltation, particles of size 0.1 to 0.5 mm diameter is carried. Saltation is caused by the direct pressure of wind on the soil particles and their collision with other particles. The particles first pushed along the ground and some of them jump a short distance and others leap 30 cm or more.

(ii) **Suspension:** Very fine particles diameters less than 0.1 mm (fine dust) are carried into air by saltation, remain suspended in the air. The suspended dust particles are carried to long distance cause the area eroded, while particles moved in saltation that creep the surface generally remain the eroded area.

(iii) **Surface creep:** Soil particles 0.5 to 1.0 mm diameter moved in saltation are pushed or spread along the surface because of weight. This is called surface creep.

Note: About 90% of total soil movement in wind erosion is below 30 cm height and about 50% within 5 cm of the ground level.

7.1.2 FACTORS INFLUENCING EROSION

soil erosion by water is influenced by

(i) Precipitation (intensity and amount)

(ii) Slope of land

(iii) Type of soil and

(iv) Nature of the ground cover and land use.

Intensity of rainfall (depth) and duration (D) over area (A) determines volume of run-off, percolation into soil. Run-off speed depends on slope of the ground, soil type (namely its structure, texture, organic matter content its infiltration capacity) and permeability effects the soil loss and run-off.

Nature of ground cover and land use determine the erosive power (presence of plants/trees decreases while bare soil increases).

Land use. If the land is left, undisturbed under a natural cover, the run-off and soil loss are the least.

The soil loss and run-off increases steeply in the absence of vegetation and land is cultivated.

Soil erosion is called " The rape of the land". Erosion destroys in days or weeks what has taken centuries to produce fertile top soil in which plants grow. It is often an irreversible reaction. It is a natural process and all our valley soils were brought about by erosion from the hills. It is ever present factor, which turns to our disadvantage by sheer stupidity or carelessness. Erosion is brought about by too little rainfall or by too much rainfall.

Erosion control-Contour ridges with a crop cover and future reforestation help to resist soil erosion.

7.2   SOIL AND WATER CONSERVATION MEASURES

Any soil and water conservation project must have the following two sets of operations.

(i) The mapping of land for classification with respect to its capability

(ii) Planning and executing measures to check erosion, improve land productivity and reclamation of waste land.

Land-capability classification is arrangement of different kinds of land based on crop produce. Land-capability or soil characteristics depend on.

(i)   The texture of the top soil

(ii) Its effective depth

(iii) Permeability of top soil and sub-soil

(iv) Land features – the slope of the land, the extent of erosion, the degree of wetness and susceptibility to overflowing and flooding.

The grouping of soils is based on their capability to produce common cultivated crops, pasture plants (without deterioration over a long period).

7.2.1 SOIL AND WATER CONSERVATION ON AGRICULTURAL LAND

Soil and water conservation on Agricultural land fall into two important classes-

(i)   Agronomic measures

(ii) Mechanical measures

Agronomic soil and water conservation help to: (i) intercept rain and reduce the splash effect, (ii) to obtain better intakes of water by soil, by improving the content of organic matter, soil structure, (iii) to retard and reduce the overland run-off through the use of contour cultivation, mulching, dense growing crops, strip-cropping and mixed cropping.

1. **Contour farming:** Farming across the slope (called contour farming), that is ploughing around the hill. It has a number of beneficial effects. When furrows are ploughed around the slope, each one serves as a micro-dam to check the flow of water. Water stands in each furrow which soaks into the ground. This decreases run-off, soil erosion and loss of plant nutrients. Contour farming conserves soil fertility and increases crop yields.

    On long slopes bunding is done to reduce the length of the slope. These binds serve as a guide for contour farming. All the cultural operations have to be done parallel to these bunds.

2. **Mulching:** Surface mulches are used (i) to prevent soil from blowing and being washed away (ii) to reduce evaporation, (iii) to increase infiltration, (iv) to keep down weeds, (v) to improve soil structure and finally (vi) to increase crop yields.

3. **Dense growing crops:** cultivated legumes provide a better cover and hence better protection to cultivated land against erosion than ordinary cultivated crops. The crops and cropping systems change from one region to region depending on the soil and climatic conditions.

4. **Strip-cropping:** strip-cropping is another form of crop-rotation. It controls run-off, erosion and maintain fertility of the soil. Strip-cropping employs crop-rotation, contour cultivation, proper tillage, stubble mulching, cover-cropping. Broad strips are cultivated on the

contour of a slope for growing row-crops such as corn, cotton, beans or potatoes. These strips alternate with strips in which cover-crops such as wheat, oats or grass are grown.

Different forms of strip-cropping: The main forms are:

(i)   Contour strip-cropping

(ii)   Field strip-cropping

(iii)   Wind strip-cropping

(iv)   Permanent or temporary buffer strip-cropping

Mixed cropping: The objectives of mixed cropping are a better and continuous cover of the land, good protection against the beating action of the rain, complete protection against soil erosion and the assurance of one or more crops to the farmer. The roots of various species in a mixed crop feed at different depths in the soil.

7.3   EROSION CONTROL-MECHANICAL MEASURES

Mechanical measures (also called engineering measures) play a very vital role in controlling erosion on agricultural land. The aim of mechanical measures for controlling erosion are: (i) basin-listing, to increase infiltration by intercepting run-off, (ii) subsoiling to reduce the velocity of the run-off and erosion by dividing long slope into several short ones and (iii) contour bonding.

Basin-Listing: It consists in making of small interrupted basins along the contour with a basin-lister. Basin-lister helps to retain rain water on retentive soil having mild slopes.

Subsoiling: This consists in breaking with a subsoiler the hard and permeable subsoil to conserve more rain-water by way of improving the physical conditions of a soil. This operation promotes greater moisture penetration into the soil, reduces run-off and soil erosion. The subsoiling is worked through the soil at a depth of 30-60 cm, at a spacing of 90-180 cm.

Contour bunding: In this method narrow embankment made across the slope on a level and along the contour. It conserves soil and water in arid and semi-arid areas with high infiltration and permeability.

Graded bunding or channel terraces: Graded bunding used in areas where annual rainfall is less than 80 cm irrespective of the soil, texture. Graded bunds may be narrow = based or broad based. A broad-based graded terrace consists of a wide-low embankment constructed on the lower edge of the channel from which the soil is excavated. The channel is excavated at suitable intervals on a falling contour with a suitable longitudinal grade.

Bench terracing: This consists of a series of platform having suitable vertical drops along contours or on suitably graded lines across the general slope of the land. The vertical drop may be 60 – 180 cm depending upon the slope and soil condition.

Note-Terracing: Terracing is used extensively to check the flow of water on steeply sloping land. A long slope is broken into a large number of short ones by forming a series of banks. The terracing graded machine is used to form flat strips on the contour of the slope. Each strip is divided from another by a bank. Drainage ditches at the base of each bank conduct the water around the slope.

Gully control: When large gullies are formed they become menace for management land resources. Gullies have been classified into four types G1, G2, G3, and G4.

G1: Very small gullies. Depth up to 3 cm, bed width not more than 18 m, Side slopes vary.

G2: Small gullies. Depth up to 3 cm, bed width more than 18 m. Side slopes vary.

G3: Medium gullies. Depth 3 to 9 m, bed width more than 18m. Side slopes uniformly sloping 8 to 15%.

G4: Deep and narrow gullies. (a) depth 3 to 9 m, bed width less than 18 m. side slopes vary. (b) Depth greater than 9 m, bed width and side slopes vary, side slopes steep or vertical.

Gully plugging (control): Gully plugs protect the gully beds by reducing the speed of run-off water, redistricting it, increasing percolation, encouraging silting and improving the soil, moisture regime for establishing a plant cover. Various materials are used for gully plugging like brushwood, live hedges, earth, sand bags, brick masonry and boulders.

7.3.1 SOIL TYPES

Soils constitute the topmost layer of the earth's crust:

Soils are formed by (i) the mechanical and chemical weathering of rocks and (ii) partly by the decomposition of organic matter. (iii) Under the action of weather and climate. More than, hundred years are required to form a few centimeters of top soil. A vertical section of soil presents a succession of natural layers, which are formed as a result of leaching and transfer of material (called profile). Soils differ in color, texture, reaction, consistency and porosity. A large portion of clay and organic matter increases the water holding capacity of a soil. The important soil varities are described below.

1. **Alluvial soil:** These are generally found along rivers and lowlying tracts. They show stratification of sandy, clay and loam.

2. **Sandy soil:** These are found along sea coast and along major river banks. They vary from sandy loam to pure sand. They are highly

porous, deficient in nitrogen, phosphorous, potassium and in some cases lime.

3. **Desert soil:** These are found in Rajasthan and adjoining parts of Gujarat. They are poor in organic matter, nutrients but they have sufficient minerals.

4. **Peaty soils:** Peat means decomposed vegetable matter. Peaty soils are found in humid regions and contain large organic matter. They are black in color, contain high clay, nitrogen and sufficient potash but low phosphorus.

5. **Saline and alkaline soils:** These are found in drier parts of North-West India, over Deccan plateau. In summer they bring dissolved salts to the surface by capillary action which form a white crust. These soils are unproductive for tree growth.

6. **Laterite:** Laterite soil is a mixture of hydrated oxides of aluminium, iron with small quantities of Manganese oxide, Titanium etc., These are found in regions of heavy rainfall zones and moist climate. They are poor in nutrients, phosphorus, potassium and calcium and deficient in nitrogen. The pH ranges 4.5 to 5.5 and base exchange capacity is low.

7. **Red soil:** They are deficient in organic matter and poor in plant nutrients. The presence of ferric oxide gives the soil red coloration. Base status low with poor exchange capacity.

8. **Black soil:** These soils are called red or black cotton soils, developed from basaltic rocks. The black colouration is due to the presence of iron salts. They are loamy to clay, expand on wetting and shrink on drying. They are poor in organic matter, phosphorus nitrogen but have sufficient calcium. They are found over most part of the Deccan plateau.

9. **Forest/mountain meadow soils:** They are rich in nitrogen and contain organic matter in surface layer. They have variable base status depending on the degree of washing by percolation (leaching).

10. **Skeletal soils:** They are found in hills of drier regions of the peninsula and in the Himalayas. These are isolated and have local importance.

   Soils under different forest vegetation differ considerably in their physical and chemical characteristics.

7.3.2 DESERT AND DESERTIFICATION

Deserts: Deserts are barren areas with scanty vegetation of thorny bushes. Rainfall is scanty an highly variable. Desserts are located both in warm and cold climates. The hot deserts are the Sahara in north Africa, Kalahari in southern Africa and Thar desert in India. Mexico, South America and

Australia also have hot deserts. The cold deserts are located in Iran and Turkey. The desert soil is rocky with sand or salt. Weather generally long periods of hot spells with extreme temperatures. Annual rainfall mostly less than 50 mm, Geophytes grow in deserts with cacti like bushes adaptation. These plants tolerate drought high temperature, alkalinity with scanty rainfall.

In 1972 there was widespread drought in monsoon region of Africa and Asia. The southwestern region of Sahara desert (in Africa) which includes Chad, Niger, Mali, Upper Volta, Mauritania and Senegal. These areas have mostly dry lands with scanty rainfall together called Sahel Desert. Sahel was stuck by severe drought in 1911, 1940 and 1968. During normal years receives rainfall about 270-290 mm. Russo, a station in Sahel received 122 mm in 1968 and 295 mm in 1969, white during 1970, 1971 and 1972 receives 149 mm, 126 mm and 54 mm respectively. The continuous years of severe drought stirred the United Nation and UN general Assembly passed a resolution in 1974 (Dec.) calling for an International Conference on Desertification, which was held in 1977. According to this Conference the desert and desertification was defined with the help of Aridity Index.

$$\text{Aridity Index } (I_A) = \frac{Precipitation}{Evapotronspiration}$$

If $I_A < 0.03$, it is termed Hyper Arid zone

If $0.03 < I_A < 0.20$ it is called Arid zone

If $0.20 < I_A < 0.50$, it is called semi-Arid zone

and If $0.50 < I_A < 0.75$, it is termed sub-Humid zone

In Hyper Arid zone, vegetation is practically absent, shrubs observed near river beds and desert condition

In Arid zone Dryland Areas. Sparse perennial and annual vegetation. Nomadic pestoration.

In Semi-Arid zone, tropical shrubs, rainfed agriculture with livestock breeding

In sub-Humid zone dense vegetation, rainfall agriculture very common, Crops adapted to seasonal drought are grown

Desert Soils: A large part of the arid region between the Indus river and Aravalli range affected by desert conditions (west Rajasthan, Haryana, Punjab) of geologically recent origin. This part is covered under a mantle of blown sand which, combined with the arid climate results in poor soil development. The most predominant component of the desert sand is quartz in well rounded grains together with feldspar and hornblende grains and fair proportion of calcareous grains. The area receives little southwest monsoon rain. The sands which cover the area are partly derived from disintegration of subjacent rocks, but are largely blown in from the coastal regions. Some of these soils contain high percentages of soluble salts, possess high pH have low loss in ignition, a variable percentage of calcium carbonate and are poor in organic matter.

There are five geographical deserts over the world.

1. Widespread bolt from Atlantic Ocean to Sahara, China including Arabian Desert, Rajasthan, Taklamakan and Gobi deserts of Magnolia and China.

2. The Sonoran desert of northwest Mexico together with southwest USA

3. The Kalahari and the Namib desert of south Africa

4. Most part of Australia

5. The Atacama desert

The above five deserts are located in the Atmospheric High pressure belts of sub-tropics, where large scale atmospheric subsidence exists with very small amounts of rainfall

Largest Deserts

Name of Deserts	Area (km^2) (About)
Sahara Desert	8.6×10^6
Australian Desert	1.35×10^6
Gobi Desert (China)	1.04×10^6
Arabian Desert (Middle east Arabia)	2.30×10^6
Kalahari Desert (Southern Africa)	0.58×10^6
Taklamakan Desert	0.32×10^6
Thar Desert (India, Pakistan)	0.45×10^6
Atacama Desert (south America)	0.18×10^6

Australian Desert = great Victoria + Great Sandy + Gibson

Australian Desert = 0.65×10^6 Km^2 + 0.39×10^6 Km^2 + 0.31×10^6 Km^2

$$= 1.35 \times 10^6 \text{ Km}^2$$

Deserts have a tendency to spread outwards because of transports of deserts sandy soil to the neighboring areas and thereby reduce soil fertility. That is why periphery of desert is prone to encroachment of desert.

The desert ecosystems are complexes of interaction between plants, animals, soils, water and energy. Environments are more hostile to vegetation (plant life). United Nations Conference aims to mitigate by combat "desertification". Desertification is largely aided by human activities like rearing goats and other live-stock in addition to low rainfall and sandy soil.

Desert Development: The best example of desert development in Rajasthan. Conservation of rain water by percolation into soil, improving plantation and crops through diversion of water from nearly rivers (irrigation channels and lift variegation channels). Harvesting crops that require less water, preserving and improving forest, rearing cattle which withstand with low water like camels, avoiding rearing goats, combating spread of desert lockust (pests)

Budyko Radiation Index of dryness (BI) as given below

$$BI = \frac{R_n}{L_r}$$

Where R_n = net radiation available for evaporation of a wet surface.

L_r = heat requirement to evaporate the mean annual precipitation

If BI < 1, it is humid area

BI > 1, it is dry

Based on BI index, the climate classification is given below

BI	Climate
> 3.0	Desert
2.0 – 3.0	Semi-desert
1.0 – 2.0	Steppe
0.33 – 1.0	Forest
< 0.33	Tundra

7.4   INTERIOR STRUCTURE OF THE EARTH

Seismology is the science of earthquakes. In order to understand the nature of seismic wave propagation and detection, a quick survey of the interior structure of the earth is essential.

The interior of the earth consists of : Lithosphere or earth's crust, Mantle and Core.

The earth's outer shell or layer is called crust or Lithosphere. Litho – means stone. Earth's crust is not monolithic but divided into strata. The top structure is made of sediments or sedimentary rocks, the second layer is made of granite rocks, and the third layer is made of basaltic rocks. The density of rocks increases, basaltic > granite > sedimentary rocks. The three crust layers, are found everywhere in continents. However under oceans the granite layer is missing (see Fig. 7.1).

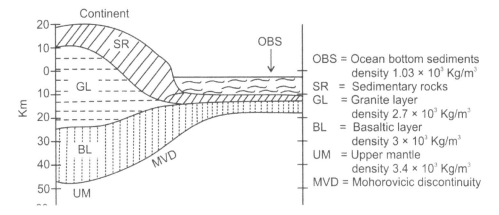

Fig. 7.1 Constituents of earth's crust

The boundary between the sedimentary and granite rocks has not been given any name, the boundary between granite and basaltic rocks is called Konard discontinuity. The boundary under basaltic (between Crust and Mantle) is called Mohorovicic or simply Moho discontinuity.

The average thickness of the crust is about 35 km, which is very thin compared to the radius of the earth (about 6400 km). The average thickness of the earth's crust under oceans is 5-10 km, while its thickness below continental mountain ranges is about 50 km. The average density of the earth is 5.5×10^3 kg/m³, Lithosphere 2.8×10^3 kg/m³, Mantle 3.2 to 5.7×10^3 kg/m³ and core 9 to 12×10^3 kg/m³. The mass of the earth is about 6.00×10^{24} kg, Lithosphere 5×10^{22} kg, Mantle 4.05×10^{24} kg (68% of the earth) and core 1.88×10^{24} kg (31% of the earth). The depth of the Mantle is about 3000 km and Core is 3000 km.

A discontinuity means an interface and indicates marked change of material property. Below the earth's crust lies Moho discontinuity (at an average depth of 35 km, where density is 3.3×10^3 kg/m³) which separates the earth's crust from Mantle (the interior of the earth). The Central Core (below the Mantle) probably consists of liquid outer core (thickness about 2000 km) and solid inner core (thickness about 1000 km). Gutenberg discontinuity separates the Mantle from the central core and is located at a depth of about 3000 km, where the density is about 5.7×10^3 kg/m³. There is still uncertainty about the composition of materials in the interior of the earth. It is generally viewed that the principal constituents of Lithosphere as oxygen (93.88%), Silica, Aluminium, Iron, Calcium, Sodium, Potassium and Magnesium (granite and basaltic rocks) all together they makeup 98.5% of the earth's crust by weight, Earth's crust mostly contains Sial (Silica, Aluminium compound) and Sima (Silica, Magnesium compound).

Mantle consists of Oxygen, Silica, Mangesium and Iron (Iron, Magnesium Silicates). The common silicate in the Mantle is probably Olivine (Formula: [(Mg Fe)$_2$ SO$_4$]). These silicates are similar to those found in stony meteorites. The core probably consists of iron, sulphur in the combined form as FeS and nickel (Fe, FeS, Ni). Core contains NiFe (Nickel, Iron compound).

The building blocks of Lithosphere are rocks namely Igneous, Sedimentary, Metamorphic rocks. Sedimentary layer is thick below the continents and becomes thin in ocean area, whereas the granitic layer almost absent (in ocean area).

The temperature of the interior of the earth increases with depth. In Lithosphere it increases about 1°C for every 30 metres of depth, however this increase will not continue to the centre of the earth. Observations indicate that the average increase in temperature is about 1°C/km. The cause of the increase of temperature with depth is suggested to be :

(i)   heat generated by the pressure of overlying rocks,

(ii) primordial heat (original heat from the time of the earth's formation), and,

(iii) by the radioactive mineral disintegration.

The interior of the earth behaves like solid elastic in respect of earthquakes and earth-tides. It is known that oceans rise and fall twice in 24 hours due to gravitational attraction of the Moon on Earth. In a similar way an earth-tide occurs in the earth's crust and Mantle which causes rise and fall about 30 cm under the action of the earth-tides. Geneva based synchrotron works only during complete rest period of the crust and it will not work even if there is slightest change in earth's crust due to ebbs and flows occuring far away from the unit. It is found, the Geneva Synchrotron operates about 30 hours a week and the rest period the surface of Switzerland is vibrating. It has been scientifically proved that the entire territory of Moscow daily rises or falls about 50 cm from a certain average level. This rise and fall is caused by the gravitation of the Moon and the Sun.

7.5 PLATE TECTONICS

According to Indian Cosmology the earth consisted at one time of seven continents joined together. They separated like the lotus petals (or leaves) from Mount Meru, the centre of the universe. Afterwards the continents floated and drifted away from the centre and were separated by seven oceans. An Amateur Russian Astronomer Y.V. Bykhanov observed a remarkable coincidence of the outlines of the American and Euro-African coastlines that they fit well without a crack if these were moved together. By this observation, in 1877, he conjectured that once a uniform continent split into parts and ever since they have been moving away. In 1910, Alfred Wagener (1880-1930), a German geophysicist propounded a theory of continent Panghela (all earth) was broken up into pieces like the lotus leaves, they separated from one another and floated away giving rise to the modern continents divided by oceans. In 1960's this hypothesis was further modified by the evidence of ocean floor structure and named tectonics of plate or global tectonics (See Fig. 16.3). According to this hypothesis, instead of continental move, plates of large areas of the earth's crust containing both continents and adjoining sections of ocean floor moved.

According to this plate tectonics hypothesis there are six major plates.

1. Euro-Asiatic plate,
2. African plate,
3. Antarctic plate,
4. Indo – Australian plate,
5 American plate and
6. Pacific plate.

In addition to these major plates there are several minor plates located between them and they move to some extent independently (see Fig. 7.2).

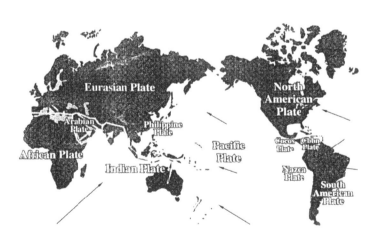

Fig. 7.2 Major global tectonic plates and average direction of motion. Subduction occurs along colliding plates. After toksoz (1975)

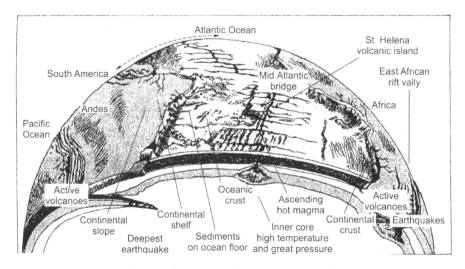

Fig. 7.3 According to the sea-floor spreading theory, the splits in the mid-oceanic ridge are deep cuts in the surface of the earth (below the sea) from which molten rock gushed up to trigger earthquakes, thus moving the sea floor away from the ridge on either side

There are three types of Plate Boundaries

1. Extension or divergent boundaries
2. Compression or Convergent boundaries and
3. Transform faults.

Extension boundaries are formed when two adjacent plates move apart. Material from below swells up and a new crust is produced at the crests of the oceanic ridges. This makes both sides of the plates are added up with mass.

Compression boundaries are formed when two adjacent plates approach each other. In this case surface is destroyed. The line along which plate destruction takes place is called trench.

A third type of boundary forms when the plates move laterally relative to each other. The line along which plates move laterally is called transform faults. In this case neither crust is formed nor destroyed. It has been found that Eurasia and American plates are converging at a rate of 2 to 4 cm per year, while American, Eurasian and African plates are increasing in size and that of Indian and Antarctican plates are not changing significantly in size.

Fault: A fracture in a rock mass or rock layer whose opposite faces move independently is called a fault. The various types of fault, movements are shown in Fig. 7.4.

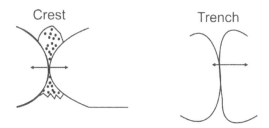

Crest   Trench

Faults

There are three types of faults – normal Reverse and Strike slip.

Normal faulting is associated with divergent boundaries ($\leftarrow \rightarrow$) (or Cristal extension).

Reverse faulting is associated with convergent boundaries ($\rightarrow \leftarrow$) (or crustal collision shorting)

Strike slip faulting is associated with lateral movement of the crust and have found at transform boundaries ($\downarrow\uparrow$)

Summary

1. Earth interior divided into 4 layers called: crust mantle outer core (liquid) inner core (solid). Temperature of mantle varies between 600 °C to 300 °C, while core temperature about 4300 °C.
2. *Lithosphere:* consists of top crust together with upper solid mantle *Asthenosphere* consists top molten layer.
3. *Tectonic plates:* Lithosphere is broken into several distinct plates which float on the asthenosphere. These plates called tectonic plates.
4. *Plate boundaries:*
   (a) Locations where two plates are moving away called *Divergent boundary.*
   (b) Location where two plate are pushing against one another in opposite directions is called convergent boundary
   (c) Location where two plates are slipping in opposite direction is transform boundary.
5. *Seismic waves:* Consists (i) Body waves (ii) Surface waves.
   (a) Body waves travel in the earth's inner layer and are two types: *Primary waves, Secondary waves.* Primary waves are longitudinal waves, while secondary waves are transverse. Primary waves toss the buildings up and down ($\downarrow\uparrow$), while secondary waves shake the building from side-to-side ($\leftrightarrow$).
   (b) Surface waves travel on the earth's surface and are two types – *love waves, Rayleigh waves.*

Love waves are surface waves move fast along the ground from side-to-side. They are named after mathematician AEH Love who worked out mathematical model.

Rayleigh waves roll long the ground like a wave that rolls across a lake or an ocean. These move on the ground up and down and side in the same direction. Shaking is felt due to Rayleigh wave.

Rayleigh waves named after William Strutt and Lord Rayleigh who predicted the existence mathematically.

Folds: During tectonic movements stratified rocks develop bending. These are called folds.

Tectonics: According to one hypothesis earth's crust has large plates. These plates move due to convective forces that emanate from beneath the crust and create rifts. It is assumed that the upper Mantle consists of Newtonian Viscous fluid and the convection currents are generated due to heating from below or insitu. The lower Mantle consists of very dense fluid which inhibit convection. It is clear in case of divergent or extension boundaries a new crust is formed at the boundaries of these plates. All such boundaries are located in the oceans. The earth's crust building up towards American plate on one side and towards African and Euroasiatic plates on the other side. The plates collide at the boundaries which leads to (convergence or) submergence or subduction of one under the other plate. Such submergence taking place in case of Pacific plate under Euroasiatic plate. It is theorised that where old crusts are buried they provoke for earthquakes and volcanic eruptions. The present scientific thinking is that earthquakes are caused by the friction on the boundaries of the plates moving together. Powerful tremors are caused by the accumulated shear stresses which periodically exceed the rock strength. From the heating of the sedimentary layers of the submerging plate volcanic cruptions take place.

The various types of fault movement are shown below in Fig. 7.4.

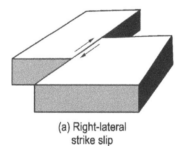
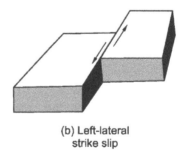

(a) Right-lateral   (b) Left-lateral
strike slip   strike slip

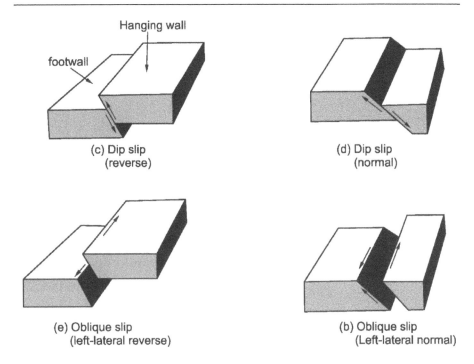

(c) Dip slip
(reverse)

(d) Dip slip
(normal)

(e) Oblique slip
(left-lateral reverse)

(b) Oblique slip
(Left-lateral normal)

Fig. 7.4 Various types of fault movement

San Andreas is (a) type of fault; Himalayas has (c), (d), (e) and (f) types while peninsular India undergoes (d) and (f) generally. Here (c) and (e) are rare. Reverse faults are due to compression (shortening of the crust), whereas the normal faults are due to tension (stretching of the crust).

Earthquake waves

A disturbance which progresses from one point to another point in a medium with transfer of energy but without the transfer of matter is called a wave motion. Elastic waves are mechanical disturbances propagated in an elastic medium. A wave is called longitudinal or compressional if the particles of the medium vibrate in the direction of wave propagation. Longitudinal waves travel through solids, liquids and gases.

A wave motion is called transverse or shear if the particles of the medium vibrate at right angles to the direction of propagation. Transverse waves travel through solids but not through liquids and gases.

An earthquake generates two kinds of waves. Primary or P-waves, which are compressional and travel through solids, liquids and gases. Secondary or S-waves, which are shear waves and travel only through solids. P and S-waves may not develop in actual displacement of a land mass on the surface crust.

Generation of P and S waves using springs as analogy shown in Fig. 7.5.

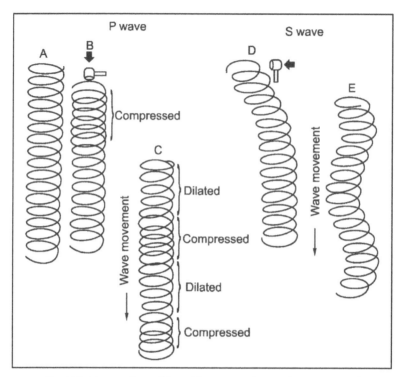

Fig. 7.5 Generation of P and S waves using spring as an analogy

The figure below shows the oscillation pattern of Rayleigh and Love waves.

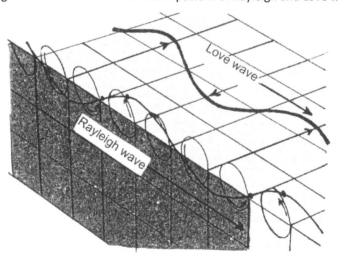

L-wave or long surface waves travel through surface layers of the earth. Seismic wave motions are shown in Fig. 7.6.

The study of seismic waves produced by earthquakes provide valuable information about the nature of the matter inside the earth in its path. Seismic waves travel deep down into the earth from earthquake site and return to the earth's surface at some distant point. Seismographs are used to detect the seismic waves on their arrival at the surface of the earth. They also provide the information of type of the wave, its intensity and time of arrival. Speed of Seismic waves partly depend on the density of material through which they pass.

The markings of seismograph is shown in Fig. 7.6.

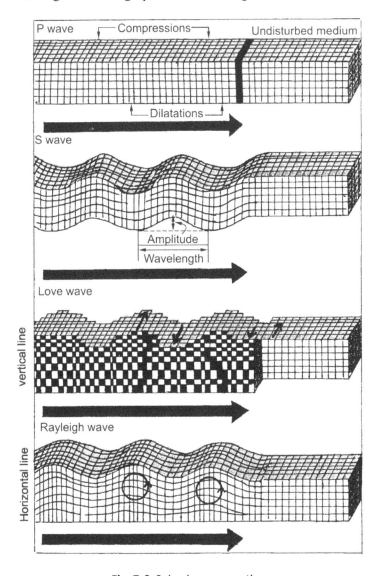

Fig. 7.6 Seismic wave motions

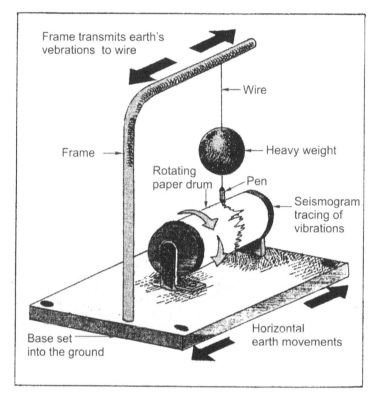

Fig. 7.7 During an earthquake, the assembly together with the frame anchored into firm ground moves to and fro. A pen attached to the weight records the tremors on the paper covering the rotating drum

Earthquake Parameters are :

  (i) **Time of origin** is the time at which earthquake has occurred.

 (ii) **Focus** is the point inside the earth where from the earthquake originated. See Fig. 7.8.

(iii) **Duration of an Earthquake** generally less than one minute.

(iv) **Epicentre :** the surface point vertically above the focus. It is expressed in latitude and longitude of the point. See Fig. 7.8.

 (v) **Focal depth** is the depth of the focus from the surface of the earth. See Fig. 7.8.

(vi) **Hypocentre** of an earthquake is the combination of epicentre and focal depth.

Classification of Earthquakes

  (i) Shallow earthquakes have focal depth 70 km

 (ii) Intermediate earthquakes have focal depth between > 70 km 300 km.

(iii) Deep earthquakes have focal depth > 300 km

    Earthquakes have not recorded focal depth exceeding 720 km.

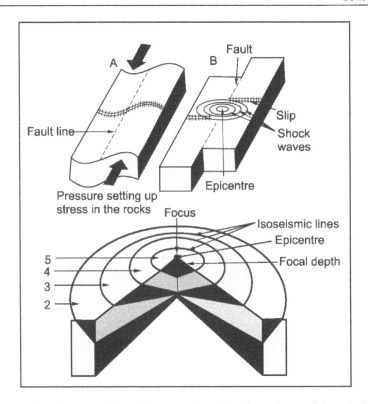

Fig. 7.8 Earthquake waves begin from a point called focus located deep below the surface of the Earth. The nearest point above the focus on the Earth's surface is called the epicentre around which the maximum damage is caused. The intensity of the quake decreases as the distance from the epicentre increases. Isoseismals are imaginary lines on a map joining points of equal intensity. These are drawn based on damage survey, after an earthquake, according to modified Mercalli (MM) scale

Magnitude (M)

The magnitude of an earthquake is a kind of instrumental measure of its size or energy (E), and is given by a linear equation

$$\log_{10} E = a + b M \qquad .....(7.1)$$

Or

$$E = 10^{(a + bM)}$$

where a and b are constants, E = Energy, M = Magnitude

The values of constants are generally found to be a = 11.8, b = 1.5

If   M = 4, we have $E_1 = 10^{a + 4b}$

         M = 6 we have $E_2 = 10^{a + 6b}$

$$\frac{E_2}{E_1} = \frac{10^{a+6b}}{10^{a+4b}} = 10^{2b}$$

or $\qquad E_2 = E_1 \times 10^{2b} = E_1.10^3 \qquad$ (since b = 1.5)

Thus the energy radiated by earthquake of magnitude 6 (E_2) is 10^3 times the energy radiated by the earthquake of magnitude 4 (E_1).

If E_3 is the energy of magnitude of earthquake 8, then

$$E_3 = E_1 \ 10^{4b} = E_1 \times 10^6$$

i.e., $\qquad E_3$ is 10^6 times E_1.

log E_s = 11.8 + 1.5 M, where E_s in ergs, M is Richter scale magnitude.

If M = 8.25, E_s = Energy = 3.68 × 10^{10} kwh or 13.248 × 10^{23} ergs.

M = 7.5 E_s = 2.86 × 10^9 kwh or 10.296 × 10^{22} ergs.

Seismic energy (E_S) yield for different magnitudes (M) given in Table 7.1.

Table 7.1

M	E_S in TNT	M	E_S in TNT
5.0	32000 tons	8	5 billion tons
5.5	80000 tons	9.0	32 billion tons
6.0	1 million tons	10.0	1 trillion tons
6.5	5 million tons	12.0	160 trillion tons
7.0	32 million tons	8.5	1 billion tons
7.5	160 million tons		

Intensity

It is based on effects of the earthquake on buildings, topography, landslide etc., i.e, on macroseismic effects.

The waves spreading from an earthquake centre pass through different rocks at different velocities. Seismic waves have a maximum velocity of 12.5 kmps. On passing through the Mohorovicic boundary the primary seismic waves speed up from 6.5 to 8 kmps and secondary waves from 3.7 to 4.5 kmps. However on passing from mantle into core the primary waves speed drops 12.5 to 8.5 kmps and the secondary waves from 7.5 to 5 kmps. The spreading waves from an earthquake centre pass through different rocks at different velocities. Since the waves are reflected, and refracted on their way, they reach observatories at different times. On an average more than one lakh (10^5) earthquakes of varying intensities are registered over the globe every year by the seismological observatories. The data is evaluated by using 12 grade modified Mercalli (M.M) scale. The scale can be divided into several groups:

1. ***Slight Intensity or First Group:*** It consists of first three grades which are weak and not imperceptible earth tremors. These are sensed by some animals. Most of domestic animals become restless, birds fly away from the place of earthquake. Cats fur stands on end. It is also said that second and third grade tremors are sensed by nervous people.

2. ***Moderate Intensity or Group Two:*** It consists of grades 4, 5 and 6. These shocks are felt by everyone. In this group objects hanging on walls move, hanging lamps/bulbs, chandeliers swing to and fro. Cracks develop in some houses. Tall factory chimneys may fall down.

3. ***Severe Intensity or Group Three:*** It consists of 7, 8 and 9 grades. The strength of this group is destructive and devastating. Tall buildings may fall down, cracks may occur in the ground and occasional human casualties are noted.

4. ***Catastrophic or Very Severe Intensity or Group Four:*** This   last group consists of 10, 11 and 12 grades. The strength is described as catastrophic. Many buildings will collapse except structures on monolithic rocks which may not be affected. The shocks and yawning cracks will be very severe. These may cause electrical fires.

Causes of Earthquakes

The most common cause of earthquake is tectonic activity. This mechanisms is shown in Fig. 7.9. The vertical dashed line (PQ) shows fissure or fault in the solid earth crust. By slow prolonged tectonic movement in the lithosphere one side X of the fault (Fig. 7.9 (a)) is displaced in relation to the other Y. This is shown in Fig. 7.9(b), by a deformation of straight lines drawn across the fault (P' Q'). This process continues until the stresses thus generated in the fault zone overcomes the friction between the two sides X and Y. Then a rupture (sudden displacement) occurs. After this the configuration is shown in Fig. 7.9(c). It is this sudden rupture constitutes an earthquake. The slow process is repeated and a new shock occurs at some time later. This mechanism of earthquake is called elastic rebound theory of tectonic earthquakes. Almost all earthquakes occur by this mechanism. However there may be some tremors by volcanic activity. Large earthquakes may occur with a volcanic explosion. The collapse of cavities can be the origin of minor tremors.

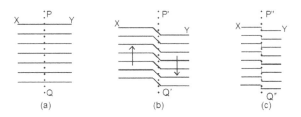

Fig. 7.9 Elastic rebound theory of tectonic earthquakes

Long ago the super continent Pangea broken and drifted like the lotus leaves. A plate containing India broke away from it and drifted towards Asian land mass. When this plate collided with Euro-Asian plate sediments of ancient sea bed squeezed together in huge folds and slowly rose to form the Himalayas.

Indian plate consisting of India and parts of Indian ocean is moving at an average speed of 5 cm / year in the north-northeast direction and colliding with Eurasian plate along the Himalayas. This resulted in faults and fractures in the Himalayas. These are responsible for some great earthquakes in the Himalayan region in the past.

Categorisation of earthquakes	*Richter scale*
Tremor or micro-earthquake	Magnitude < 3.0
Slight earthquake	Magnitude 3 to < 5.0
Moderate earthquake	Magnitude 5 to < 7.0
Great earthquake	Magnitude 7.0 or more

Important Seismic Belts
1. First belt, the most important Seismic belt runs along the Pacific and includes western coast of South America, North America, eastern coast of Asia, the Island of south coast Pacific and New Zealand.
2. Second belt runs from south Pacific Islands through Java, Sumatra and Central Asia mountains, further passing through Caucasus mountains to Greece, Italy and Spain.
3. Third belt which is not so important runs from north to south in the middle of the Atlantic ocean.

7.6   SEISMICITY OF INDIA

IMD maintains a catalogue on earthquakes in India and neighbourhood from available historical records and also instrumental data (See Fig. 7.10 and 7.11). This catalogue is continuously updated. From this records it is observed that moderate to great earthquakes have occurred all along the Himalayan region, the Rann of Kutch, Manipur, Myanmar (Burma) belt and further continuation to Andaman and Nicobar islands. Scattered earthquakes also occurred but less frequent in peninsular India with magnitude less than 6.5. Latur (Osmanabad district) earthquake had magnitude 6.3. The catatrosphic earthquakes of magnitude 8 or more is given in Table 7.2.

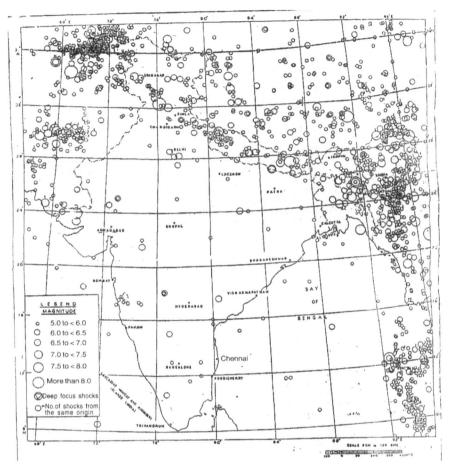

Fig. 7.10 Map of India showing epicenters (upto 1994)

Source : India Met Dept.

Table 7.2

Date	Place	Magnitude
12.06.1997	Assam	8.7
15.08.1950	Arunachal Pradesh	8.5
15.01.1934	Bihar-Nepal border	8.3
26.06.1941	Andaman islands	8.1
04.04.1905	Himachal Pradesh	8.0
16.06.1819	Rann of Kutch	8.0

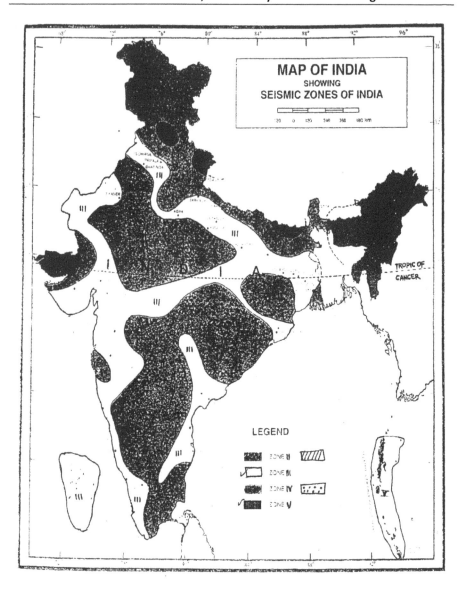

Fig. 7.11 Map of India showing seismic zones (source IMD)

Research: The upgraded seismological network gave rise to useful and unique digital broad band and strong motion data sets for several significant earthquakes in the last decade including the recent great Sumatra earthquake of 26th December 2004. This helped in understanding about the earthquake process in the inter-and-intra-plate seismic regions. The Crust and upper Mantle structure of Peninsular shield region generated by the regional events. For the first time ground motions expected from future scenario earthquakes have been estimated from the Jabalpur (1977) and Bhuj (2001) earthquake data.

Role of IMD

IMD is the nodal agency of Government of India, responsible for monitoring seismicity in and around India. IMD rendered more than 100 years of seismological service to the nation. The first seismological observatory of the country was set up at Kolkata in 1898. The operational task of the (IMD) department is to quickly determine the earthquake parameters immediately after the occurrence of an earthquake and disseminate the information to all the concerned State and Central Government, agencies responsible for rendering relief and rehabilitation. The information is also broadcast to public on AIR and DD and press etc.

National Seismological Network

It consists of 47 permanent observatories and 4 observatories in Northern India for special studies (See Fig. 7.12). Of the existing observatories 10 of them located at Ajmer, Bokaro, Bilaspur, Bhopal, Bhuj, Chennai, Karad, Pune, Thiruvananthapuram and Visakhapatnam, recently upgraded with global seismograph Network standard digital broadband seismograph system and 14 also been upgraded with Broad Band Seismograph systems of different makes. All 24 systems are of the state-of-art type having broadband sensors, high dynamic range (24-bit) digitizers, Global Positioning System time synchronisation and facility to access the data remotely through telephone mode or satellite communications.

A central Receiving Station has been set up at IMD HQ in New Delhi, which has the operational responsibility of keeping round the-clock watch of seismic activity, downloading the waveform data from remote field stations through dial up facility, analyse and disseminate the earthquake information to user agencies (See Fig. 7.13).

National Seismological Data base centre at IMD HQ in New Delhi, supplies a number of seismicity related reports for specific regions for establishment of Industrial units, power houses etc, and provides consultancy services to various State and Central Government Agencies on earthquake related matters.   The other important activities of the division – all correspondence related to earthquake prediction, disaster management, supply of seismological data to various national and international organisation including research and academic institutions, river valley projects etc.

The Seismological Division publishes a monthly National Seismological Bulletin which contains the phase data and the processed information on epicentral parameters of all earthquakes located by the National Seismological Network. The bulletins are periodically sent to International Seismological center (ISC) to incorporate into the global network.

IMD is a permanent Member of the International Seismological Centre UK, earthquake information, publications, bulletins are regularly exchanged by IMD and ISC.

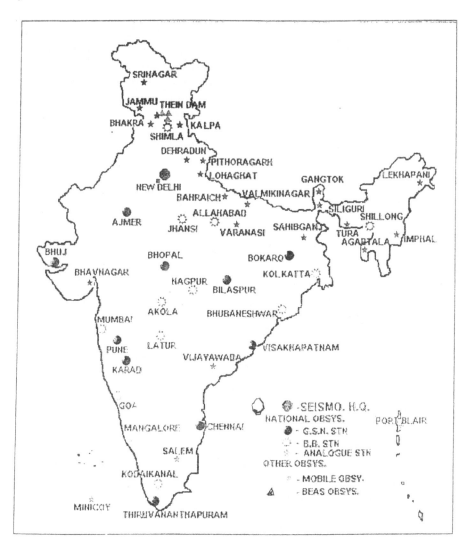

Fig. 7.12 Seismological observatories of India meteorological department

Seismic Zoning of India: Based on Bureau of Indian standards (IS-1893 - part 1 : 2002) collection of scientific inputs from a number of agencies, the country has been divided into four Seismic zones ; Viz. zone II, III, IV and V. Of these zone V is the most Seismically active region intensity on Richter scale 9 or more, zone IV 8 to 9, zone III 6 to 8, while zone II is the least 5 or less.

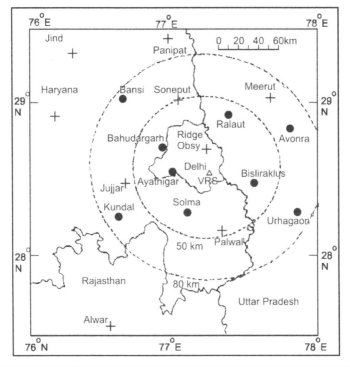

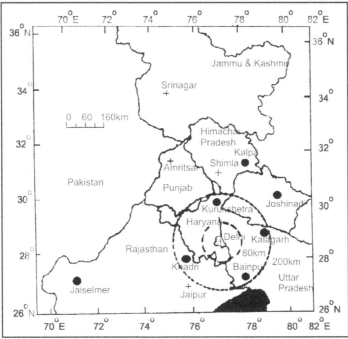

Fig. 7.13 Seismic telemetry network in and around Delhi

Important Services of Seismological Division

Micro-earthquake surveys for monitoring after shocks, Swarm type seismic activities and site response studies by deploying portable seismographs in the affected area.

After shocks in Bhuj area is monitored at Rajkot, Surendranagar, Jamnagar for studying the explosion like blast sound reported in the region.

As part of Tsunami warning system for the Indian ocean region, a real time Seismic monitoring and dissemination of data is in progress.

Major event in 2004: A great earthquake of magnitude 9.3 occurred on 26 December 2004 at   06 hrs 29 minutes IST off west coast of Sumatra Island region, lat 3.3° N long 96.1° E. It generated destructive Tsunamis, took a toll of more than 3 lakh people. The Seismic belt extended from Sumatra to North Andaman Islands and caused several earthquakes of large magnitude. The main shock was followed by intense aftershocks. To study these an array of five temporary field observatories set up in Port Blair and observatory upgraded deploying state-of-art digital broadband Seismograph system. The five field Seismological observatories are located at Port Blair, Baratang, Havelock, Hut Bay and Great Nicobar.

CSO Shillong (Central Seismological Observatory Shillong) was established in 1952 where Seismology, Meteorology and Radiation units are functioning with the round-the-clock watch of seismology is functioning from 24-4-1989. All earthquakes of magnitude 4 or more within epicentral distance of 25 degrees transmitted to HQ New Delhi. Recording of plate motion with the help of Geographical Position System is continued from December 2001.

Earthquake Risk Evaluation Centre (EREC) was established in 2004 by Government of India to guide national effort in mitigating disastrous impact of earthquake and to evaluate earthquake risk. Microzonation of Jabalpur, Guwahati   and Delhi is undertaken in collaboration with Global Seismological Institute (GSI), National Geophysical Research Institute (NGRI), IIT Roorkee and   Central Building Research Institute (CBRI) Roorkee.

According to National Disaster Management Authority (NDMA) all new constructions must be earthquake resistant, particularly in cities located in seismic zones. The guidelines are selective seismic strengthening and retro-fitting of existing priority structures located in high-risk areas. Compliance of fresh building codes, revised town planning that will make it mandatory for all builders to incorporate earthquake-resistant features in their construction plans.

Earthquake Safety

It is said that earthquakes do not take lives but the ill constructed structures kill the people. Bhuj (Gujarat) earthquake of 26 Jan 2001, intensity 6.5 on Richter Scale took a toll of thousands, while the earthquake of higher magnitude (about 6.5 - 7.0) shook Seattle in February 2001 without any casualitiy and there were no house collapses, because the houses were constructed with earthquake safety guidelines. During an earthquake, structures (buildings) will experience two kinds of seismic wave forces. One is lateral longitudinal waves or horizontal to and fro and the other uplift (vertical or transverse waves). The lateral or shear forces which are horizontal they shook the buildings back and forth, while the transverse waves cause the buildings move up and down. Keeping these movements in mind, using metal connectors, shear walls and fasteners, the structural stability of the building may be strengthened. This is the basis of earthquake proofing.

Retro-fitting process is reinforcing the foundation of existing buildings to bring them to the level of the present norms of earthquake resistence. The cost of retro-fitting may take about 10% more than the actual building cost. It must be noted that a single storey or a multi-storied building are equally hazardous to earthquakes but a poorly designed structure may crumbel while a well-designed multistoried structure may only shake.

When mild earth tremors are felt rush out of the house to an open area, avoid taking shelter near walls or structures which may collapse. Even if you are in sturdy house observe the safety rule of Drop-Cover and Hold.

CHAPTER - 8

Environmental Pollution

INTRODUCTION

According to WHO air pollution is world's top environment health risk. It causes one-in-eight deaths globally. Death due to air pollution increased four folds across the global over the past decade and the number of death due to air pollution in plugged at 8 million every year. Of this 3.7 million (globally) are from outdoor or ambient air pollution. About 88% of premature deaths attributed to air pollution exposure occurred in low and middle income countries and greatest number in the Western Pacific and south-east Asia regions. WHO studies indicate besides development of respiratory diseases, cardiovascular diseases as strokes and ischaemic heart disease and cancer occur due to air pollution. Latest resolution passed during 68th World Health Association called for all countries to develop air quality monitoring system.

The UN conference on Human Environment held at Stockholm in June 1972 stressed to take appropriate steps for the protection and improvement of human environment. According to UN Report, at present mankind is persisting with thoughtless and extravagent consumption of natural resources in an unprecedented manner. The UNEP (October 2007) outlook report says, at present consumption rate it requires 21.9 hectares per person while earth's capacity 15.7 hectares per person. As a result of this 116550 square km of forest area being lost across the world each year. According to WHO report, about one fourth of global deaths are caused by environmental risks, like polluted air, dirty water, hazardous work places and dangerous roads. The WHO estimates 12.6 million deaths in 2012 or about 23% of total world deaths were attributed to such environmental factors. The burden is greatest on the poor and the youngest: Majority from environmental risk is highest in sub-Saharan Africa and low and middle income countries in Asia.

It is the fundamental duty of every citizen to protect and improve the natural, environment including air, forests, lakes, rivers and wild life. It is in this context, the following Government of India Protection Act came into existence.

1. The Environmental Protection Act 1986

2. The Air Pollution Act 1981

3. The Water (Protection and Control) Pollution Act 1974

4. The India forest Act 1927 and Forest Protection Conservation Act 1980

5. The Wild life (Biodiversity) Protection Act 1972

6. The Public Liability Insurance Act 1991.

8.1   AIR POLLUTION

The presence of solid, liquid and gaseous contaminants in the atmosphere in such quantities which damage materials and property, and cause injury to human, animal or plant life or interfere unreasonably with normal human activity and enjoyment of life is called air pollution. In short any undesirable concentration of contaminants in air may be called air pollution.

Natural Contaminants

SO_2 (Sulphur dioxide), H_2S (Hydrogen sulphide), HF (Hydrogen fluoride), HCl (Hydrogen chloride) which emanate from volcanoes, due to thunderstorms, putrefication. These are considered harmful only when the concentration exceeds certain limit. Toxic gases like Nitrous oxide (N_2O), Ozone (O_3) form or evolve due to electrical discharges in the atmosphere. Natural particulate matter, such as dust from the deserts and volcanoes, ashes from forest fires, salt particles from sea, meteoritic dust, pollen from flowers are important.

Artificial Pollutants

Emission from chimneys, exhaust from automobiles, aeroplanes and other man made activities. Agricultural insecticidal, pesticidal dusting and spraying, domestic and municipal incinerators, burning of vegetation introduces pollutants into the atmosphere. Thermo-nuclear explosions introduce radioactive pollutants.

The Important Air Pollutants

CO (Carbon monoxide), Hydrocarbons, SO_2, Oxides of nitrogen, Aldehydes, Ammonia etc., photochemical process lead to further generation of harmful gases and aerosols. Hydrocarbons formed by the incomplete combustion of liquid petroleum products. Pollutants that enter the atmosphere by human activities are called Anthropogenic pollutants.

Some basic definitions are given below

A gram molecule or mole (kilogram molecule or kilomole)

It is the amount of chemical substance whose mass expressed in grams (kilogram) is numerically equal to its molecular weight.

The volume of one mole of a substance is called its molar volume.

A gram atom (kilogram atom)

It is the amount of a simple chemical substance (an element) whose mass expressed in grams (kilograms) equals to its atomic weight.

Avogadro number

For all substances,

number of molecules in a gram molecule

$$= \text{the number of atoms in a gram atom} = N_A$$

This number N_A is called Avogadro number.

$$N_A = 6.023 \times 10^{23} \text{ mol}^{-1}$$
$$= 6.023 \times 10^{26} \text{ k mol}^{-1}$$

Molar ratio (or volume ratio): The number of moles of the compound per mole of air.

Mass ratio: The mass of the compound per unit mass of air.

Mass Concentration: The mass of compound per unit volume.

Concentration (or molecular concentration): The number of moles of the compound per unit volume.

$$\text{Mass ratio} = \text{molar ratio} \times \frac{M_j}{M}$$

where M 29 = the average relative molecular mass of air

M_j = Relative molecular mass of the compound considered.

ρ = density of air

Mass concentration = Mass ratio $\times \rho$

$$\text{Concentration} = \frac{\text{Mass concentration}}{M_j}$$

ppm = parts per million (10^{-6})

ppb = parts per billion (10^{-9})

ppt = parts per trillion, (10^{-12})

At temperature 15 °C, pressure p = 1013 h Pa, the air density = 1.225 kg/m^3

Notation: (m) in bracket denotes mass ratio, (v) in bracket denotes volume ratio.

Examples

1. Conversion

(i) Given molar ratio of oxygen in dry air is 21% (v)

$$\text{Mass ratio of oxygen} = \text{Molar ratio} \times \frac{M_j}{M}$$

$$= 21 \times \frac{32}{29}$$

$$= 23.17 \text{ (m)}$$

(ii) Given mass ratio of $SO_2 = 1.1$ ppb (m)

Molar ratio or volume ratio of SO_2

$$= \text{Mass ratio of } SO_2 \times \frac{M_j}{M}$$

where   mol wt. of S = 32

mol wt. of O = 16

mol wt of $SO_2 = 64$

$$= 1.1 \text{ ppb (m)} \times \frac{29}{64}$$

$$= 0.5 \text{ ppb (v)}$$

Mass concentration and molar concentrations of minor constituents of air are expressed in units $\mu g\ m^{-3}$ (micrograms per cubic meter) or $\mu \text{mole}\ m^{-3}$.

2. Consider x molecules cm^{-3}.

$$x \text{ molecules } cm^{-3} = x \frac{\text{mole}}{N_A} cm^{-3}$$

$$= x \frac{\text{mole}}{N_A} 10^6\ m^{-3}$$

$$= \frac{x}{N_A} 10^6\ \mu \text{ mole } m^{-3}$$

$$= \frac{x}{N_A} 10^{12}\ m \text{ mole } m^{-3}$$

3. Consider 145 ppb (v) of CO as mass concentrations and molar concentration

$$145 \text{ ppb (v)} = 145 \times \frac{28}{29} \text{ ppb (m)} \quad \text{at } T = 15\ ^\circ C,\ P = 1013 \text{ h Pa}$$

$$= 140 \times 10^{-9} \times 1.225 \text{ kgm}^{-3}$$

$$= 140\ \mu g \times 1.225\ m^{-3}$$

$$= 171.5 \ \mu g \ m^{-3}$$

$$= \frac{171.5}{28} \ \mu \ mole \ m^{-3}$$

$$= 6.12 \ \mu \ mole \ m^{-3}$$

$$= 6.12 \ \frac{N_A}{10^{12}} \ molecules \ cm^{-3}$$

$$= \frac{6.12 \times 6.03 \times 10^{23}}{10^{12}} \ molecules \ cm^{-3}$$

$$= 3.69 \times 10^{12} \ molecules \ cm^{-3}$$

During 20th Century (1900-2000) Global industrial activity increased 20 to 25 folds, Global population grown from about 2.5 billion to about 6 billion. More land has been cleared for cultivation during this period than in all preceding human history, fossil fuel consumption increased 30 to 35 folds. Global demand for water increased enormously due to growth of industries, irrigation and domestic use. This resulted in water withdrawal from existing source was more than six fold, which was more than double the population growth rate during the same period. All this resulted in destruction of green lungs of the earth (plant kingdom), burning of fossil fuels in motor vehicles, furnaces, factories, electricity producing plants. The later added carbon dioxide, Sulphur dioxide and oxides of nitrogen emissions to the atmosphere together with toxic chemicals like lead, mercury. Some of these are responsible for the destruction of ozone layer in the stratosphere, acid rain, photochemical smog. Increased agricultural activity added methane, toxic pesticides to the atmosphere. In addition to these greenhouse gases (namely CO_2, N_2O, CH_4, CFCs), tropospheric ozone have been added into the atmosphere which are responsible for the Global warming of the atmosphere.

If N_2 and O_2 were the only two constituents of the atmosphere, the average air temperature near the earth's surface would have been −18 °C, that is 33 °C colder than the at present 15 °C. The gases CO_2, N_2O, O_3, CH_4, CFCs, water vapour warm the earth's surface like that of a glass house by allowing solar radiation coming to the earth and preventing the terrestrial radiation escaping from the atmosphere. These gases are called green house gases. The principal sources of green house gases are man-made, such as fossil fuels, aerosol sprays, refrigerants, agriculture, deforestation. Natural sources are volcanoes and forest fires which are insignificant. During last 100 years barring water vapour all other green house gases increased in the atmosphere. In totality during last 150 years the green house gases increased in the atmosphere by an amount radiatively equivalent to a 53% increase in CO_2, although the gas (CO_2) itself increased by only 26%. The main impact of the green house gases are climate change, sea level rise and adverse effect on health.

Except the CFCs the green house gases cycle (through atmosphere, biosphere-earth system) with sources and sinks are adding to or subtracting from the atmospheric concentrations. On the whole man has achieved in reducing the sinks and adding to the sources. The life time of the green house gases in the atmosphere vary from few hours or weeks to more than 100 years. Tropospheric ozone has a life period of few hours or days, CFCs have about 75 to 110 years, nitrous oxide (N_2O) 150 years, methane 7-10 years, CO_2 50-200 years. The concentration of CO_2 increased from 275 ppmv before industrialisation to 354 ppmv by the end of 20th century. The major sources of emission of this gas is burning of fossil fuels, deforestation and change in land use. The concentration of tropospheric ozone increased 15 ppbv to 35 ppbv by the end of 20th Century. Ozone formed from vehicle exhausts and industrial pollutants in sunshine (photo chemical action). Ozone contribution to the green house effect is about 8%. The increased effect of CFCs to the green house effect during last 100 years is about 25%. It was non-existent in atmosphere before industrialization and it is found to the order of 0.25 to 0.45 ppbv by the end of 2000 AD. The sources of CFC's are man-made chemicals used as solvents, spray can propellants making of foam and refrigeration. Nitrous oxide contributes 7-8 percent in green house effect. It is increased 228 ppbv before industrialisations to 320 ppbv by the end of 2000 AD. It (N_2O) is found in fossil fuels and burning of biomass, fertilizers and land use change. Methane contributes about 13% to the green house effect. Its concentration roughly trebled (0.7 ppmv to 2.0 ppmv) during last 100 years. The major methane producing sources are : Swamps, rice paddies, ruminants (animal cud-chewing) and fossil fuel extraction. Overall man has achieved to add green house gases and reduced the sinks.

According to WMO, No. 735 (1990) the annual carbon fluxes were: photosynthesis on land removes 100 Gt [Gt (gigaton) = one million metric tons] of carbon from the atmosphere annually in the form of CO_2, plant and soil respiration together returns 100 Gt (each 50 Gt). Fossil fuel burning and deforestation releases carbon into atmosphere 5 and 2 Gt respectively. Physicochemical processes at the sea surface releases about 100 Gt into the atmosphere and absorb about 104 Gt. The net gain by the atmosphere is about 3 Gt annually.

Of the green house gases increased by human activities 70% accounts to energy sector. Vegetation is called the green lungs of the earth is also a carbon sink. Deforestation reduced one third of forested land on the globe (six billion hectares to four billion hectares).

In addition to release of green house gases, many metals and chlorinated hydrocarbons are also released into the atmosphere by industrial activities. Some of these substances are toxic in high concentration to human beings, animals and plants. According to WMO estimates the anthropogenic toxic substances that are introduced into the air in kilotons per year are: lead 332,

Cadmium 7.6, Copper 35, Nickel 56, Zinc 132, Arsenic 18 while the natural process introduces these substances 19, 1.0, 19, 26, 46 and 8 Kilotons/year respectively.

Group of Experts on the Scientific Aspects of Marine Pollution (GESAMP) established that all oceanic areas over the globe are affected by man-made pollutions. Though concentrations are not high, but these toxics are deposited in the seas and on land over long periods and bio-accumulate in the food chain. From the atmosphere to sea, fluxes of metals such as Pb, Hg, Cd, Zn are estimated to be greater than or comparable in magnitude to direct discharges and transport by rivers.

By burning Coal, oil and metal smelting industry introduces mainly SO_2 and oxides of nitrogen into the atmosphere, which are largely responsible for acidic rain. Acid rain is mainly confined to local, regional scale than global. Rain and snow are acidic and have a global pH of about 5.6.

8.2 pH VALUE

Concept of Acids and Bases

A substance which can donate or give a pair of electrons to form a coordinate bond is called a base. A substance which can accept a pair of electrons to form a coordinate bond is called an acid. Thus base is an electron donor while an acid is an electron acceptor. pH of a solution is the negative logarithm of the concentration (in moles per liter) of hydrogen ions which it contains.

$$pH = - \log [H].$$

A neutral solution has pH $=7$

If pH is less than 7 the solution is acidic

If pH is greater than 7 the solution is alkaline.

The pH of lemon juice is about 2.2, Vinegar 3.0, pure rain 5.6, Distilled water 7, Ammonia 12.

Buffer Solution

A solution which resist change in its pH value on addition of an acid or a base is called Buffer solution.

Le Chatelier's Principle

The state of equilibrium is affected by concentration, temperature and pressure. According to Le Chateliers principle if a system at equilibrium is subjected to a change of concentration, temperature or pressure the equilibrium shifts in the direction that tends to undo the effect of the change. Thus in a chemical equilibrium, when concentration of reactants is increased the equilibrium shifts in favor of the products. Conversely when the concentration of the products is increased the equilibrium shifts in favour of the reactants.

Thus a liquid changes into vapour by absorption of heat. The vapour pressure of a liquid increases with increase of temperature. The boiling point of a liquid increases with increase of pressure. The solubility of substance increases with increase of temperature.

Methods of measurements of important pollutants are given below.

SO_2	:	Measurements are made by UV-pulsed fluorescence, flame photometry, dilution or permeation tube calibrations.
CO	:	Measured by Non-dispersive IR tank gas and dilution calibration.
O_3	:	Measured by ozone UV-generators, UV-spectrometers, Gas Phase Titration (GPT) calibrators.
NO_2	:	Measured by Chemiluminescence method, permeation, or GPT calibration.
Pb (lead)	:	Measured by High Volume Sampler and atomic absorption analysis.
Sulphates and Nitrates:		Measured by High volume Sampler and chemical analysis.
Hydrocarbons	:	Measured by flame ionization method or Gas Chromatograph, *calibration* with methane tank gas.

TSP (Total Suspended Particulate) Measured by High Volume Sampler and weight determination method

8.3   POLLUTIONS STANDARD INDEX

Based on pollutants effects, pollutions standard index (PSI) is determined. Corresponding to PSI air quality, pollutants concentration are given in Table 8.1.

Table 8.1

PSI	Air Quality 24-hr	Pollutants concentration				
		TPS 24-hr $\mu g/m^3$	SO_2 24-hr$\mu g/m^3$	CO 8-hr $\mu g/m^3$	O_3 1-hr $\mu g/m^3$	NO_2 1-hr $\mu g/m^3$
0		0	0	0	0	0
50	50% NAAQS	75	80	5	80	–
100	NAAQS	260	365	10	160	–
200	Alert	375q	800	17	400	1130
300	Warning	825	1600	34	800	2260
400	Emergency	875	2100	46	1000	3000
500	Significant harm	1000	2620	57.5	1200	3750

NAAQS : National Ambient Air Quality Standards. μg = Micrograms

NAAQS – 1970 given in Table 8.2.

Table 8.2

Pollutant	Period of measurement	Primary standard $\mu g/m^3$	Ppm
CO	8–hr	10,000	9
	1–hr	4000	3
Hydrocarbons(non-methane)	3–hr	160	0.24
NO$_2$ (Nitrogen oxide)	1–yr	100	0.05
O$_3$	1-hr	240	0.12
So$_X$	1–yr	80	0.03
	24-hr	365	0.14
	3–hr	None	None
TSP	1–yr	75	None
	24–hr	260	–
		–	
Pb	3–months	1.5	–

The Table 8.3 gives PSI, corresponding health effects and precautions

Table 8.3

PSI	Air quality	Health effects and Precautions
200	Alert	Generally unhealthy. Irritation symptoms in healthy persons. Mild aggravations of symptoms in susceptible persons. Persons suffering from heart and respiratory ailments reduce physical exertion and outdoor activity.
300	Warning	Persons suffering from heart and lung diseases symptoms of significant aggravation and decreased exercise tolerance. Healthy persons lose exercise tolerance. Elderly persons with heart and lung ailments should stay indoors and lessen their physical activity.
400	Emergency	In ailing persons aggravation of symptoms. In eldely persons premature onset of certain diseases. In healthy persons decreased exercise tolerance Ill and elderly persons stay indoors, avoid physical exertions. Others avoid outdoor activity.
500	Significant harm	In ill and elderly persons onset of premature death. Healthy persons experience adverse effect in their normal activity. All persons should remain indoors. Doors and windows should be closed. Minimise physical exertion. Avoid traffic.

A. The effects of Air Pollutants

The effects of air pollutants depends on the toxicity of the substance its concentration, period of exposure and individual personal/animal/plants internal resistivity.

B. Effects of Gases

Compounds

Effects of some important air pollutants are given below.

SO_2: Affects health, damages to materials and ecosystem, aids acid precipitation and affects climate indirectly.

CO: Directly affects health and indirectly affects climate.

CO_2 and CFCs: Directly affects climate and indirectly affects stratospheric ozone.

O_3: Directly affects health and climate and damages materials and ecosystem.

Hydrocarbons: Directly affects health and indirectly affects climate, causes damages to materials and ecosystem.

NO, NO_2: Directly and indirectly affects health, damages to materials and ecosystem, aids acid pptn, directly affects stratospheric ozone layer and indirectly affects climate.

NO_2: Indirectly affects stratospheric ozone layer and directly affects climate.

C. The Effects of Particulate matter

Aerosol particles directly affects health and climate. Sulphate and nitrate containing particles directly affects health and climate and damages to materials, aids acidic rain.

SO: it affects climate. Particulate matter of heavy metals and radionuclides directly affects health and causes damage to ecosystem.

The above discussed pollutants are directly introduced into the atmosphere by industrial activities. However a part of N_2O and CO_2 are let into the atmosphere by agriculture and forestry activities. The primary species, like SO_2, NO and NO_2 aid for the secondary formation of tropospheric ozone, sulphate particles and some particulate matter. It may be noted here that all of these compounds barring CFCs and some radionuclides are natural constituents but they should be considered pollutants only when their anthropogenic emissions effect their concentration in air significantly.

It has been estimated that on average in a year the atmosphere is polluted by more than 200 million tons of CO, more than 50 million tons of various hydrocarbons, about 146 million tons of SO_2, 53 million tons of oxides of nitrogen, 200-250 tons of dust and 120 million tons of ash and soot. Daily

10000 tons of cosmic dust falls on the earth. Its origin is still a mystery. It may be the result of the activity of the sun or originate in zodiacal nebulae.

Suspended Particulate Matter

On an average a man breathes 22000 times in a day and inhales about 15-22 kg of air. The daily respirable particulate matter (RSPM) should be 100 Mg/m^3. If on any two consecutive two days RSPM levels cross 100 mark, it is categorised as high pollution. Delhi is the most polluted city in India and fourth in the world.

It was mentioned earlier that particulate matter in the atmosphere effects human beings and also weather system. In high concentration many kinds of air borne particles were found to be toxic. The hazards of Chernobyl radioactive fallout and Bhopal isocyanide gas leakage were well documented. The accident of Chernobyl (USSR), nuclear power plant radioactive fallout moved with the winds to Scandinavia. The fallout was scavenged by rains to render lichen and the reindeer browsing on lichen radioactive. Animals were found to carry too high, a burden of radioactivity to be consumed by the people of the region. The radioactive debris were carried over the whole of northern hemisphere by westerly wind circulation.

Dust and ash particles that spewed out of volcanoes or industrial smoke stack has cooling effect on the atmosphere. The eruptions of Tambora (1815 which took a human toll of 12000 lives) spewed out about 80 cubic kilometers of ash and rock pieces into the atmosphere. Some particles entered into stratosphere and remained there. This caused cooling. As a result there was no summer in 1816 in northern parts of US and it caused extensive crop failure. The volcanic dust, ash, water vapour reached high altitudes and were carried by winds around the globe. The precipitation of this dust back to the earth took years. The ejected particles into the atmosphere, dispersed sunlight and temporarily increased the earth's albedo. The Krakatoa (in East Indies) eruption on August 27, 1883, was the most powerful volcanic eruption in the history of mankind till that date. The eruption of clouds reached the heights about 30 km. It turned day into night in Batavia, about 150 km away. A Tsunami wave, caused by the fall of broken away part of Krakatoa island travelled around the world. Rare optical phenomena were observed in Europe in the end of November 1883. The sky remained purple for several hours during sunsets. This was caused by the dispersion of sunlight by a layer of dust particles (size about 2 mm) injected by the volcano into the stratosphere. The dust/ash particles did not settle down on earth for several years. Successive years of weather over the entire earth was cooler than the normal. This was due to the increase of earths albedo and partly due to the mixing of ocean waters by Tsunami waves. The violent eruption of volcano Gunung Agung in Bali in 1963 injected particulate matter into the upper troposphere and stratosphere which gave rise to the spectacular sunsets throughout the world. The eruption caused the rise in stratospheric temperature by about 5 °C in the height range of 18-20 km (60-80 hPa). The volcanic particles consisted of fragments of lava and crystalline material (Calcium and Ammonium sulphate). The ash and pumice discharges of a few renowned volcanic eruptions are given in Table 8.4.

Table 8.4

Name of volcano	Year	Estimated amount of discharge (ash and pumice) (Cubic kilometers = km^3)
Mt. Mazama	4600 BC	42
Mt. St. Helens	1900 BC	4
Vesuvius	79 AD	3
Tambora	1815	80
Krakatoa	1883	18
Mt. Katmai	1912	12

Airborne particulate matter acts as a cloud condensation nuclei and freezing nuclei. It scatters and absorbs solar radiation and this influences the weather. Sea dust (spray) caused by the surf rises into PBL and on evaporating in dry air leaves sea salt and material particles in it. Since major part of the earth's surface is covered with oceans, the concentration of salt particles in the atmosphere are found to be about 10^8 particles /m^3 over oceanic area and 10^6 particles/m^3 over land. The fallout of salt in coastal areas and islands range 10-150 kg per acre per annum with extremes as high as 1400 -1800 kg per acre per annum. Atmospheric concentrations of aerosol particles vary from less than 0.1 µg/kg in polar regions to 100 µg/kg or more in dusty continental air. On an average 1 µg/kg have been noted in the middle and upper troposphere.

Airborne particles inhaled deposit in lungs passage and lungs. This may effect the tissues and cause or aggravate lung diseases. Carcinogenic substance present in the air cause lung cancer. Airborne particles may stick to any surface and cause soiling or corrosion. In addition to the scattering of sun light, air-borne particles reduce insolation reaching earth when present in bulk and under favourable atmospheric conditions (inversion of temperature close to the surface of the earth). It causes haze and smog. This deteriorates the visibility and becomes hazardous in airports and other surface transports. Airborne particles are subjected to all sorts of chemical reactions such as photochemical action, oxidation, reduction, polymerization, condensation and catalysis. It is also possible that some of these reactions lead to the formation of substances which are more toxic.

Note: *Precipitation scavenging* means removal of any particulate or gaseous matter from atmosphere and depositing on the earth's surface by cloud droplets, rain, snow flakes, fog droplets (atmospheric hydrometeors).

E.g., In Hyderabad (India) normal day TSPM at important junctions found to be 220-290 $\mu g/m^3$. After continuous 3 days raining (13-15 July 2004) it was found to be 150-200 $\mu g/m^3$ due to rain washout.

Cycles denotes the circulation of the element between the reservoirs, that is individual atoms after sometime return to the same reservoir again.

Resident time is the length of time that an atom or molecule spends in each reservoir. Resident times of aerosol particles depend on the size of the particles and on meteorological factors.

Reservoir is defined by its physical characteristics such as the atmosphere, the oceans, the biosphere etc. However it also refers to the chemical form in which the element occurs. For example sulphur occurs in the atmosphere in the form of SO_2 [Most of the elements or compounds in different environments are transferred by physical or chemical transport processes between different reservoirs].

Airborne particles (aerosols) may have different size, shape, chemical properties and density. Consequently their atmospheric residence time and removal processes differ widely. Normally sea spray particles, fine dust particles size any from 2 to 20 μm, larger particles (called coarse type particles) raised by gusty strong winds have smaller residence time–minutes to hours because they are heavy and have rapid sedimentation. Coarse type particles are removed from the atmosphere mainly by sedimentation, dry deposition at the surface or by precipitation scavenging. Fine airborne particles have very low rate sedimentation and dry deposition. Their residence time in the atmosphere is governed by precipitation scavenging. Residence time of aerosol particles is a function of the size of the particles and precipitation frequency. On average the residence time of these particles vary from a few days to a few weeks. In polar regions the average residence time extend to many weeks because of low rate of precipitation and stable stratification of atmosphere. Similarly is the case with the residence time of aerosol particles in the upper troposphere.

Non-radioactive trace gases pollutants contain mainly Carbon, Sulphur, Hydrocarbons, Nitrogen and Ozone. On global scale the natural sources of many of these trace gas pollutants and particles immensely exceed the anthropogenic emissions. However in case of industrial and urban community air pollution, anthropogenic emission are much more than natural ones. Thus hazardous air pollutants are confined to these population areas. Trace pollutant substances may be of organic or inorganic in nature. As regards radionuclids, they have three sources; (i) Natural substances from the soil and oceans, (ii) Cosmic dust, (iii) man-made nuclear bomb-debris.

Compared to anthropogenic non-radioactive trace gas pollutants, radioactive pollutants are very few but they are potentially health hazards. The consequence of nuclear radiations can be realised from the world war nuclear explosions over Hiroshima and Nagasaki. However radioactive bomb debris are used in the study of understanding stratospheric processes (they affect the electrical conductivity).

The important natural sources (earth's crust) of radio nuclides found in the atmosphere are: Uranium-U^{238}, Radon-Rn^{222}, Thorium Th^{232}, Radium - Ra^{226}, Thoron-Tn^{220}. These have half life period 4.5 × 10^9 yr, 3.83 days, 1.65 × 10^{10} yr, 1.58 × 10^3 yr and 54.5 sec respectively. Tropospheric residence time of U^{238}, Ra^{226}, Th^{232} is 5-30 days, Rn^{222} has 3.8 days and Tn^{220} has 50 sec. The important cosmic dust radionuclides are Beryllium – Be^{10} & Be^7 Carbon-C^{14}, Sulphur S^{35} and phosphorus- P^{32}. Their half-life period is 2.7× 10^6 yr and Be^7 53 days, 5.7 × 10^3 yr, 87 days and 14.3 days respectively. Tropospheric residence time of Be^{10} and Be^7 is 5-30 days and that of C^{14}, S^{35} and P^{32} is not known.

Important man-made sources of radionuclides in the atmosphere are nuclear weapons testing, waste from nuclear reactors and nuclear power fuel cycle. The principal emissions are Strontium-Sr^{90}, Cesium- Cs^{137}, Carbon-C^{14}, Zirconium – Zr^{95}, Barium- Ba, their half-life period is 27.7 yr, 28.8 yr, 5.7 × 10^3 yr, 65 days, 12.8 days respectively and tropospheric residence time 5-30 days for all except Ba which has 13 days. In addition to the above, Tritium-H^3 (t), Krypton K_r^{35} - also have anthropogenic sources and transuranium species Plutonium, Americium and Curium.

Note:

1. Radioactive form of carbon is called Carbon-14 (C^{14}). C^{14} is readily detectable with Geiger counter and hence it is used as a tracer. C^{14} combined with oxygen form a tell-tale carbon dioxide which is used in biological research.

2. Radioactivity effects of radiation on living tissue are very complex. It mainly destroys the blood cells when exposed to it. White blood cells form the marrow of bone. Consequently radioactive emission destroys bone and marrow. Radioactive emission destroys/influences gases and cause mutation. Human mutations are fetal. It causes death of a baby before it born. People who were exposed to nuclear radiation due to atomic explosion over Nagasaki and Hiroshima became the victims of Leukemia, Cancer of Thyroid, breast, lung and stomach. It caused damage to bone marrow, the spleen, the gastrointestinal tract lining and the central nervous system. Mental and physical damage was found in unborn and born infants. It was found radiation exposure causes destruction of reproduction cells, sudden increase in white blood cells count and then drop, slight drop in Red Blood Cells count, fatigue, fever, sore throat, vomiting, diarrhea, and cataract of eye.

3. Radiation energy received by an organism is called the absorbed dose. It is expressed in units of rads.

Radiation shows effects in man when it exceeds 100 rads. Radiation commonly measured in millirads. When a group of people exposed to a single dose of 350 to 450 rads, about half of them lose their lives. This is expressed as LD 50 (Lethal Dose-50) for humans. Another unit rem (roentgen equivalent man) is defined as the product of rad and a qualifying factor, where qualifying factor is a function of tissue exposure, ionising radiation and energy associated with radiation. In general radiation absorption expressed in millirems. An average man receives about 1 m rem per year from energy production.

Absorbed radiation dose =

$$\frac{\text{Radiation energy absorbed}}{\text{The mass of the substance exposed to radiation}} (J/kg)$$

$1 \text{ rad} = 10^{-2} \text{ J/Kg}$

It was estimated that in America, an average man receives total whole body radiation doses from all sources (natural /environmental, diagnostic and other sources) is about 180 m rem per year per person. Human body normally recovers from a maximum of an acute absorption of 250 rems.

4. Geiger-Muller counter/detector is used for radiation measurement.

   $1r = 2.580 \times 10^{-4}$ Coulomb Per Kilogram   (where r = roentgen)

   The roentgen is the exposure dose which produces a total of 1 esu (esu = electro static unit) ions of each sign in one cubic centimeter of air under standard conditions.

5. *Radioactivity* is defined as the spontaneous disintegration (or transmutation) atoms of unstable isotopes of one element into isotopes of another element. The disintegration is accompanied by the emission of certain particles (helium, α, β or neutron particles, or gamma radiation).

8.3.1   HAZARDS OF NUCLEAR FALL-OUT

Ionizing Radiation: Radiation like X – rays, Radioactive wastes of atomic and nuclear reactors cause ionization of gas. Natural radioactive elements and cosmic rays are called background radiations. Radioactive elements emit ionizing radiation consist of α – particles, β – particles and γ – rays. The primary cosmic ray particles have high amounts of energy.

Biological effects due to ionising radiation: Ionizing radiation exerts both short-range and long-range effects. In short range (or acute) effects may take place within a few days/weeks after exposure to radiation while long range effects or delayed effects over a long period. Short range effects include physiological, morphological that lead to physical disability or instant death of organisms on exposure. Long range (or delayed) effects include genetic changes, mutation like tumors, cancer, shortening life span, hazardous body growth and changes.

The interaction of radiation with living tissues are very complex. Excessive exposure to radiation including sunlight, x-rays and nuclear radiations can cause destruction of tissues. In mild cases the destruction reflect as burn (like sunburn). Long exposure may cause severe illness or even death. For example the destruction of the components in bone marrow that produces red blood cells. However exposure to natural sources of radiation including cosmic radiation to the extent 10 to 100 times very rarely harmful.

Natural background radiation causes low doses of ionizing radiation. Long period background radiation might have caused mutation in wild populations for new evolution. Man-made ionizing radiation is causing great harm to environment which is a matter of great concern to life on earth. The physical harm of radiation is directly proportional to the damage. It is assumed any radiation is harmful.

Hazards of Nuclear fallout

At present the explosion of nuclear devices, atomic, (hydrogen) bombs etc are greatly adding to the amounts of background radiation. Hiroshima atomic bomb explosion on 6 August 1945 released energy 8×10^{10} ergs. [1 Joule = 10^7 ergs]

1Megetion (TNT) nuclear bomb = 4×10^{22} ergs.

North Korea tested Hydrogen Bomb on 3 September 2017. It is the biggest ever nuclear detonation. It is five times bigger than Hiroshima explosion. It caused 6.3 magnitude tremor. Since 1945, there had been close to 2056 nuclear tests conducted, of these only 10 tests conducted after the comprehensive Nuclear Test-Ban Treaty in 1996. After the NTB treaty was signed, India and Pakistan conducted two nuclear tests each in 1998 and then announced unilateral moratorium. Only North Korea conducted by defying understanding six nuclear tests.

Nuclear fallouts: The main radioactive nuclear fall-out are 1. iodine–131 (I–131) and strontium – 90 (Sr – 90). Iodine is required in the formation of thyroxine which is essential for the proper functioning of thyroid glands in animals, I– 127 is stable, while the radio isotope I–131 is unstable. Plants contain a small amount of I (iodine) and animals get their iodine requirement through food chain. The concentration of iodine increases along food chain pyramid- from producer level (plant) to the top. Biological system do not distinguish I–131 and I–127. Hence I-131 enters the food chain at all levels and becomes concentrated in the top (end) level. Man mostly lies in the last link of many food chains. As a result I–131 generally gets concentrated in the human body. I-131 concentration in human body can damage white blood cells, bone marrow, spleen and lymph nodes. This may result in lung tumors, skin cancer, sterility (unproductive) defective eye sight etc. 2. In plant and animals Sr-90 can replace calcium. In plant calcium is needed for the formation of middle lamellae of cell walls and in animals it is required for bone formation. Radioactive Sr-90 may also enter the food chain and get concentrated (in the terminal food chain pyramid) in human body like I-131. This accumulation leads to bone cancer, tissue disintegration in most animals and man.

Effect of radioactive elements on aquatic life

It is best explained by the following two examples

Example-1: Radioactive isotopes biologically magnified in food chains are P-32 (phosphorous-32) and Zn-65 (zinc-65). In addition to these Co-58 and Co-60 are also produced in radioactive wastes in nuclear fission process. Absorption of radioactive elements as nutrients by aquatic algae leads to their accumulation in fish beyond-safe or permissible levels.

Safety measures from nuclear radiation

There is no straight method to stop back-ground ionizing radiation, but its dose is very low. Hence it poses no serious hazard to life. In case of nuclear reactors, effective safety measure is the solution, where the leakage of radioactive elements from nuclear reactors is prevented. The second step is to store the radioactive wastes in places of dump where they gradually decay to their final stable product without causing significant harm to biological life.

Nuclear technology, its knowhow of methods used for handing and safe disposal of radioactive materials and radiation are matters of National Security and National Choice. According to the IAEA (International Atomic Energy Agency), April 1986 Chernobyl Nuclear disaster (rated level-7) and Fukushima (Japan) Nuclear disaster (noted level-7) explained as major release of radioactive materials with widespread health and environmental effects require implementation of planned and extended counter measures.

Note:

1. A UN Treaty to ban atomic weapons appears to be an exercise in futility. Several Resolutions of the UN General Assembly have affirmed that use of Nuclear Weapons constitutes a Crime against Humanity.

2. Radiation effects the well being of individual himself is called *somatic damage* and that which affects his ability to reproduce or the well-being of his progeny is called *genetic damage.*

3. **Isotopes elements:** Elements which have same atomic number (z) but different mass numbers (A) (or atomic masses) are called isotopes of that element.

$$Z = \text{Protons} = \text{electrons} = p + e$$

$$A = \text{Protons} + \text{Neutrons} = p + n$$

Isotopes hydrogen	Z	A	E	P	n = A – Z
Protium	1	1	1	1	0
Deuterium	1	2	1	1	1
Tritium	1	3	1	1	2
Isotopes of oxygen					
$^{16}_{8}O$ or ^{16}O	8	16	8	8	8
$^{17}_{8}O$ or ^{17}O	8	17	8	8	9
$^{18}_{8}O$ or ^{18}O	8	18	8	8	10
Carbon isotopes					
$^{12}_{6}C$	6	12	6	6	6
$^{13}_{6}C$	6	13	6	6	7

Nuclear Fission: A nuclear reaction in which one heavier nucleus is split up into two lighter nuclei of almost equal size with release of a huge amount of energy is called Nuclear fission or atomic fission

Nuclear Fusion: A nuclear reaction in which two lighter nuclei are combined or fused together to form a heavier (hence stabler) nucleus in called nuclear fusion.

Nuclear fusion is the opposite of nuclear fission in which one heavier nuclear is split up into two lighter nuclei.

Uses of Nuclear Fission: The enormous energy liberated in nuclear fission and the occurrence of chain reaction have been used in Atomic Bomb (Fission Bomb) and Atomic (or Nuclear) Pile or Nuclear Reactor.

Uses of Nuclear Fusion: Nuclear fusion adequately explain the energy of the solar system, called Stellar energy and the basic principle of Hydrogen bomb (Fusion bomb).

8.3.2 NUCLEAR ENERGY AND RADIATION HAZARDS

In the universe nuclear energy is a common form. At present the importance is given to generation of energy from nuclear reactors to replace diminishing fossil fuels. Nuclear energy is produced by a fission or a fusion reaction. All nuclear reactors are working on fission process in which energy is released by electrostatic repulsion.

A nuclear reactor consists of a modulator, control rods with provision of cooling along with the fissionable materials. The fission products are radioactive, emit beta radiation or beta and gamma rays.

A nuclear weapon may be based on fission or fusion. The energy availability from fusion type weapon is almost unlimited.

Biological (radiation) damage is caused mainly due to ionization of molecules in the tissues, which may be somatic (affecting the individual) or genetic. Living organisms are affected by nuclear radiations. In testing of nuclear weapons and in nuclear war, radiation fallout cause considerable damage. Alpha radiation damage is negligible, beta radiation damage is slight but gamma radiation damage is very considerable.

Radiation from radioactive fallout is a major hazard of nuclear weapons. A single explosion of nuclear weapon requires prompt evacuation or radiological shelter over a large area of thousands of square kilometers. Adequate radiation shelters could reduce to a very large extant causalities from a nuclear attack.

Foods must be protected from nuclear fallout. Personal safety from radiological contamination involves complete bathing and removal (discarding) of outer clothes. The radioactivity fallout decreases with time. Radiation from a fallout from nuclear weapon depends on the size and type of the weapon, the elevation at which it is detonated, the physical topography and weather conditions, particularly wind speed and direction.

8.4   CARBON AND ITS COMPOUNDS

The second most pollutant of atmosphere after dust particles is carbon and its compounds. The majority of earth's living matter is contained in the green plants (or the green lungs of the planet earth) which entrap solar energy and prepare food for itself (carbohydrates and some proteins) by photosynthesis process. In this process plants absorb CO_2 and water and give out oxygen. This forms the basis for life on earth.

Photosynthesis

$$x\ CO_2 + y\ H_2O \rightleftharpoons C_x(H_2O)_y + xO_2.$$

respiration

In this process one gram molecule of CO_2 absorbs 112 k cal of energy. The first photosynthetic organism on earth probably blue-green algae. It is estimated that about 248 billion tons (metric tons) of O_2 is released into the atmosphere every year by the green lungs of the earth. Earth's biosphere contains 1.44×10^{18} tons of water, 1.18×10^{15} tons of O_2 and 2.33×10^{12} tons of CO_2.

The first geochemical history of carbon begins with volcanic eruptions of CO_2, CO and CH_4 from the mantle of the earth. Most of the atmospheric carbon occurs as CO_2 and in smaller quantity as CO, CH_4. Any living organism of the earth is mainly composed of carbon, oxygen, hydrogen and nitrogen which are also the basic chemical elements of water and atmosphere.

CO_2 (Carbon dioxide)

It enters the atmosphere through fuel combustion, biological decay, deforestation, volcanoes and release from ocean waters. The mass of CO_2 in global oceans is estimated to be about 1.4×10^{17} kg while its mass in atmosphere is 2.33×10^{15} kg (the farmer is about 60 times the later). The annual fluxes of Carbon globally are as follows : Plant and soil respiration adds (100 GT) 10^{14} kg to the atmosphere, fossil–fuel burning, deforestation adds 5×10^{12} kg, 2×10^{12} kg respectively. Physio-chemical processes at sea surface releases (carbon source) about 10^{14} kg to the atmosphere while it absorbs (carbon sink) about 1.04×10^{14} kg from the atmosphere. Photosynthesis removes about 10^{14} kg of carbon in the form of CO_2 from the atmosphere. Thus in totality 3×10^{12} kg of carbon per year is added to the atmosphere which contributes to the global warming. It is estimated that during last 250 years (from middle of eighteenth century) 26% of CO_2 increased in the atmosphere which is radioactively equivalent to 53% increase in CO_2 in the atmosphere. The life time of CO_2 in atmosphere varies 2-4 years to 50-200 years. The pre-industrial concentration of CO_2 was about 275 ppmv while it was 354 ppmv during the last decade of 20th century. The background concentration of CO_2 is 320 ppm and water solubility 1.64.

CO (Carbon monoxide)

It enters the atmosphere mainly through auto exhaust and incomplete combustion, forest fires. In small amount from oceans and volcanoes. Anthropogenic pollution mass is about $(4.5 \text{ to } 6.4) \times 10^{11}$ kg per year while total emission into atmosphere is about $(2.2 \text{ to } 6.5) \times 10^{12}$ kg per year. The global sink is $(3.3 \text{ to } 5.6) \times 10^{12}$ kg per year. Atmospheric background concentration is 0.1 ppm and residence time about 3 years. As compared to natural emission, anthropogenic emission of CO is much less.

CO is colorless and odourless gas does not support combustion, sparingly soluble in water (Water solubility is 0.00284). CO has more affinity with hemoglobin of the blood than oxygen (about 250 times) and forms Carboxyhemoglobin (COHb) and leads to carbon monoxide poisoning. COHb prevents oxygen meeting with body tissues and thus starves tissues from oxygen and causes tissue suffocation. CO poisoning is dangerous as symptoms are not readily seen. When COHb level exceeds 5% level it starts effecting the body. It starts with burning of eyes followed by headache, dizziness, throbbing of temples, weariness, nausea, ringing in the ears, fast heart beating, buckling of the knees. Some victims show irritability, obstinacy, soon paralysis sets in and death fallows due to tissue suffocation. When first symptoms are noticed victim should be shifted and oxygen be given immediately. Heart attacks likely with people suffering from angina pectoris when CO level exceeds 5%.

Hydrocarbons (CH$_4$)

Gasoline evaporation is the main source of hydrocarbon pollution. The other sources of hydrocarbon emission into atmosphere are combustion, biological decay, agriculture and release from ocean waters. The estimated emission of anthropogenic origin is about 8.8×10^{10} kg per year while natural emission is about 4.8×10^{11} kg per year. Atmospheric background concentration of CH$_4$ is about 1.5 ppm and non-methane hydrocarbons less than 1 ppb. Atmospheric residence time of methane is about 16 years. It is removed slowly from atmosphere by photochemical reactions with oxides of nitrogen and ozone.

8.5 SULPHUR (S)

According to one estimate of global sulphur cycle (mainly gaseous H$_2$S, dimethyl sulphate and SO$_2$) about 6.5×10^{10} kg per year emitted into the atmosphere by anthropogenic sources while the natural emission is about 4.0×10^{10} kg, from volcanoes (3×10^9 kg), volatile sulphur from water logged soil (3×10^9 kg) and biological decay (3.4×10^{10} kg). Sea spray emission is about 4.4×10^{10} kg. This shows that other than sea spray emission anthropogenic emissions of sulphur is more than the natural sources. The atmospheric background concentration of SO$_2$ is about 0.2 ppbv, residence time 1-4 days, water solubility is 11.3. It is removed from the atmosphere by photochemical oxidation by ozone. Sulphur is removed from the atmosphere by wet and dry deposition by adsorption (7.0×10^{10} kg) and gaseous adsorption (3.5×10^{10} kg).

SO$_2$ is colourless and non-combustible gas and has bleaching properties. Most of sulphur in the atmosphere is found as SO$_2$. Its toxic effects are: irritates the mucous membrane of nose, throat and eyes. Sometimes causes emphysema and swelling of throat. In acute cases it causes paralysis of respiratory track. It destroys green plantation.

8.6 ATMOSPHERIC NITROGEN AND ITS COMPOUNDS

Nitrogen is present in the atmosphere about 78% by volume and by mass 3.865×10^{18} kg. Its atmospheric life time is about 2×10^7 yrs, consequently its global variation is negligible. In combined forms it occurs in the atmosphere as NO (Nitric oxide), N$_2$O (Nitrous oxide), NO$_2$ (Nitrogen peroxide) and NH$_4$ (Ammonia). Nitrogen is an essential constituent of animal and vegetable matter. It is non-poisonous, does not support combustion and respiration. It does not combine with metals and non-metals. However it combines with oxygen in (lightning) the atmospheric electrical discharges and forms oxides of nitrogen.

Fixation of Nitrogen

Combining atmospheric nitrogen to form commercial compounds nitric oxide (then nitric acid and nitrates), ammonia (ammonium formate, bicarbonate, and sulphate), cyanides and cynamides, nitrides (finally obtain ammonia) is called fixation of nitrogen.

Natural emission of oxides of nitrogen consists of decomposition of nitrite in soils, fixation by lightning and conversion from ammonia, while anthropogenic emissions involve mainly high temperature combustion processes which are associated with transportation and energy production. A large part of ammonia emission is through urea from domestic animals.

Nitrogen Cycle

Atmospheric nitrogen is a chief source for numerous nitrogen compounds. Free ammonia occurs in small quantities in air which is washed down along with rain water. Nitrous and nitric oxides are formed in the atmosphere by photochemical reactions in sunlight between nitrogen and oxygen and also during electrical discharges in air. These oxides form nitrous and nitric acids with rain. When they come down to earth they form calcium nitric and calcium nitrate in the soil which is absorbed by the plant as food. There are some bacteria in the roots of leguminous plants which directly take atmospheric nitrogen. Azotobacter (one kind of bacteria found in soil) algae, fungi and mosses also use atmospheric nitrogen directly. The dead plants give out ammoniacal compounds to the earth by nitrisifying and nitrifying bacteria. Animal wastes and urea on decomposition give out ammonia and ammonium compounds. These are formed in the animal body by the process of peptic and tryptic digestion of the nitrogenous plant proteins taken by the animals as food. There are some denitrifying bacteria which breaks up soil nitrates and nitrites and ammoniacal compounds to nitrogen and thus completes the nitrogen cycle, By photochemical and bacterial processes ammonia to nitrite, nitrite to nitrate and again nitrate to nitrogen completes the nitrogen cycle.

N_2O

It enters the atmosphere through natural process of biological action in soil. Its estimated emission of mass is about 5.9×10^{11} kg per year, background concentration is 0.25 ppm, residence time in atmosphere is about 4 years. Its water solubility is about 0.121. It is removed from the atmosphere by photodissociation and biological action in soil. When inhaled with oxygen causes nervous excitement. Pure N_2O causes unconsciousness and may be fetal.

NO/NO_2

These gases enter the atmosphere by combustion and bacterial action in soil. The estimated mass of pollution of NO is 5.3×10^{10} kg and background concentration 0.2 to 2 ppb, while the estimated pollution mass of NO_2 is 5.3×10^{10} kg and background concentration is 0.5 to 4 ppb. However their natural emission is 4.36×10^{11} kg (for NO), 6.58×10^{11} kg (for NO_2) per year, atmospheric residence time about 5 days. The water solubility of NO is 0.00618. It is removed from the atmosphere by photochemical reactions and oxidation.

N^+ (Active Nitrogen)

It exists in metastable state and not an allotropic modification of nitrogen. It shows band spectrum. Active nitrogen is produced under electrical discharges and shows bright luminescence after electrical discharge stopped. It has short life and emits phosphorescence in the presence of sulphur and sodium.

Note: Both natural and anthropogenic trace pollutant gases are removed from the atmosphere by hydrometeors viz cloud droplets, raindrops, snowflakes, fog droplets. Removal process are: Wet removal process, dry removal process and chemical transformation.

8.7   OZONE (O_3)

In the composition of the atmosphere it was briefly discussed about stratospheric ozone. Here we shall consider about its formation in stratosphere and near the earth's surface. Incoming solar radiation consists (about 99%) in the wave length (λ) range of 0.15-4.0 µm with peak energy at ($\lambda = $) 0.5 µm while terrestrial radiation consists (about 99%) in λ- range of   4-80 µm with peak at 10 µm. Visible radiation consists in the range 0.4-0.7 µm. Ozone absorbs UV- radiation (UVB) between range 0.2- 0.3 µm and IR- radiation at $\lambda = 9.6$ µm.

Ozone is found in the atmosphere between altitudes 10-45 km with a maximum concentration (one ozone molecule to 10^6 molecules of normal oxygen) between altitudes 18-20 km in a globe encircling ozone layer. The temperature of top ozone layer varies $-$ 20 to $+$ 10 °C (at altitude of about 45 km) due to powerful absorbing of UVB by this layer. Ozone is formed in the atmosphere in two stages. In the first stage NO_2 is dissociated by short wave UVB [$NO_2 + h\nu$ $NO_2 + h\nu \rightarrow$ $+NO + O$ where $h\nu$ is the energy of a solar photon] into NO and atomic oxygen. In the second stage a recombination of atomic oxygen and oxygen molecule of air ($O + O_2 \rightarrow O_2$). Ozone is also generated in small amounts in the atmosphere by the electrical discharges (thunderstorms). In free atmosphere the life period of ozone varies from a few weeks to months. According to WMO assessment (1989) there is a decrease in stratospheric ozone content with the evidence of substantial decrease in stratospheric ozone over Antarctica in spring time (the later one is described as Antarctica ozone hole). Pre-industrial period background ozone concentration was 15 ppbv which shot up to 35 ppbv in 1990. Further it was estimated that the ozone contribution to the greenhouse effect was about 8%.

Ozone shows diurnal variation in concentration in air, over land with maxima in the day time and minima at night. Near the surface of the earth ozone is found in urban polluted atmosphere as photochemical smog which is a complex mixture of ozone and other pollutant gases. In smog, ozone is produced by the oxides of nitrogen and hydrocarbons in the presence of sunlight. These gases are emitted from vehicle exhaust and other industrial plants. Thus anthropogenic pollutants are the cause of smog and its adverse effect on nature.

Ozone has a special characteristic that it absorbs the sun's harmful UVB and prevents it from reaching the surface of the earth. UVB induces skin cancers, cataract of the eyes, suppression of immune system in humans and effects the productivity of aquatic and terrestrial ecosystems. It has been estimated that a decline of 1% ozone content in stratosphere causes 3% rise in the incidence of skin cancers in the humans. If ozone concentration exceeds $800\mu g/m^3$ in one hour it is harmful, particularly to elderly and ailing-persons. Ozone and photochemical smog attacks and damages rubber and various synthetic materials. Ozone concentration of 20 ppm or more is poisonous. Ozone causes impairment of lung function and acceleration of aging. The vertical distribution of ozone is changing due to the effects of man-made emissions of oxides of nitrogen and hydrocarbons. As a result there is a net decline in total ozone column which is a danger signal of decreasing protection to the life matter on earth from the lethal effects of UVB. The chlorofluorocarbons (CFCs) are powerful green house gases and they are the main cause for the decline of the stratospheric ozone.

In pursuant to the global convention for the protection of the ozone layer (Vienna,1985), the Montreal Protocol held in 1987 to control substances (and related compounds) CFCs that deplete the ozone layer in the first global agreement for the protection of earth's atmosphere. In amending the protocol, in 1990, signatory countries have committed themselves to a complete phase out the use of chemicals that deplete the stratospheric ozone layer by the year 2000 AD.

8.8   ELEMENTARY WAYS OF POLLUTION CONTROL

Pollution control may be accomplished:

(i) by the use of substitute materials which are less pollutants,

(ii) by diminishing the use of pollutant sources, and

(iii) by treatment for the removal of pollutants.

For example we can use natural gas instead of coal. We can use natural gas/electricity driven autos instead of mineral oil autos. We can convert the pollutant to a harmless substance such as exhaust catalytic unit to convert CO to CO_2.

Control of Suspended Particulate Matter

Incomplete burning is the cause of carbon particulates (smoke and soot), and flyash. Dust and other particulate matters are produced in industrial manufacturing processes such as grinding, crushing, cement/asphalt plants, foundaries, construction and demolition works.

Particles from combustion and dust can be filtered by air cleaning equipments such as cyclones, scrubbers, baghouses and electrostatic precipitators which are briefly given below.

In *cyclone* dust collector, dust-laden air is introduced at the top outer edge to whirl around and around inside the cylinder. Centrifugal force flung the dust particles against the outside wall where it slides down to the bottom and removed periodically. The dust cleared air escapes from a duct located in the centre of the cylinder. By this process 50 to 80% of the dust particles of size 10 μm or more are removed. Small size particles still remain in the cleared air.

Scrubbers are simple screens of water/liquid spray from which air moves up. It removes large particles of air. Scrubbing towers are used to remove particles and gaseous pollutants. Water is used if the gas is soluble in it otherwise a liquid is used in which the gas is soluble.

Baghouses consists of bags or long sleeves of fabric which withstand against high temperatures. As the air passes through the fabric to the otherside particles in the airstream are filtered out. The sleeves are periodically cleared. To prevent condensation the temperature of the bag house is maintained to be higher than dew point temperature of water. Bag houses are very effective for removal of fine particles of air.

Electrostatic Precipitator: It consists of positively charged wires centred between plates which are negatively charged and grounded. When high electric charge is passed on wires, they create charge on air stream particles which are attracted and settled on grounded plates. Settled particles are removed periodically. This system is efficient when particles in the air are electrically charged.

Control of Oxides of Sulphur

Oxides of sulphur mostly emitted into the atmosphere by smelters, oil refineries, paper industry and burning of fuels containing sulphur. Combustion of fuels containing sulphur evolves sulphur dioxide (SO_2) and sulphur trioxide (SO_3), the farmer being in bulk. Parts of SO_2 converts into SO_3 by photochemical processes. Moisture converts SO_3 into sulphuric acid (H_2SO_4). Sulphur can be undressed from oil but with the declining of oil reserves, it is wise to use coal fuel or hydropower for power generation plants. Coal contains sulphur in organic form as iron pyrite. By crushing and washing about 30% of iron pyrite can be removed. Rest of sulphur is chemically bound in organic form to the coal. The later can be removed by gasification. Large scale coal use requires flue gas desulfurization.

Control of CO

Most of CO enters the atmosphere through internal combustion engines. A diesel engine provides more complete combustion and emits less CO as compared to gasoline engine. Autos operating on combustion engines are to be fitted with catalytic converters to reduce CO and hydrocarbons exhaust.

Control of Oxides of Nitrogen, Hydrocarbons and Ozone

Reduction of oxides of nitrogen in the atmosphere is very difficult. Most of the hydrocarbons enter the atmosphere through evaporation of gasoline which can be reduced by proper usage of floating roofs on storages. In autos it can be reduced by the use of (PCV) Positive Crank case Ventilation systems.

8.9 ACID RAIN

Man's industrial activities changed the composition of atmosphere and precipitation both locally and regionally. In industrialised region the chemical composition of precipitation is found to be acidic with pH value around 4.5.

The gas phase chemical equations of atmospheric acidic rain are given below.`

$$O_3 + h\upsilon \rightarrow O_2 + O(D') \qquad\qquad .....(8.1)$$

where $h\upsilon$ = energy of solar photon.

$$O(D') + H_2O \rightarrow OH + OH \qquad\qquad .....(8.2)$$

O (D') atomic oxygen (by photo dissociation)

$$NO + O_3 \rightarrow NO_2 + O_2 \qquad\qquad .....(8.3)$$

$$NO_2 + OH (+ M) \rightarrow HNO_3 (+ M) \qquad\qquad .....(8.4)$$

(+ M) indicate molecule

$$SO_2 + OH (+M) \rightarrow HSO_2 (+M) \qquad .....(8.5)$$

$$HSO_3 \text{ is oxidised to } H_2SO_4 \qquad .....(8.6)$$

In liquid phase oxidation processes of conversion of nitrogen species into nitrate is not known clearly. Gas phase conversion to nitric acid (HNO_3) is significant. However HNO_3 (gas) $\square$ HNO_3 (liquid).

Oxidation of SO_2 to sulphate (SO_4) in the liquid phase proceeds via dissolved compounds, particularly ozone and hydrogen peroxide (H_2O_2).

Most of the acidity in precipitation may be attributed to oxides of sulphur and oxides of nitrogen which enter into the atmosphere mainly from the combustion of fossil fuels and industrial processes. SO_2 oxidized to sulphuric acid in rain drops.

Sulphuric acid and nitric acid accounts most of the acidic precipitation, the former accounts two-thirds while the later one-third.

Atmospheric acid is causing erosion of priceless statues, marvellous buildings through dry deposition of sulphate particles (which become acid with rain). Similarly aquatic and terrestrial ecosystem (relationship of organisms to its environment) is being damaged through long term deposition of acidic rain and snow. A number of fresh water lakes without natural buffer are suffering from acid rain or deposition. It was found during 1960 the pH of rain water was 6 to 7 (neutral value is 7). By 2000 AD it changed to 4.5 to 4.0 in some lakes in Europe and America. This acidification resulted in heavy losses of commercial fish population, and gradual disappearance of fauna. Fat head minnows, some zoo plankton disappear at pH 5.9, algal forms, lake trout disappear at a pH value of 5.6 or less. Ground water in shallow wells and aquifers also affected due to contamination by toxic metals such as mercury, cadmium. On land forests were reported dying in Europe and north America due to acid rain, high ozone concentration and smogs. Acidity of the soil increased by ten times in Europe to a depth of about one meter.

8.10   BAPMoN

In the mid-1960s, WMO established Background Air Pollution Monitoring Network (BAPMoN) to measure the changing background chemical composition of the atmosphere near the ground, away from cities and strong polluting sources (industries/plants). A complementary Urban network is coordinated by the World Health Organisation (WHO). This network of observatories collect samples for measuring the chemical composition of rain and snow. Under Global Atmospheric Watch (GAW) it provides information on (i) global increase of greenhouse gases (namely CO_2, CH_4, Nitrous oxides, CFCs, tropospheric ozone), (ii) regional distribution of sulphur, nitrogen compounds - which result in acid rain, fog, smog and other changes in the lower level (PBL) chemical composition and (iii) radiation transmitted through the atmosphere and aerosols that were ejected from volcanoes, factories to assess their contribution in cooling the atmosphere. At a few locations toxic metals (mercury, lead etc.) and toxic organic compounds (pesticides, herbicides) are also determined. Complementary data on air pollution within cities and by the side of heavy polluting sources are coordinated through WHO and United Nations Environment Programme (UNEP). In mid 1990's there were about 350 GAW stations operating in 70 countries. 164 stations collect sampling for measurements of precipitation chemistry (rain and snow), 95 BAPMoN stations record turbidity (transparency of air), 78 stations suspended particulate matter, 52 stations CO_2, 22 stations surface ozone, 9 stations methane (CH_4), and   5 stations measure CFCs. The atmosphere is sampled according to well defined criteria, instruments and procedures by trained personnel. Through the global ozone observing system continuous measurements of atmospheric ozone in total column, its vertical distribution and changes are made, This network has about 140 stations over the globe-complemented during 1980's by satellite remote sensing.

UNO, through its agencies WMO and UNEP, urged to monitor the quality of air and environment. WMO coordinating with national services for setting the network of air pollution monitoring stations to monitor both in time and space. Horizontal scale ranges 100 to 1000 km, vertical scale ranges 1 to 10 km and time scale few hours to a year or more. Each individual station is expected to represent space radius 5000 km, time one year. A regional station may provide hourly values representative over a circle of radius 50 km. Station should be located away from Urban areas (cities/industries), high ways, power generation station etc ; however it should experience frequent natural phenomena like heat/cold waves, cyclones, forest fires, sand/dust storms, volcanoes etc. Local influence should be as least as possible. When such conditions are satisfied the monitoring station represent the background values. An example is given below.

The Table 8.5 and 8.6 gives emissions of Singrauli thermal power plant of 2000 MW capacity, situated at Shakthi nagar, Uttar Pradesh (India), during September 1992, measured over 11 days. The thermal power plant burns about 1000 tones of coal per hour and emits about 9680 kg SO_2 per hour from four smoke stacks (height about 225 m). The plant is equipped with eletrostatic precipitators for removal of particulates from the plume emission. Measurement of trace gases $-SO_2$, NO_2, NH_3, O_3 and TSP were made upwind of the power plant at a distance of one km north of the power plant. Daily three samples of trace gases for 3 hrs duration were collected (morning, afternoon and evening), one sample of TSP 24 hrs duration collected on Whatman 41 filter paper using high volume air sampler with a flow rate of 1.2 m^3/min.

Table 8.5 Average concentrations ($\mu g\ m^{-3}$) of trace gases and TSP at Shaktinagar

Sept 1992	SO_2	NO_2	NH_2	O_2	TSP
	33.2	8.5	22.5	7.5	134

Table 8.6 The average chemical (ionic) composition of TSP ($\mu g\ m^{-3}$)

Cl	SO_4	NO_3	NH_4	Na	K	Ca	Mg
1.16	2.47	1.30	0.41	0.64	0.61	2.55	0.86

The conclusions drawn based on the above data was that the average concentrations of NO_2 and O_3 are of the order of background values. The maximum ground level concentration of SO_2 was due to the burning of coal in the thermal power plant. The concentrations of TSP and its water soluble components (SO_4 and Ca) are high at Shakthi nagar.

A number of WMO member countries provide data to the centralized data collection centres for GAW. Canada operates the WMO World Ozone Data Centre and publishes ozone data every other month. Amercia provides a base for data on precipitation chemistry analyses, acid rain and atmospheric turbidity measurements. Russia (earlier soviet union) is responsible for the collection of solar radiation and atmospheric electricity data and Japanese Government to operate the World Data Centre for Greenhouse gases.

8.11   SMOG

In the Meuse Valley (Belgium) a mysterious yellow-brown fog was formed in December 1930. It caused irritation and many become ill the next day. In 1944, an air pollution episode in Los Angeles area damaged vegetation and when it became severe it caused eye irritations and tearing. It was later found that auto exhaust gases oxidised by the sun's rays (photochemical oxidation) formed complex compounds which were responsible for irritation.This phenomena later termed of smog. Smog is a combined word of smoke and fog, represents the mixture of smoke and fog. In 5-9 Dec 1952 another air pollution episode over London took a death toll of 4000 people. This attracted the world scientist to find the cause. It was found that there was temperature inversion in the atmosphere close to the ground (PBL) and SO_2 concentration was six times than usual. Subsequently it was confirmed that all air pollution death cases were associated with temperature inversion close to the ground (that is warm air above and cold air near the ground) and smog. Further it was found that these incidence also affected the vegetation.

Smog is of two types. (i) Photochemical smog and (ii) Coal-burning smog. Photochemical smog results from the action of sun rays on contaminated organic vapours particularly unburned gasoline, and oxides of nitrogen. Coal-burning smog is a combination of coal smoke and fog. Present day smog in most cities is a combination of these two types in varying proportions. Dense coal-burning smog contains carcinogenic compounds, polynuclear hydrocarbons is one which is effective carcinogen; and 3,4-benzypyrone is another carcinogen. This exists in particles.

In photochemical smog the following reactions take place,

$$NO_2 + h\upsilon \rightarrow NO + O \text{ (atomic oxygen or nascent oxygen)}$$
$$\text{(photochemical dissociation)}$$
$$O + O_2 (+ M) \rightarrow O_3 (+ M)$$
$$O_3 + NO \rightarrow NO_2 + O_2$$

Atomic oxygen reacts with various hydrocarbons (such as olefins of auto exhaust) forms free radicals

$$O + olefins \rightarrow R + R'O$$

or

Where   R, R' are free radicals

$$O_3 + olefins \rightarrow products$$
$$R + O_2 \rightarrow RO_2$$
$$RO_2 + O_2 \rightarrow RO + O_3$$

Thus organic compounds such as aldehydes, acrolein or peroxyacetyl nitrate are formed. Probably these compounds are responsible for eye irritation in smogs. Smog is notorious in corrosive action. It corrodes plant cells, rubber fabrics, nylon stocking, buildings and monuments. In India Taj mahal, Red fort and other historical buildings in Agra are being targeted by smog. In 1966 Tokyo was alarmed for smog 154 days. After air pollution death episodes smog alarms are introduced in some countries.

8.12   TOXIC AIR POLLUTANTS

Air pollutants that are health hazards are: asbestos, arsenic, antimony, cadmium, copper, lead, nickel, zinc, mercury, vinyl chloride, benzene, radionuclides, polycyclic organic matter, ethylene chloride, methyl chloroform, toluene, trichloroethylene, benzopyrene, pesticides/chlorinated hydrocarbons. Some are carcinogens. They are found in urban run-off, industrial discharge of waste water. EPA restricted the use of DDT, DDD, heptachlor, diendurine, lindane, chlordane; 3, 4- benzopyrene is a carcinogen, asbestos when inhaled may cause cancer. Toxic metals and chlorinated hydro carbons/pesticides are released into atmosphere as byproducts from industries and agriculture. During 1970's and 1980's it was found minute quantities of pesticides in melted snow, in the tissues of polar bears and in breast milk at the Arctic and in birds at the Antarctic.

Toxic some trace metals of natural processes and anthropogenic sources with global values of emission are given in Table 8.7.

Table 8.7

Metals	Sources		Remarks
	Natural in 10^6 kg/year	Anthropogenic 10^6 kg/year	
Lead	19	332	White lead is very poisonous
Cadmium	1.0	7.6	
Copper	19	35	
Nickel	26	56	
Zinc	46	132	
Arsenic	8	18	Arsenious oxides are drastic poisons, a dose of 0.125- 0.25 gm is fetal.

In the Arctic haze it was found high concentration of aerosols of sulphate, man-made pesticides like lindane, dieldrine, DDT and toxic heavy metals lead and mercury. WMO Marine pollution studies shows the presence of a number of chlorinated hydrocarbons over marine areas around the world but their concentrations are not high. According to its estimates 80% of the global chlorinated hydrocarbons in air are spread over oceanic areas. The contribution to the sea water pollution through atmosphere is of comparable magnitude (if not more) with that of contamination of sea water by highly polluted river waters particularly of fluxes of metals such as lead, mercury, cadmium and zinc whose sources are located far away from the sea areas. Examples are the pollution of Mediterranean sea, Great lakes of North America. It was found that fish in some lakes contained PCBs and were unfit for human consumption. Water pollution transmitted through atmosphere is difficult to address as pollution sources involved several countries and States that are far away.

8.13   DISPERSAL OF AIR POLLUTANTS

Dispersal of pollutants in air depends on wind and environmental air stability and turbulence. Precipitation acts as scavenger. Thus weather effects the dispersal of air pollutants and inturn air pollutants effect the weather (particularly precipitation, visibility, acid rain, fog, smog, insolation).

Wind direction and speed at surface and also at the chimney stack level has a great bearing on the dispersal of air pollutants from stacks. Stronger the wind speed quicker the dispersal and the pollutants are carried away farther before they are settled. When the wind is light or calm pollutants concentration near the chimney stack is dense and not carried away to great distance. Gustiness dispersal covers a wide area of the chimney plume. Stable conditions of environmental air causes concentration of pollutants locally and unstable environmental air conditions causes pollutants dispersal over a large area leading to less concentration of pollutants. Based on environmental stability/instability (lapse rates) we observe six types of plumes. Looping, coning, fanning, lofting, fumigation and trapping. These cases illustrated in Figs. 8.1 and 8.2.

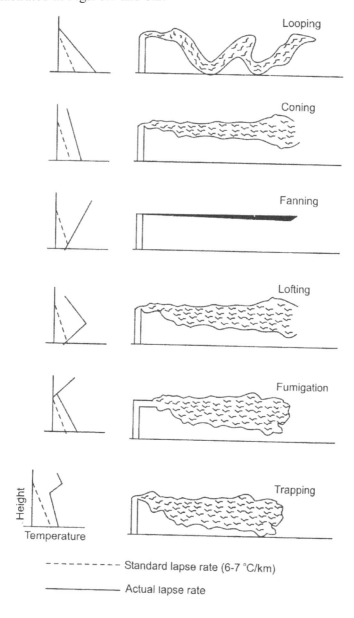

Looping

Coning

Fanning

Lofting

Fumigation

Trapping

Height

Temperature

‑ ‑ ‑ ‑ ‑ ‑ ‑ ‑ ‑ Standard lapse rate (6‑7 °C/km)

————— Actual lapse rate

Fig. 8.1 Plume classes under different stability conditions

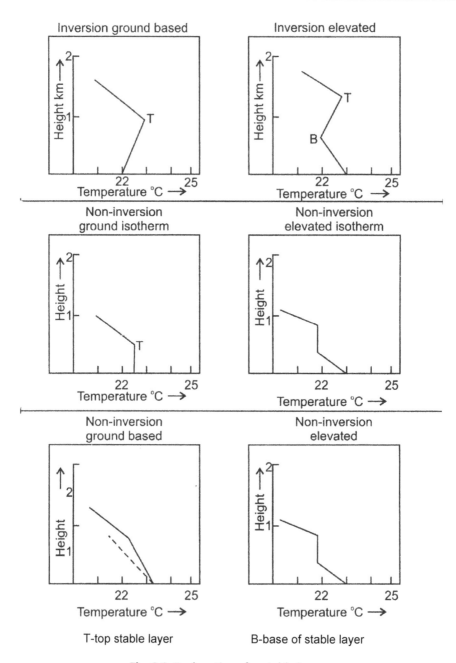

Fig. 8.2 Explanations for stable layers

Precipitation and Humidity Effects

Natural scavenging of air pollutants occur by gravity fall out, rain out and wash out. In the absence of turbulence fine particles settle down by impaction with objects. Heavy particles settle down by gravity. In rain out small particles form condensation nuclei. On condensation they fall out to ground with rain drops. In wash out, rain drops while coming, sweep the pollutants in its way.

The Main Objectives of Air Monitoring

  (i)   Collection of basic data of air pollutants (quantitatively and qualitatively),

 (ii)   meteorological factors aiding pollution,

(iii)   topographic influence,

 (iv)   climate study ,

  (v)   population.

The Table 8.8 gives the methods of measurement of air pollutants.

Table 8.8 Measurement of Air pollutants

Types of Pollutants	Sampling equipment	Analytical Method
Dust fall	Dust fall Jar	Gravimetric
Suspended particulates	High volume sampler	Gravimetric
Total sulphur compounds	Lead-candle	Gravimetric
Sulphur dioxide	Air sampling kit	West and Gaeke method
Oxides of nitrogen	Air sampling kit	Jacob and Hochhneiser method
Hydrogen sulphide	Air sampling kit	Methylene blue method
Any other gaseous pollutants	Air sampling kit	–
Wind direction and speed	Wind vane recorder D.P.T recorder	Recording chart
Temperature and humidity	Stevenson screen with thermometer and evaporimeters	Instrument reading Thermographs Hydrographs

8.14   METHODS OF ESTIMATION OF PARTICULATE MATTER IN AIR

Dust-fall Jar

It is an open mouthed polyethylene jar or glass. It contains water at the bottom which is exposed to the atmosphere for a period of one month. The contents of the jar for total deposits, from which deposits per unit area calculated. Dust-fall jar method used to measure the amount of settleable particles such as soot, fly ash, smoke particles in the air. Dust fall is expressed as tonnes per square kilometer per month (tonnes/km^2/month) (see Fig. 8.3).

High Volume Sampler

It contains a 20 cm × 25 cm glass fibre filter, which collects sample over a period of 24 hrs. The glass fibre filtrates 100% of all particulate matter of 0.3 μm diameter or more. Air is drawn from vertically upward flow at the average speed of 19 m/min. It sucks all particulate matter up to diameter of 100 μm. Filtered particles over a period of 24 hrs is measured by weight and expressed as $\mu g/m^3$ of air. Generally air is drawn at the rate of 1-1.5 m^3/min.

Methods of Estimation of Sulphur Compounds in Air

Lead Peroxide Candle Method

Gaseous Sulphur compounds + Solid lead peroxide = lead sulphate

$$PbO_2 + SO_2 \rightarrow PbSO_4$$

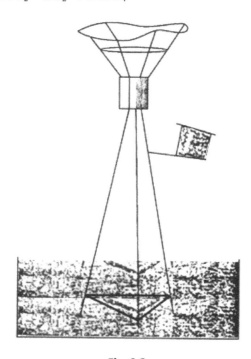

Fig. 8.3

A piece of tapestry cloth size 10 cm × 10 cm is wound round on a glass cylinder or polyethylene tube of 100 cm^2 curved surface area. Tragacanth (gum) and lead peroxide paste is made and is evenly applied on tapestry cloth surface. This is dried in a desiccator to make it a candle. This lead peroxide candle is erected in a wooden stevenson screen like box, which protects it from rain but allows the air to circulate over it. In this way the candle is exposed to air for a period of one month. Then the lead sulphate formed over the candle is estimated gravimetrically after converting into barium sulphate. The results expressed as milligrams SO_3/100 cm^2/per day.

The above lead peroxide candle arrangement generally attached to one of the legs of dust fall jar tripod.

Multi-gas Sampling Kit

It is used for sampling measurements of different gases simultaneously. Selective absorbants are kept in different impingers (generally four) to trap respective gaseous pollutants. A vacuum pump operates to suck atmospheric air which circulates over the impingers containing different absorbents and the purified air leaves the kit. Gaseous pollutants react with the absorbents. Generally air is sucked into the kit at the range of 0.1 to 3 liters per minute. Using this kit O_3, SO_2, H_2S, oxides of nitrogen, ammonia can be sampled. The results are expressed as ppm (ml/m^3) or as $\mu g/m^3$ (μg = micro grams).

8.15 GLOBAL WARMING EFFECTS

Climatic records indicate rise in global temperature by 0.3 to 0.7 °C during last one hundred years see Fig. 8.5. International Scientific Community agreed that there is compelling evidence of climate change and human beings are largely responsible for this. This requires adaptation to actual or expected impacts of climate change. Planning for adoption is important and which should reduce the adverse impacts of climate change and enhance beneficial effects. It is said African nations might have to spend 5 to 10% of GDP on adopting to climate change. Climate change likely to increase the frequency and intensity of extreme weather events. It is therefore necessary to increase awareness among communities, and policy makers about the implication of climate change. See Fig. 8.5(a) and 8.5(b). Awareness campaign should focus school children and village youth in addition to normal tools like local media, street wall posters, Cinema film shows.

According to UN Climate Panel (Oslo / London) global warming may melt the Himalayan glaciers by 2030s, 40 animal and plant species face extinction due to rising temperatures which destroy the ecosystem that supports them. It warns that the poorest nations are likely to suffer most. Heat wave incidence may increase in US and damage to corals like the Great Barrier Reef of Australia. The rise in temperatures are blamed to the increase in greenhouse gases in the atmosphere emitted by human beings from burning fossil fuels. This would cause desertification, droughts and rising in sea levels. At the current warming rate, the Himalayan glaciers melt and shrink from the present 500,000 km^2 to 100,000 km^2 by 2030. The UN Climate Panel reports that 29 × 10^9 tonnes of CO_2 released into the atmosphere every year, which is acidifying the oceans. As a result it is likely to destroy coral reefs, plankton and many commercial fish species. Further it noted that the area in the equatorial belt (in Africa) may affect soil and reduce the crop yields. This may result in hundreds of millions of people hungry. The Global warming may destroy snow cover by about 70% by 2050 and destroy Alpine Ski resorts and increase in sea level water. Half a meter rise in sea level water may result in submergence of some island States and a considerable part of Bangladesh.   However some good effects of global warming likely that cause Canada, Russia, New Zealand, Scandinavia become warm and help in raising crops. The UN Climate Panel study squarely blames humans for global warming saying that 90% of recent global warming attributed to human cause. See Fig. 8.6(a), 8.6(b) and 8.6(c).

The UN Climate Panel on 6 March 2007, issued the starkest warning about impact of global warming based on the finding of 2500 scientists.

According to Intergovernmental Panel on Climate change:
   (i)   20 to 30 species face extinction if temperatures rise 2°C above average in the 80s and 90s.
   (ii)   Heat waves, floods, storms, fires and droughts will cause more deaths and harm.

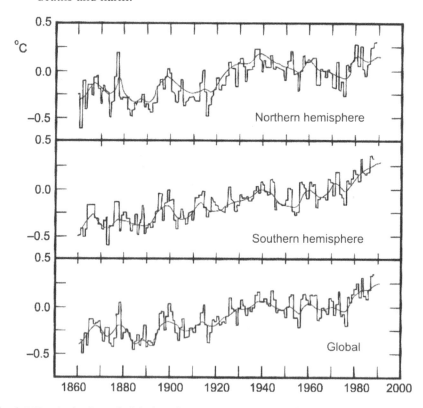

Fig. 8.4 Hemispheric and global surface temperature trends based on world climate observation networks

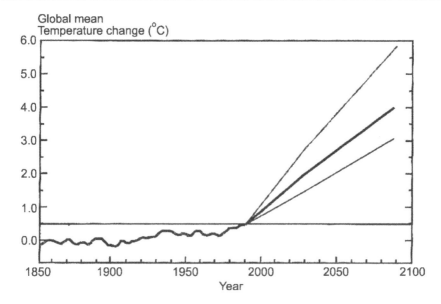

Fig. 8.5 Global mean surface temperature (According to IPCC)

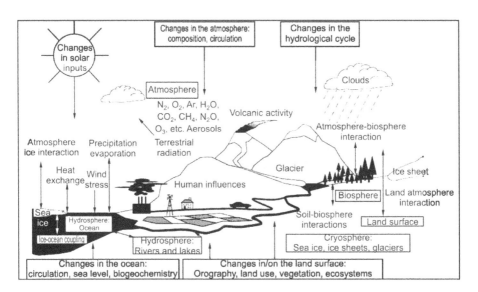

Fig. 8.5(a)

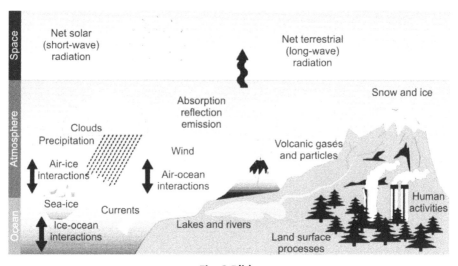

Fig. 8.5(b)

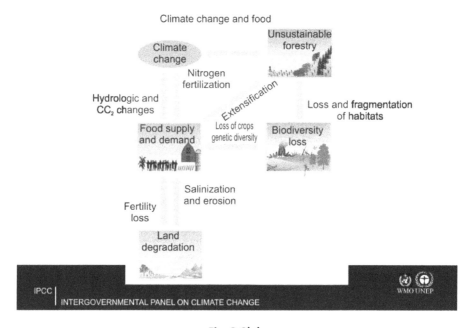

Fig. 8.6(a)

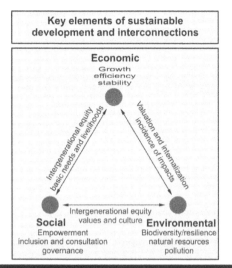

Fig. 8.6(b)

Fig. 8.6(c)

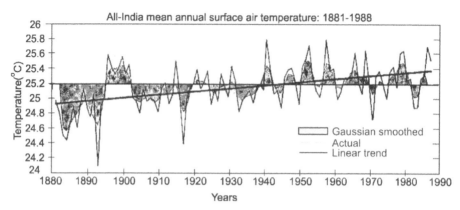

Fig. 8.7 All-India mean annual surface air temperature : 1881-1988

(iii)   Glacial retreat in the Himalayas will affect billions of people.
(iv)   Millions in coastal areas will be at risk from sea level rises.
(v)   Production of wheat, maize and rice in India and China will drop.
(vi)   Over one billion (10^9) people may face shortage of fresh water by 2020.

UNs Kyoto Protocol aims to cap greenhouse gas emissions by 2012. US pulled out of Kyoto in 2001 but planned to tackle limiting CO_2 emissions on its own.

India has recorded weather data of past 150 years. Study of these records indicate that there is no change in monsoon rainfall and the extremes are in natural variation (such as floods, droughts, heatwaves etc) 120 years rainfall data indicate, 72% (86 years) south west monsoon was normal (within ± 10%), Excess of normal monsoon for the country as a whole 4% (about 17 years). And below normal 14% (about 17 years). 86% of occasions southwest monsoon was normal or above normal. It is found that there is rise of 0.5°C during last 100 years see Fig. 8.6. Model simulation shows rise in temperature by 3 to 4°C during 21st century. Night temperatures are increasing faster than the day temperatures. Extreme precipitation appears to be on rise over large area.

Coming decades may witness warmer weather, rising sea levels, intense storms, changes in ocean currents. These will wipe some idyllic destinations off the tourist map. Global warming models indicate that the Himalayan Ski resorts may be affected due to higher temperatures. The hazards of landslides would increase. Beaches from Bondi in Sydney to Fiji, Bali, Thailand, the Philippines and Maldives are also under threat. In Europe the Alpine Ski resorts likely to be affected with losing about 70% of snow cover by 2050 (see Fig. 8.4). Global farm potential might increase with a rise of 3 °C in temperature before sinking worldwide. Crops might grow better in nations far from the tropics (such as Canada, Russia, New zealand, Scandinavia). Mediterranean region might become arid. In the US, rising seas and storm surges could severely affect transportation along Gulf, Atlantic and Northern Coasts.

According to Andhra Pradesh pollution monitoring units: Air pollution in Hyderabad, AP,India become doulbed during 2002-2007. In 2004 the average reading of suspended particulate matter in the air was less than 150 units/m^3 of air. In 2007 it was 267 units/m^3 of air. The accepted (normal) level is 100 units /m^3 of air.

According to Dr.Yashwant Oke (pediatrician) noise pollution may rupture the eardrum, induce cardiac and cardiovascular changes, fatigue and cause sleep disturbances, headache and insomnia. According to Environmental Protection Act of 1986, which notified noise as a pollution. It laid down a fine of up to Rs. 100,000 or imprisonment up to five years or both.

Permissible noise level in residential areas	:	55 decibels
Loud Speakers	:	75-115 db
Drilling Machine	:	90-100 db
Vehicle - Horns	:	80-84 db
Typewriter	:	50-60 db

In 1970 there were about 80,000 vehicles plying in Hyderabad, which increased 10 lakhs in 2001 and 18.5 lakhs by the end of 2007 in the city roads.

As of January 2007 there were over 18 lakh vehicles plying in Hyderabad. The following are the pollution measurements at some important junctions in city of Hyderabad.

According to A.P pollution Control board oxides of nitrogen 26 mg/m^3 on 2 June 2007; 38 mg/m^3 on 2 June 2008.

1. **Noise pollution in the city (As on 2 June 2008)**

Permissible level	65 dB
Panjagutta	86.23 dB
Charminar	85.88 dB
Paradise	85.84 dB

2. **CO (Carbon monoxide)**

Permissible	4 mg/m^3
Panjagutta	26 mg/m^3
Charminar	23 mg/m^3
Paradise	23 mg /m^3

3. **Total suspended particulate matter**

Abids	242 µg/m^3
Panjagutta	321 µg/m^3

Paradise   347 µg/m³
Permissible   200 µg/m³
Langar House   448 µg/m³
Uppal   358 µg/m³
Balanagar   338 µg/m³

Notation :   µg = micro grams

mg = milligrams

m³ = cubic meter or meter cube

dB = deci Bels

Consequences of Global Warming

Climatic fluctuations are caused by large scale aberrations of atmospheric circulation. The likely factors are :

(i)   air-sea interaction,

(ii) injection of large scale ash, dust into the atmosphere,

(iii) changes in the composition of atmosphere particularly water vapour, carbon dioxide, ozone (which are selective absorbers of radiation that could modify the heat balance of the earth).

Climatic fluctuations means large departures from seasonal or annual averages, while climatic change stands for longer time scales, decades to centuries, when the average temperature of a whole hemisphere or whole globe increases or decreases significantly. Climate changes have far reaching consequences.

The study of past civilizations show how climatic changes had major impacts on human society. The last glacial period terminated about ten thousand years ago, with this the boundary of glaciers moved northwards and higher in mountains, the ocean level slowly elevated and reached the present level at about 6000 years ago. People who lived in near tropical zone moved northwards and populated over vast territories of Europe. The most significant migration was from Hindustan and Iranian Highland. The languages of the majority of European peoples have common roots with the ancient Hindustan, the Sanskrit. This is the reason why one of the major human races is called Indo-European. The ever changing earth's climate has warmed up by 0.5 °C during previous 100 years, which is attributed to the human activities.

Atmospheric pollution, emission of greenhouse gases and the resulting impacts (on global scale) are:

(i)   warming of the climate,

(ii)   reduction in stratospheric ozone layer,

(iii)   contamination of food chains over land and sea,

(iv)   acid rains,

(v)   regional photochemical smog.

(a)   **General circulation,** statistical or numerical climate models take into account of air-sea interaction (at least the upper layers of the oceans), biological feedback loops, the ice and snow fields (their fluctuations) and hydrological cycle (precipitation – runoff – soil moisture–evaporation – cloud formation and back to precipitation again).

These models outputs are: For doubling of carbon dioxide, the global mean temperature will increase range 2 °C to 5 °C. More warming towards poles (particularly in Northern Hemisphere but not significant change in precipitation distribution. On global basis precipitation will increase with increase in evaporation as a consequence of global warming. Intensity of tropical cyclones, extreme events of heat wave, cold waves, will increase. The inter-governmental Panel on Climate Change estimates a warming of about 3 °C by the end of (year 2100) 21st century (with range of estimates 2 °C to 5 °C with the present level of increase of greenhouse gases). The warm climate would result in 10 to 30 cm rise in mean sea level water (due to melting of glaciers, Greenland ice cap and thermal expansion of the sea water). Sea surface water temperature likely to remain the same. Island nations Maldives, Vanuatu etc. greatly reduce in size.

(b)   **CFCs are powerful greenhouse gases** and they are the main destroyers of stratospheric ozone. Stratospheric ozone layer is a protecting sheild against incoming solar harmful UV-radiation (UVB). Ozone is present in the atmosphere between 10-45 km altitude with maximum concentration at altitude around 18-20 km in a globe encircling ozone layer. The total ozone column in the atmosphere sheilds the earth's surface from harmful UVB. UVB induces skin cancer, damages eye, causes suppression of immune systems in human beings. It effects the productivity of aquatic and terrestrial ecosystems. One percent decrease in ozone concentration causes three precent rise in skin cancer in humans. Stratospheric ozone decline causes stratospheric cooling, which will effect the global circulation, climate, however which is not understood. WMOs global ozone observing system concluded that use of chemicals containing Chlorine, Bromine can lead to a significant depletion of stratospheric ozone. This is the cause of ozone hole over Antarctica in spring time.

(c)   **Contamination of seas, lakes and land:** Contamination of atmosphere by toxic metals and organic compounds infecting all parts of the earth. The biological communities of the water and land absorb, and the bio-accumulate these contaminants through the food webs. By the process of food chain when these substances found in higher level mammals, including humans, the trace concentrations deposited would be in harmful amounts. Arsenic, lead are known to be carcinogens. Mercury, Lead attack the central nervous system. Toxic-persistent chemicals carried through the atmosphere to land, lakes, seas and into sediments may be viewed as "time bombs". Metals and organic contaminants deposited and absorbed initially may not show any impact on the environment but after a few decades by natural process (such as microbiological methylation of mercury, and other metals) become highly toxic. The slow release of "time bomb" in soil and water can rapidly multiply in concentration at higher levels in the food webs. These affect seriously on high level predators. For example DDT caused decimation of North American eagle population during 1960s and 1970s and death and deformation of humans, the 'Minamata' condition created by ingestion of fish poisoned by methylated mercury.

(d)   **Acid rains** - Discussed in Air pollution.

(e)   **Smog** - Discussed in Air Pollution.

Radio activity

The accident of nuclear power plant at Chernobyl USSR opened the eyes of scientists about its very harmful effects. They found the nuclear traces in long range transport or airborne pollutants. The cloud of radioactivity from Chernobyl accident moved with wind towards Scandinavia where rains scavenged in sufficient amounts. This rendered lichen and the reindeer browsing on lichen radioactive. The animals found to be unfit for human consumption as they were carrying very high radioactivity. After Scandinavia, the cloud trajectory moved southwards to central Europe, where radioactive particles found deposited in foodstuffs, which were banned for short period for health hazard reasons. Subsequently the radioactive debris were moved by westerly wind circulation throughout the northern hemisphere. After Chernobyl event the International Atomic Energy Agency and WMO established an international warning service forecasts by meteorological agencies regarding transport, dispersion and deposition of radioactive particles.

In 1979 the first World Climate Conference established world climate programme of four components – Data, Applications, Impact studies and Research.

(i) *Data*: Measurement, collection and exchange of data on climate and factors affecting it.

(ii) *Application*: Applying climatic information to improve the efficiency of many economic activities.

(iii) *Impact Studies*: Evaluating the Socio-economic impacts of projected changes in climate.

(iv) *Research* : Undertaking research on the climate system (atmospheric - oceans - land - biota) and the factors affecting it including predictions of effects of increasing greenhouse gas concentrations.

It is estimated that one to two billion tonnes of carbon per year could be removed from the atmosphere for each 100 million hectares of forest planted or green lungs of the earth. Under FAO/UNEPs Tropical Forest Action plan, aggressive restocking and replanting programmes in the temperate and boreal zones could restore the major biological sink for carbondioxide.

8.16 SOLID WASTES

Solid wastes include any garbage, refuse, sludge, air pollutants, any discarded material from domestic houses, industrial operations, mining, agriculture, construction and demolition of buildings. Solid wastes may be biodegradable or combustible. Industrial wastes include organic and inorganic chemicals, pesticides, explosives, paints and allied products, petroleum refining materials, rubber, plastic, waste oil. Some of these are toxic in nature. Municipal wastes include garbage, fats, paper, leaves, grass, wood, plastics, rubbers, rags, ash, glass, ceramic etc.

Generation of solid wastes cannot be stopped but they can be reused or recycled for use. Community refuse may contain reusable materials such as steel, aluminium, glass, paper. These materials may be collected separately and reused, which helps in reducing the cost of manufacture.

Solid waste disposal mainly carried by: (i) land fill, (ii) incineration, and (iii) compost.

Toxic solid waste disposal by land fill may create health problems. Incinerators reduce the volume of solid waste by burning combustible materials. In this case also one must take care of air pollution. Land fill consists of depositing refuse in a low place or excavated trench. The refuse after dump is covered by earth about one feet depth.

Incinerators generally burn 85 to 90% of combustible material. Heat generated in incinerators are used for many purposes and now considered as a source of energy.

Dumping of food and agricultural waste, sewage sludge in pits to decompose by biological action is called composting. The residue of composting material is used as manure in agriculture. Micro-organisms break down the degradable material into powdery material called compost.

8.17   FIRE SERVICES (PREVENTION, CONTROL AND FORECASTING)

Fire is one of the Pancha Bhutas. The others are Air, Earth, Water and Sky. Fire is considered as the enemy of forests, because it may rage an inferno in a short time and reduce the forest (its earned centuries savings) wealth into ash and scarred black remains. Not only vegetation but also flora and fauna destroyed and heavily damaged and make the forest land infertile for long time (by destroying top humus soil). It is a known fact that forests control soil erosion, avalanches, landslides and runoff of water which causes flash floods. Forests regulate water supply and aids in improving ground water table. All these favours are destroyed by the fire. Besides forest fires, it is a common feature, during heat wave conditions, burning of huts, haystacks, thatched houses, electrical short circuits.

In Russia, which has about 25% of world's forest area, fire danger ratings are published and warned of dangers regularly. Fire forecasters compute fire danger ratings based on maximum temperature, humidity and wind direction or speed.

In a day to day life common man uses water and sand for fire fighting. In government offices, factories, industries, fire fighting arrangements are made by use of water, sand, carbon dioxide extinguishers and making use of fire fighting services where available.

In aviation, the principal fire extinguishing agents are water and foam. The amounts of water for foam production and the complementary agents required depend on fire fighting category, which is determined from Tables 8.9, and 8.10 as appropriate. The discharge rate of the foam solution shall not be less than the rates given in Table 8.10 as appropriate. The discharge rate of complementary agents are selected for optimum effectiveness of the agents used.

For example at an elevated heliport, at least one hose sprary line capable of delivering foam in a jet spray pattern at 250 l/min.

Table 8.9 Heliport fire fighting category

Category	Helicopter overall Length
H_1	< 15 m
H_2	≥ 15 m but < 24 m
H_3	≥ 24 m but < 35 m

Table 8.10 Minimum usable amounts of extinguishing agents for
surface level Helicopters

(1) Category	(2) Water in liters	(3) Foam Solution l/min	(4) Dry chemical Powders (kg)	(5) Or Halons(kg)	(6) CO_2 (kg)
H_1	500	250	23	23	45
H_2	1000	500	45	45	90
H_3	1600	800	90	90	180

Fire fighting vehicles are used with hydraulic platforms that can extend up to 54 meters high-rise buildings.

Modern Fire Control Technology

In Classical approach firefighting is done using water sprinklers, fire alarms. Modern technology aims at prevention of fire accidents, which is named fire control technology. In this (later) system sensory devices introduced to target corporate houses and industrial units. It is aimed to fight off the fire before it breaks out. The devices heat-sensing cable, nitrogen injection and wireless system developed with the purpose of detecting and extinguishing fire at the beginning itself and eliminating the chance of casuality. For example take the case of heating cable. The optic fibre cable is made of quartz, which can detect hot gases and at the same time it is free from electromagnetic disturbances. Temperatures are recorded along the sensor cable. Depending on temperature, warning signals are delivered immediately. This helps in putting out the fire at the start itself. A German company 'Detectomat' manufacturing microprocessors with software tools for fire control panel to detect fire at any stage. The advanced electrical cables focus exclusively on detecting and preventing fire in the corporate and industrial units.

CHAPTER - 9

Water and its Pollutions

INTRODUCTION

Water (Prevention and Control of Pollution) Act 1974. Aims to provide safe water, sustainable use by preventing pollution.

All life on earth is sustained by water. It is the second most vital matter for life after air and then follows food, clothing, shelter etc. Man can survive without food for a few weeks but without water he cannot survive even a few days (a week) and without air even a few minutes. It is estimated nearly 70% of man by body weight is water. Water is inorganic and does not provide energy to tissues. Much of water is in body – protoplasm and in spaces between cells. Plasma (fluid part of the blood) contains 91-92 percent water. Water is essential for digestion of food, blood circulation and removal of waste products from body. Man's average daily loss of water is 2500 CC. Of which 1500 CC as urine, 500 CC in perspiration, 400 CC in exhaled air and 100 CC in faeces. Any loss of water more than 250 CC (10%) is fetal. On an average man drinks about 1500 CC of water and the remaining he gets it from food. Fruits and vegetables contain about 80% water. Kidneys remove excess water along with cellular waste.

On an average man requires 2.5 to 3 liters of water per day for drinking and cooking or one cubic meter per year. However to improve his living conditions he is using 100-200 liters per day for his household needs and a thousand times for industrial consumption. For example, production of one ton sugar requires about 100 cubic meters (m^3) of water, one ton of paper requires 250 m^3 of water, processing of 1000 liters of milk requires 5000 liters of water. Production of one ton steel requires 150 m^3, Nickel 800 m^3, Aluminium 1500 m^3 of water. In recent times man has been mining ground water for agriculture, domestic and industrial use. This is causing depletion of ground water resource which was reliably dependent through the ages. Unknowingly man has made many blunders in water use and conservation. The worst part is pollution of lakes and rivers. Pollution is the main factor that is threatening the exhaustion of fresh water resources. One cubic meter of sewage dump contaminates more than 12 cubic meters of clean water in rivers and lakes. Use of polluted water is a severe health hazard. Pollution of water is the foremost concern of mankind. Experts are of the opinion that water famine may arise only through inefficient use of natural water resources rather than lack of water in the world. Further they concluded that the water resources on the earth fully sufficient to meet all the growing needs of man for an indefinite time period provided man must practice to avoid water pollution, extend recycling of suitable water resources.

According to recent World Water Forecast, a grim prognosis is that among the worlds ten most endangered rivers include Ganga, Indus, Nile and Yangtze. Ganga basin covers one third land area of India and its fertile soil is the cradle of Indian civilization and the home of millions of people. It is a sorry state of affair that indiscriminate extraction of water with table wells from river and its basin, damming of its tributaries for irrigation purpose seriously eroded its reproduction which lead to the free flow arrest. It is said Ganga is dying due to pollution over extraction of water and serious climatic changes. According to World Water Forecast, glaciers account for as much as 30 to 40% in the case of Indus. Ganga flows through India but also drains from Nepal and China. It covers about 2500 km length and 30% of land area. As regards Indus, it flows through Pakistan, but also drains from Afghanistan, India and China. It covers 2900 km length. The main warrant threats of death is the over water extraction, pollution and drains. The endangered species in the Ganga basin besides human beings are 140 species of fishes, dolphins, fresh water shark and 90 amphibian species. In Indus the endangered species are Dolphins, about 22 species of fishes, 25 amphibian species. From source in the Himalayas to the sea Bay of Bengal the story of Ganga is India's civilization. The World Water Forecast limits a warning the death of Ganga is not merely the river, but includes the great Hindu Civilization, Ravi singh Secretary General and Chief Executive Officer, World Water Forecast India, pleads for conservation of rivers and wet lands which are part and parcel of National Security, health and economic success and warns, people must change their mindset atleast now, lest they have to pay the heavy price in the not so-distant future.

In 1950's the world population was about 2.5 billions which shot upto 6 billions by the end of 2000 AD and it is projected to be 9.2 billions by 2050. India would become the world's most populus country with 1.7 billion people according to UN report of 2005. Half the increase in the global population between 2005 and 2050 would be on account of rise in the over 60 years population, while the number of children under 15 will begin to decline. A 2004 UNICEF report of India said that only a third of population has access to potable water and adequate sanitation facilities. Poor water quality, sanitation and waste management lead to plethora of maladies. In 1950s about one third of the population lived in cities which rose to one-half by the end of 20th century and it is projected to rise to two thirds by 2025.

Water is essential to all lives of animals and plants. It has crucial role in industries and steam generation, plays important role in engineering. Water is used in large quantities in the production of steel, sugar, paper, rayon, chemicals, textiles, ice, drinking, bathing, washing, irrigation, fire fighting, power generation and atomic energy.

Water in the World

The volume of the global water is estimated to be 1455×10^6 cubic kilometers (km^3). About 1370×10^6 km^3 or 94% of the global water is estimated to be in the seas, which occupies 361 million km^2 (square kilometers) of the global surface area or 70.8%. 85×10^6 km^3 or 6% of global water is fresh water, which accounts water of rivers (1200 km^3 or 0.0001% of global water), lakes (750×10^3 km^3 or 0.05% of global water), glaciers (2.4×10^6 km^3 or 4.11% of global water). Water in the atmosphere is 14×10^3 km^3 or 0.001% of global water.

It is estimated that about 4.5 million km^2 or 3% of land area is occupied by inland water bodies and 16.5 million km^2 or 11% (land) area occupied by glaciers. The volume of fresh surface water is about 751200 km^3 which is very small amount compared to the volume of the oceans and seas water, yet it plays a vital role in the life of man.

9.1   THE WATER POLLUTION ACT 1974

After industrial revolution the growth of industries lead to the urbanization, in turn pollution of rivers, streams become a severe problem. It had become essential to ensure that the domestic and industrial effluent's not to the discharged into the water courses without adequate treatment. Otherwise the water world be unsuitable for drinking purposes, unsupporting for fish life and irrigation. Pollution of rivers, streams a serious problem that cause damage to country's economy. A committee was set up in India to look into the problem of water pollution in 1962. The committee after considering relevant local provisions recommended for maintaining/restoring wholesomeness of water courses and laid stress for controlling the existing and new discharges of domestic and industrial waste into water flows. Consequent to recommendations, the water (prevention and control of pollution) Bill passed by both the Houses of Parliament and received the assent of the President on 23rd march, 1974. It came on to the Statute Book as the water (prevention and control of pollution) Act. 1974. Further it was Amended in 1978 and 1988.

Urban growth, water demand and extreme weather events

The total world population in 1950 was 2.5 billions, of which about one third lived in cities. By the year 2000 the population was about 6.2 billion, of which half the population living in towns and cities. The ratio is expected to increase to two-thirds by the year 2025 (with projected population of 8.0 billion).

Between 2015-2050 (over the world) the number of people living in cities will grow 3.9 billion to 6.3 billions and urban dwellers increase from 54% to 67% of the world population.

Along side this increase is the emergence of mega cities in developing countries where some 80% of the worlds urban residents will live. Statistics show that extreme weather events (both meteorlogical & hydrological events) account for about 70% of all natural disasters.

The global demand for water has increased dramatically over the past century. Between 1900 and 1995, water withdrawal from existing sources has increased more than six fold. The increase is more than double the rate of population growth during the same period. The demand is a result of increasing irrigation, urban agriculture, industrial growth and rising water consumption per capita for domestic and sanitation purposes.

Significant proportion of urban dwellers, particularly in developing countries have limited (or no) access to a safe potable water supply. Estimates put this figure at 16% worldwide, 21% in southeast Asia, 22% in the eastern mediterranean and up to 48% in Africa.

Explanation of some terms

Outlet: Any conduit pipe or channel, open or closed, carrying sewage or trade effluent or any other holding arrangement which causes (or likely to cause) pollution.

Water pollution: Contamination of water or such alteration of the physical, chemical or biological properties of water or such discharge of any sewage or trade effluent or any other liquid, gaseous or solid substance into water as may or is likely to, create a nuisance or render such water harmful (or injurious) to public health or safety or to domestic, commercial, industrial agricultural or other legitimate uses or to the life and health of animals or plants or of aquatic organisms.

Sewage effluent: Effluent from any sewage system or sewage disposal works and includes sullage from open drains.

Sewer: Any conduit pipe or channel open or closed, carrying sewage or trade effluent.

Stream: Includes- 1. River, 2. Water course 3. Inland water (natural or artificial) 4. Sub terranean waters, 5. Sea or tidal waters to such extent or as the case may be, such point as the state govt. may by notification in the official gazette, specify in this behalf.

Trade Effluent: Any liquid, gaseous or solid substances which is discharged from any premises used for carrying on any industry operation or process, or treatment and disposal system, other than domestic sewage.

Central and state water pollution control boards are constituted similar to Air Pollution Control Boards.

However some executive measures are different which will be briefly given.

The Board Meetings: A Board shall meet at least once every three months.

9.2   GLOBAL AND INDIA WATER USES

Global water: Earth's water 1386 million cubic kilometers. Only 2.5% of earth's water (34.65 m km^3) is fit for consumption. Of this fresh water 68.7% present in glaciers and ice caps, 30.1% groundwater and 1.2% surface and other fresh water. Of this groundwater is fast depleting and other water sources are also threatened. Of the global water (1386 m km^3) oceans contain 96.5% (1337.49 m km^3) and other saline water 0.9% and fresh water 2.5% (34.65 m km^3).

Of the global fresh water, 69% ground ice and permafrost 20.9% in lakes, 3.8% soil moisture, 2.6% in swamps & marshes, 0.49% is rivers and 3% (1.0395 m km^3) in atmosphere and 0.26% in living things.

In India fresh water resources = 4000 b m^3

1b = 10^9, m^3 = cubic meter, 1m^3 = 1000 liters.

Of the fresh water (4000 b m^3)

1047 bm^3 lost in evaporation,

1084 bm^3 non available water and

1869 bm^3 available water.

Of the available water 1123 bm^3 usable water.

Of the usable water 395 bm^3 ground water, and

725 b m^3 surface water.

India's water demand

In 2006 it is 829 bm^3

In 2025 it would be 1093 bm^3

And in 2050 it would be 1047 bm^3

Note: Water demand decreases as population declines and water use efficiency increases

Water use share in %	India	World	Europe
Agriculture use	83	69	33
Industry use	12	23	54
Domestic use	5	8	13
Total	**100**	**100**	**100**

India consumes most of its usable water for agriculture (83%) purpose, while households get only 5%.

9.3 FUNCTIONS OF CENTRAL / STATE WATER POLLUTION CONTROL BOARDS

The main functions of the central water pollution control board is to promote cleanliness of streams and wells in different areas of states.

1. Providing technical help and guidance to the State Board to carry out and sponsor investigations and research relating to problems water pollution, prevention, control or abatements (P, C/A) of water pollution.
2. Plan and organize the training of persons engaged in the programmes of water pollution, P, C/A.
3. Organize through mass media a comprehensive programme regarding water pollution, P, C/A.
4. Collect, compile and publish technical and statistical data relating water pollution and effective measures; prepare manuals, codes and guides for treatment and disposal of sewage and trade effluents.
5. Lay down the standards for a stream or well depending on the nature of use of water in such streams or wells.
6. Plan a nationwide programme for water pollution, P, C/A.
7. Establish or give recognition to a laboratory/Laboratories to enable the Board to perform its functions efficiently which include analysis of samples of water from stream/well or of samples of sewage or trade effluents.
8. State Boards encourage, conduct and participate in investigation and research relating problems of water pollution, P, C/A.
   Organize mass education programmes of water pollutions. Organize training to person engaged in programmes related to water pollution.
9. Inspect sewage or trade effluents, works and plants for treatment of sewage and trade effluents, review plans already made, make specifications to plants to set up for treatment of water, works for the purification thereof and the system for the disposal sewage or trade effluents. Lay down, modify or annual scrap effluent standards for

sewage for the quality of receiving water resulting from the discharge of effluents and classify waters of the state.

10. Evolve economical and reliable methods of treatment of sewage/trade effluents, keeping is view of the conditions of soils, climate and water resources of different regions, prevailing flow characteristics of water in streams and wells which render impossible to attain even the minimum degree of dilution.

11. Evolve methods of utilization of sewage and suitable trade effluents in agriculture.

12. Lay down standards of treatment of sewage and trade effluents to be discharged into any particular stream taking into account the minimum fair weather dilution in that stream and the tolerance limits of pollution permissible does not exceed in the water of stream after discharge of such effluents.

13. If required construct new systems for disposal of sewage and trade effluents or modify, alter any existing system to adopt remedial measures to prevent water pollution

14. State water pollution control boards make survey of any area and gauge and maintain records of the flow or volume and other characteristics of any stream or well, maintain rainfall measurements, install and maintain water gauges and carryout stream surveys.

9.4   PROHIBITION ON USE OF STREAM OR WELL FOR DISPOSAL OF POLLUTING SUBSTANCES/MATTER

No person is allowed to let any poisons, noxious or contaminating substance into any stream, well or sewer on land or cause to abstract flow of the water of the stream that will aggravate contamination/pollution.

No person is allowed to establish any industry, operation or process/any treatment and disposal system or extension thereto which may discharge sewage or trade effluent into a stream or well or sewer or on land.

Bring into use any new or altered outlet for the discharge of sewage.

9.5   EMERGENCY MEASURES IN CASE OF POLLUTIONS OF STREAM OR WELL

When any poisonous, noxious or contaminating substance/matter is present in any stream or well or on land by way of discharge or has entered into the stream or well due to some accident or unforeseen act or event it is necessary to take immediate action which are necessary-

1. Remove the contaminating substance from the stream or well or on land and dispose it appropriately

2. Mitigating any contaminating material by its presence in the stream or well

3. Instant orders may be issued restraining or prohibiting the person concerned from discharging any poisonous, noxious or contaminating substance or from making insanitary use of the stream or well.

9.6 RIVERS IN INDIA

There are four systems of rivers in India. They are: (a) The Himalayan, (b) The Deccan plateau, (c) The central and (d) Inland drainage basin.

The Himalayan rivers (Ganga, Brahmaputra) are perennial are fed by rainfall and snow melt and are flood prone. The Ganga is the largest basin, has five tributaries.-The Yamuna, The Kali, The Ghagra. The Gandak and the Kosi.

They originate in the Himalayas and fall into the Bay of Bengal. However, the three tributaries-the Chambal, the Betwa, the Son rise in the Vindhyas flow northward and join, the Ganga or Yamuna. Brahmaputra has Teesta and Manas tributaries, while Indus (Sindhu) has five tributaries Jhelum, Chenab, Ravi, Beas and Satluj, which falls into the Arabian sea.

Narmada and Tapi run east to west direction in Vindhyas and fall into the Arabian Sea. Mahanadi flows in Chhattisgarh, Odisha states and fall into the Bay of Bengal. These rivers have fertile alluvial soil.

The Godavari and Krishna rivers originate in Western Ghats, flowing eastward fall into Bay of Bengal. These two rivers cover most part of the Deccan Plateau are fed by rain water. Together with their tributaries virtually dry up in summer. Streams in western Rajasthan are short lived, which fall into salt lakes like the samber or lost in the sands. Only Luni flows into Rann of Kutch.

9.7 HYDROLOGICAL CYCLE

Definition

Hydrology may be defined as the science that deals with the processes governing the depletion and replenishment of surface water and ground water resources of the earth. Consequently it may be viewed as part of physical geography and hydrometeorology.

Hydrometry is the technology of water measurement spanning all aspects of water movement within hydrological cycle.

The hydrological cycle is a continuous process of movement of water from surface of earth to atmosphere (as evaporation), from atmosphere to ground (as precipitation), then to rivers, lakes, underground reservoirs and to the sea (see Fig. 9.1). Like lithosphere, the bulk composition of hydrosphere is oxygen. The composition of water has 88.9% oxygen and 11.1% hydrogen by weight. Waters of oceans, rivers, lakes have very small amounts of almost all elements of the earth's crust. Sea water has 3.5% dissolved minerals, of which NaCl (about 2.6%) is the most abundant. Consequently sea water has salty taste.

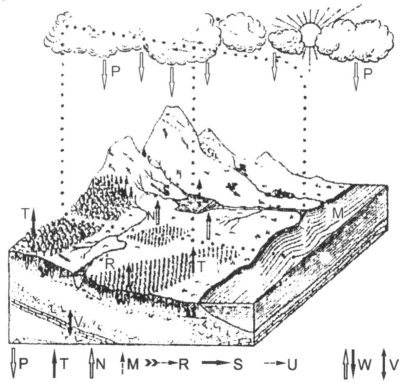

P = precipitation, T = Transpiration, N = Soil evaporation, M = Water surface evaporation, R = Total river flow, S = Surface (flood) runoff, U = Ground water runoff, W = Soil moisture, V = Exchange with ground water.

Fig. 9.1

About 80% of the total water vapour in the atmosphere comes from the evaporation of sea water.

Note: Steam occupies a volume of 1600 times greater than that of the liquid water. In gaseous state molecules are about 12 times apart than they are in liquid state.

Thus hydrological cycle is a God's gift to all living beings because fresh water resources on land are formed from rain water. In a sense it is a gigantic desalination plant which converts saline sea water into water vapour and then into fresh water that falls on earth as precipitation. It may be noted that all minerals/matter transforms to other forms but water remains as water. In Sanskrit water is called Amrita, meaning deathless. At any one time the lakes and rivers contain 751200 km^3 of water which is little more than 0.05%

of global water. An estimate of annual water balance of the world by M.I. LVOVICH is given in Table 9.1.

Table 9.1

	Volume in km^3	Average depth in mm
A. Over periphery of land area $(116.8 \times 10^6 \text{ km}^2)$		
Precipitation	106000	910
River discharge	41000	350
Evaporation	65000	560
B. In land area $(32.1 \times 10^6 \text{ km}^2)$		
Precipitation	7500	238
Evaporation	7500	238
C. World ocean $(361.1 \times 10^6 \text{ km}^2)$		
Precipitation	411600	1140
Inflow river water	41000	111
Evaporation	452600	1251
D. World total $(510 \times 10^6 \text{ km}^2)$		
Precipitation	525100	1030
Evaporation	525100	1030

Global volume of rain water = $4 \pi R^2 h$,

$$= 4 \times \frac{22}{7} \times (6370)^2 h$$

$$= 510 \times 10^6 \text{ km}^2 \times h$$

$$= 510 \times 10^6 \times \frac{1030}{1000} \times \frac{1}{1000} \cong 525.100 \text{ km}^3$$

where R = Mean radius of the earth 6370 km.

$$h = \text{height of pptn} = \frac{1030}{1000} \times \frac{1}{1000} \text{ km.}$$

9.8 WATER RESOURCES

Water resources can be divided into two categories. (i) Surface water and (ii) ground water.

Surface water resources are Rivers, lakes, swamps, glaciers and sea water.

Underground water resources: are wells and spring water.

Rain water is the purest form of natural waters. However it contains traces of dissolved material, gaseous compounds. Rivers, lakes, swamps waters are all accumulated rain water. These have some dissolved minerals and suspended matter. Sea water contains dissolved salts about 35 gm/kg or 3.5% by weight. Six ions of sea salts together will be more than 99%. They are: ions of Chloride (55%), Sodium (30%), Sulphate (8%), Magnesium (4%), Calcium (1%) and Potassium (1%).

Water, in the form of vapour, is exchanged between sea and air. More than 80% of water vapour in the atmosphere comes from the evaporation of sea surface water. Along with water vapour energy is also transferred from sea to atmosphere in the form of latent heat of water vapour. On condensation it releases latent heat. Condensed water returns to the sea in hydrological cycle but energy remains in the atmosphere. Part of this heat energy is converted to mechanical energy (wind). Two-thirds of the precipitation on landreturns to the atmosphere by soil evaporation and plants. The remaining one-third precipitation either percolates into soil or runs off on the land surface. The bulk of the fresh water on land is stored in the form of ice in glaciers (volume 24×10^6 km^3) which is about 20000 times the worlds river waters. Ice melted water flows into rivers and lakes. Water stored in lakes (750×10^3 km^3) is about 625 times the water in rivers. This water is of great importance in many parts of the world. They are the fresh water source for cities and agriculture. Because of importance artificial lakes are created by damming rivers.

Percolation of water through large pores into the soil is called gravity water. This gravity water accumulates at some depth over rocky layer and fills all cracks and pore spaces. The level below which soil and rocks are saturated with water is called water table. Water below water table is called ground water. The depth of water table varies from place to place and with season. In dry season the depth will be more than in wet season. During good rains water table slowly increases and this results in ground water flow into nearby streams. When there are persistent heavy rains stream channels will be overflowing into adjacent areas. This causes floods. Surface run-off of water depends on precipitation and varies with the intensity of precipitation. Ground water flow on the other hand is more steady. The ground water (volume $= 60 \times 10^6$ km^3) is about 50,000 times the volume of river water.

Artesian Wells

When rain occurs on mountain slopes water passes through the porous rock and then flows underground. If this water flows between two layers of low permeability artesian system (like water flow through a pipe) develops. The underground aquifer (water saturated zone) flow carry large quantities of water. This can be drawn through digging wells on plane area, where water comes to the surface by pressure. This type of well is called Artesian well.

Water impurities can be categorized as:

(a) suspended impurities,

(b) dissolved inorganic impurities and

(c) organic impurities.

(a) suspended impurities can be removed by filtration or settling. Suspended impurities include clay, silt, bacteria, algae, protozoa.

(b) dissolved inorganic impurities include (i) calcium, magnesium and sodium carbonates, bicarbonates, sulphates, chlorides, fluorides (ii) Metal and oxides e.g. magnesium, iron oxide, lead, Arsenic (iii) gases e.g., Oxygen, CO_2, H_2S.

(c) organic impurities include (i) Suspended (e.g., vegetables, dead animals). (ii) Dissolved (e.g., vegetables and animals).

The essentials of drinking water standard in India are given in Table 9.2.

Table 9.2

Item	Desirable	Maximum permissible limit (in the absence of alternate source)
pH	6.5 – 8.5	- - -
Alkalinity (mg/l)	200	600
Total hardness(as $CaCo_3$)	300	600
Dissolved solid material (mg/l)	500	2000
Calcium (mg/l)	75	200
Chlorides (mg/l)	250	1000
Chlorine (mg/l)	1.2	- - -
Fluroid (mg/l)	1.0	1.5
Nitrate (mg/l)	45	100
Iron (mg/l)	0.3	1.0
Copper (mg/l)	1.5	1.5
Manganese (mg/l)	0.1	0.3
Zinc (mg/l)	5	15
Arsenic (mg/l)	0.5	- - -
Cyanide (mg/l)	0.5	- - -
Lead (mg/l)	0.5	- - -
Total Coliforms	nil/100ml	10 counts/100ml
Faecal coliforms	nil/100ml	- - -

mg = milligrams   l = liter

9.9   HARDNESS OF WATER

Water that does not produce lather easily with soap but produces white curd like form is called hard water, while water that produces lather easily with soap is called soft water. Hardness of water is of two types :

1. Temporary hardness and

2. Permanent hardness

Temporary hardness is caused by the presence of dissolved bicarbonates of calcium, magnesium and heavy metals and the carbonates of iron in the water. Temporary hardness can be removed by boiling of water which decomposes bicarbonates into insoluble carbonates and carbondioxide. Carbonates can be removed by filtration while carbondioxide escapes out.

Permanent hardness of water is caused by the presence of dissolved chlorides, sulphates of calcium and magnesium, iron and other heavy metals in the water and it cannot be removed by boiling.

Hardness of water is generally measured as parts per million (ppm). Parts per million is the parts of $CaCO_3$ equivalent hardness per 10^6 parts of water.

1 ppm = 1 part of $CaCO_3$ equivalent hardness in 10^6 parts of water.

Equivalent of $CaCO_3$ is defined as below.

Equivalent of $CaCO_3$ =

$$\frac{\text{Mass of hardness causing substance} \times 50}{\text{Chemical equivalent of hardness causing substance}}$$

where 50 = Chemical equivalent of $CaCO_3$.

Hardness of water classified as in Table 9.3

Table 9.3

Hardness in ppm	0-50	50-100	100-150	150-200	200-250	more than 250
Nature of hardness	Soft	Moderately soft	Slightly hard	Moderately hard	Hard	Very hard

Example - 1

What is the temporary hardness in ppm and total hardness in ppm of a sample of water which have the following composition. Ca $(HCO_3)_2$ = 16.2 mg/l, Mg $(HCO_3)_2$ = 7.3 mg/l, $MgCl_2$ = 9.5 mg/l, $CaSO_4$ = 13.6 mg/l.

Solution

Where molar mass of $Ca(HCO_3)_2$ = 40 + 2 (1 + 12 + 48) = 162 etc

Substance	Mass mg/l	Molar Mass	Multiplying Factor	Equivalent of CaCo$_3$
Ca(HCO$_3$)$_2$	16.2	162	100/162	$16.2 \times \dfrac{100}{162} = 10$
Mg (HCO$_3$)$_2$	7.3	146	100/146	$7.3 \times \dfrac{100}{146} = 5$
Mg Cl$_2$	9.5	95	100/95	$9.5 \times \dfrac{100}{95} = 10$
CaSO$_4$	13.6	136	100/136	$13.6 \times \dfrac{100}{136} = 10$

Temporary hardness caused by dissolved bicarbonates of Ca and Mg.

∴ temporary hardness = 5 + 10 = 15 ppm.

Total hardness = 10 + 5 + 10 + 10 = 35 ppm.

Example - 2

What is the total hardness and permanent hardness in ppm of a sample of water containing Ca(HCO$_3$)$_2$ = 16.2 mg/l, Mg (HCO$_3$)$_2$ = 7.3 mg/l, MgCl$_2$ = 9.5 mg/l and CaSO$_4$ = 13.6 mg/l.

Solution

Total hardness = 10 + 5 + 10 + 10 = 35 ppm.

Permanent hardness = hardness of MgCl$_2$ and CaSO$_4$ = 10 + 10

$= 20$ ppm

Substance	Mass mg/l	Molar Mass	Multiplying Factor	Equivalent of CaCo$_3$
Ca(HCO$_3$)$_2$	16.2	162	100/162	$16.2 \times \dfrac{100}{162} = 10$
Mg (HCO$_3$)$_2$	7.3	146	100/146	$7.3 \times \dfrac{100}{146} = 5$
Mg Cl$_2$	9.5	95	100/95	$9.5 \times \dfrac{100}{95} = 10$
CaSO$_4$	13.6	136	100/136	$13.6 \times \dfrac{100}{136} = 10$

There are several disadvantages in use of hard water for domestic purpose, industries and in steam generation.

In Domestic Use

In washing and cleaning, bathing, cooking, drinking hard water causes lot of wastage of soap, skin becomes dry and dark, consumption of more fuel, gives unpleasant taste, pulses and beans etc., are not properly cooked and digestive system may be affected.

In Industrial Use

It casts bad effects in industries such as textile, sugar, dyeing, paper, laundry, concrete etc. Use of hard water in steam generation causes boiler scale, sludge formation, corrosion, priming and foaming, caustic embrittlement, boiler corrosion etc. These cause adverse effects on industrial production. Some of these terms are explained below.

Boiler Scale

When hard water is boiled, dissolved material in it forms a hard deposit on the inner surface of the boiler. This is called boiler scale.

Sludge

It is loose and slimy soft precipitate formed in the boiler, which can be removed by scrapping with brush.

Boiler Corrosion

It is the decay of boiler material caused by chemical or electro-chemical attacks by environment.

Priming

When hard water steam produced rapidly, some liquid water particles jump out along with steam. This is called priming.

Foaming

When boiling hard water contains some oily substances, it produces foam or bubbles which do not break.

Caustic Embrittlement

It is a form of boiler corrosion formed by alkaline water.

Softening of Water

To avoid ill effects of corrosion, boiler scale etc., water used in industries must be sufficiently soft. Methods used for removal of dissolved salts from water is called softening.

Lime Soda Process

In this method lime $Ca(OH)_2$, Magnesium hydroxide $Mg(OH)_2$, Soda Na_2CO_3, Calcuim carbonate $CaCO_3$ are added to the hard water. Soluble salts of Calcium, Magnesium are converted into insoluble compounds which are removed. By this method water with 10-15 ppm hardness is obtained.

Zeolite

Zeolites are complex hydrated silicates of Al, Ca, Na, K or Fe.

Permutit or Zeolite Process

In this method hard water is slowly percolated through Zeolite bed of sodium-alumino silicates. This bed converts calcium, magnesium salts into calcium, magnesium-zeolites and sodium salts. Filtered water becomes soft water with 10 ppm hardness and contains soluble sodium salts, while Ca, Mg salts are removed.

9.10   DRINKING WATER

Water for drinking purpose must be soft and clean. It should be colourless, odourless and have pleasant taste. Turbidity should be less than 10 ppm and dissolved solids should be less than 500 ppm. It should be free from micro-organisms. Natural water from rivers, lakes and underground do not satisfy these qualities and requires purification

Suspended impurities are removed by screening (passing water through screens) and by sedimentation (keeping water in big tanks without disturbing to settle down suspended particles at the bottom by gravity). Sedimentation with coagulation is used to remove suspended finer clay particles, colloidal matter. In this process chemicals, called coagulants (Alum, Ferrous sulphate etc) are added to the water before sedimentation. Coagulants are also useful for removing colour and odour and to provide pleasant taste. Alum is most widely used as coagulant. Sodium aluminate, ferrous sulphate are also used as coagulants depending on water alkalinity.

With the application of screening, sedimentation and coagulation all impurities are not removed, some will be left over which will be removed by special filtration beds. Special filteration bed removes colloidal matter, microorganisms and most of the bacteria. These filters remove 98% of bacteria and almost all suspended impurities. A typical filtered bed is shown in Fig. 9.2.

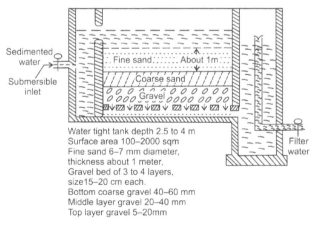

Fig. 9.2 A typical filter bed

Disinfection and Sterilization

Nearly killing or destroying of all micro-organisms pathogenic bacteria to make water potable is called disinfection, while sterilization means total destruction of all living organism in water.

Removal of Micro-organisms from Water

After filteration the residual pathogenic bacteria is killed by disinfection. Water is disinfected by adding bleaching powder, by chlorination, by using chloramine, ozone, potassium permanganate and by ultraviolet light.

Chlorination

Generally level of 0.2 mg/liter of free chlorine will kill bacteria. However to kill viruses it requires 3 to 4 mg/liter and to kill protozoa it may require a level of 500 mg/lit. Disinfection by chlorination is widely used and considered safe.

Sewage

All waste waters or liquid domestic waste, industrial waste, ground waste, storm waste, human waste is called sewage. Sewage emanates gases, such as hydrogen sulphide, ammonium sulphide, phosphine etc. Some of these gases have dirty odour. Sewage contains aerobic and anaerobic bacteria which cause oxidation of organic compounds in it.

Aerobic

Living only in the presence of free molecular oxygen gas or dissolved oxygen in water.

Anaerobic

Living in the absence of free oxygen as gas or dissolved in water.

Prototrophic bacteria takes food from minerals (like nitrites, carbonates etc) present in the sewage.

Metatrophic bacteria takes food from organic compounds, nitrogenous and carbonaceous.

Important sewage characteristics: are Physical, Chemical and Biological.

Physical characteristics include colour, odour, temperature and turbidity of sewage.

Sewage contains colloidal matter, dissolved gases and suspended matter. Fresh sewage is odourless and has gray colour. In about 4 hours time it becomes stale (oxygen being exhausted) and starts emitting offensive gases like hydrogen sulphide, ammonia, methane etc. Its colour becomes dark. Temperature of sewage in general slightly higher than the water supply. Sewage contains about 99% water and the remaining solid matter such as faecial solids, matches, bits of paper, twigs, grease, vegetable matter.

Chemical Characteristics

Fresh sewage is little alkaline while stable sewage is acidic in nature. Solid sewage consists of organic (about 40%) inorganic (about 55%) matter and it contains dissolved gases like CO_2 H_2S, CH_4 etc. It is estimated that sewage consists of 0.045% solids of which 0.0225% in dissolved form, 0.0112% in suspended form and 0.0113% in settleable.

Biological Characteristics

Sewage contains large quantity of bacteria such as algae, fungi, pathogen, protozoa and other microorganisms.

Biological Oxygen Demand (BOD)

BOD of a sewage is the amount of free oxygen required for biological oxidation of the organic matter in aerobic condition at 20 °C for a period of 5 days. BOD is expressed as mg/l or ppm. An average sewage has BOD of 100-150 mg/*l*.

Sewage Treatment

To dislodge its harmful effects sewage must be treated to public health or aquatic life before it is letoff into natural water course (rivers/ponds or on land). Sewage is passed through bar screens or mesh screen which separate the large suspended floating matter and coarse solid silt, gravel etc. It is then let off into sedimentation tanks where it will be treated with chemicals before settlement. Generally alum, ferrous sulphate are used to coagulate, which also removes colloidal matter. It is then aerated or subjected to aerobic biochemical oxidation. In this process sewage is passed through special sprinklers to maintain aerobic conditions. In aerobic condition organic carbon material is converted into CO_2, the nitrogen into NH_3/nitrites/nitrates (this forms salts). Filtering media removes micro-organisms. Treated waste sewage is removed from the bottom which is used for agriculture fertiliser etc. See flow chart in Fig. 9.3.

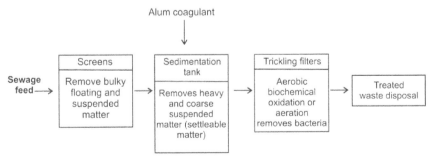

Fig. 9.3 Flow chart of sewage treatment

Desalination of Brine

Sea water contains about 3.5% of dissolved salts and it is called brine or brackish water. Brackish water is completely unfit for drinking purpose. The removal of salts from brackish water is called desalination. Over sea areas and in coastal areas desalination is essential for getting potable water. See Fig. 9.4.

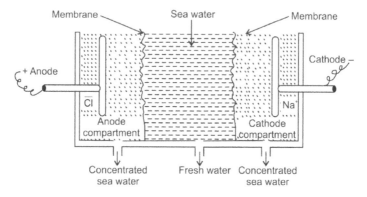

Fig. 9.4 Desalination of brine

An age old method of desalination is boiling of sea water and then condensing the water vapor to obtain fresh water. Where electricity is easily available, the desalination is achieved by electro-dialysis. In this method direct electric current is passed through brine. By this sodium and chlorine ions are attracted at cathode and anode respectively. Fitting special compartments of permiable memebrane around cathode and anode, the central compartment gets accumulated with desalinated water, which is removed.

Reverse Osmosis

If two solutions of different concentrations of a solute are separated by a semi-permeable memebrane, by Osmosis process, dilute solution flows into concentrated solution through the membrane. This flow continues till the two

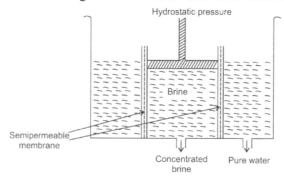

Fig. 9.5 Reverse osmosis

sides of the membrane attains equal concentrations. If a hydrostatic pressure in excess of osmotic pressure is applied on the concentrated solution, the solvent flow reverses, that is, flow starts from concentrated to less concentrated solution across the membrane. This is called reverse osmosis. Reverse osmosis method is used for desalination of brine. This method is shown in Fig. 9.5.

CHAPTER - 10

Forests

10.1 DEFINITION

A track of land covered with trees, shrubs, bushes or woody vegetation whether of natural growth or planted by human agency, existing, maintained with or without human effort or such track of land on which such growth is likely have an effect on the supply of timber, fuel, forest-produce or grazing facilities, or on climate, stream flow, protection of land from erosion, or other such matters and include.

(a) A land covered with stumps of trees of a forest, (b)Such posture land, water-logged or non-cultivable land lying within a adjacent to a forest as may be declared to be a forest by State Government, (c) A land covered by forest or intended to be utilized as a forest, and prescribed by rules.

Forest is one of the resource of renewable energy in nature. World over, forests occupied about 10% of the land area and they are regarded as crops by Adivasis and Tribal people. State Governments and local people/ Agencies pay attention for the conservation of forests. The world's biggest dense tropical forests are located in Amazon basin in south America. Forests are being destroyed worlds over for the purpose of Agriculture and industries.

Annual consumption of wood world over estimated to be about 3.7×10^9 (3.7 billion) metric tons. About 1.5×10^9 people in different countries of world depend on fuel wood as primary source of energy. Forests are valuable resource for human welfare and green plants are regarded as the lungs of earth. Acts of deforestation like grazing, burning, cutting tress etc, have to be stopped to protect the forest wealth. The clearing of forest for farming purposes in often a self-defeating exercise. In Morocco 40000 hectares of land are destroyed each year by erosion. Protective plantations are the only effective defiance against erosion by wind, dust storms and the encroaching desert. Forest is a reviewable resource. It provides us timber, fruits, nuts, seeds, fuel wood or charcoal and other forest products include fibers, flosses, grasses, bamboos, essential oils, tans & dyes, gums, resins, drugs, spices, edible products, leaves and annual/mineral/miscellaneous products.

10.2 FORESTS IN INDIA

Palaeobotanical evidence suggests there were dense forests in India in the Permian period (270-350 MY ago) 270×10^6 years ago. Main was evolved in the Pleistocene Age (1-10 MY ago). ($MY = 10^6$ = million, Y = years ago), a million years ago. At that time India had thick forests, and Rajasthan, Punjab was under a swamp-the remnant of the Tethys Sea. In historical times (4000-5000 BC) Mohen jo daro civilization flourished, which disappeared before 3000 BC. It is believed Aryans came to India about 2000 BC, who used iron axes, javelins, ploughs etc. Aryans were pastoral people who worshipped Nature, constructed shelters for themselves and for their domesticated animals by clearing forests in Gangetic Basin. Their parnasalas and Ashrams (educational centres) were spread in woody surroundings and inspiring landscapes. In such environment, the small population composed the Vedas, the Upanishads and the Aranyakas, which glorified the Nature and lay down precepts of righteous living. When Valmiki Maharshi wrote the epic Ramayana, there were dense forests in Naimisharanya, Chitrakoot, Dandakaranyam and Panchavati, which abounded with forests and wildlife, thick greenery. By the time Mahabharata compiled, city Kingdoms existed with much, less forests compared to the earlier epic period.

When Alexander the great invaded India (327 BC) still there existed impenetrable forests along Indus river. By the time of Emperor Ashoka, forests were dwindled. According to rock edict record, Ashoka ordered-plantation of trees along the roads. By the time of Mughals came to India, Sher Shah Suri planted trees along Delhi-Patna High way. When Aurangzeb marched to Deccan, he found such forests near Burhanpur a little north Aurangabad. During British rule in India, they found teak and certain other Indian timbers very-suitable for building ships for British Navy, for railway steepers and thus valuable timber trees became targets.

Scientific forestry began in India with Sir Dietrich Brandis, a German who took charge as first Inspector General of Forest India, in 1864. Under his able guidance, survey and maps of suitable for protected forests prepared and enacted as Indian Forest Act, 1865.

10.3 FOREST POLICY

1. State Forests are administered for public benefit.
2. Forests situated on hill slopes maintained as protected forests, to preserve the climate and physical condition of the country and to protect the cultivated plains that lay between them.
3. Forests are store houses of timber and to be managed as a source of revenue to the state
4. For agricultural demand, forest land can be given under special considerations

5. Forests that yield only inferior timber, fuelwood or fodder or used for grazing may be managed in the interest of local people.

10.4   FORESTS AND MAN

Man was born in dense forest. To protect himself man burnt forest. The increased demand for wood and other forest products, the forests became the targets of abuse. The agricultural demand of fuel and fodder and medicinal plants made people to think of conservation of forests and forestation in large scale denuded forest areas. All these reflects the importance of forest to man. The land use statistics of India 1987- is given below

S. No	Category	Area m. ha	m. million ha = hectares % of total Area
1.	Forests	6406	19.4
2.	Agricultural use	152.27	46.3
3.	Uncultivated land	42.18	12.9
4.	Non-agricultural land use	26.76	8.1
5.	Barren and uncultivable land	43.45	13.3

According to FAO 1985, world geographic area 13390 m ha, of which Forest Area 4077 m ha, which is30.4%. In case of India, the geographic area 329 m ha. Forests occupy 64 m ha or 19.4% of geographic area.

10.5   INDIAN FOREST ORGANIZATION

Forest services head : IGF = Inspector general of forests of the country.

Head office located in Dehra Dun.

Dy IGF (General)   [Dy = Deputy]

Dy IGF (Wildlife)

Dy IGF (R & T)

Secretary Central Forestry Commission

Director (Project Tiger)

Asst. IGF (Industries)

Asst. IGF (Wildlife)

Asst. Director (Forest Statistics)

Tech, Officer (Plan)

Tech, Officer (Forest industrial)

The IGF at Dehra Dun has under him:

The ICFRE-Forest Research Institute, Computing.

ICFRE Head Quarters:

(i) Forest Survey of India, and

(ii) Training Institution

Forest Development [Planning Industries and Delhi Zoological Park]

At state & union Territory (level):

(i) Principal Chief Conservator of Forests (PCCF),

(ii) Director General of Forests

Dy CF = Deputy Conservator of Forests

D FO = Divisional Forest Officer

Ro = Range Officer

FG = Forest Guards

ICFRE = Indian Council for Forest Research and Education, HQ : Dehradun

FRI = Forest Research Institute at Dehra Dun

10.5.1 MAIN FOREST PRODUCTS

(i) Cellulose and paper (using bamboo for paper making pulp)

(ii) Chemistry of Forest Products (oils and Tanneries)

(iii) Timber Mechanics (Determination of strength, properties of Timber)

(iv) Timber Engineering(Research bark utilization of wood)

(v) Composite Wood (ply wood study)

(vi) Wood working and Saw Milling (study of working properties, woods)

(vii) Wood Seasoning (air seasoning kilns) and

(viii) Wood Preservation (patented preservatives)

10.6 ICFRE (INDIAN COUNCIL OF FORESTRY RESEARCH AND EDUCATION) HQ DEHRADUN

Objectives

1. To undertake, aid, promote and coordinate forestry education, research and its application

2. To maintain research and reference library

3. To act as a clearing house of research and general information relating to forests and wildlife

4. To develop forest extension programmes and propagate the same through mass media, audio-visual aids and extension machinery

5. To provide consultancy services in the field of education, research and training in forestry and allied sciences, and

6. To do other things considered necessary to attain the above objectives

10.7 FOREST FIRES

The worst enemy of forests is fire. Forest fire destroy-1. Standing timber, 2. Burns the seeds, 3. Burns young trees, 4. Forests rich humus is burnt down by large fires 5. Forest fires also cause death to animals.

Main causes of forest fires:

1. Incendiarists set fire deliberately

2. Debries burners, who let the brush fires get out of control

3. Smokers (by negligence) throw the lighted cigarettes and matches from automobiles

4. Atmospheric lightning during thunderstorms

5. Campers-who leave burning line coals in a campfires

6. Rail roads

7. Lumbering, and

8. Other miscellaneous events

Fire Prevention: Place/install fire towers at strategic points in forest. Rangers survey forest from at least two towers. Fire fighters must be equipped with trucks, water tankers, chemical extinguishers.

Fire lines penetrate the forest at regular intervals. Arrange dropping of fire-fighters where necessary.

Safety rules to the observed:

1. Do not ever throw lighted sigarett or maches

2. Build camp fires only in protected areas and put off the fire completely

3. Watch for fire while driving and report if found immediately to forest authorities

4. Never burn a field or debris close to a forest

5. Get to know your nearest firefighting warden or Forester (authority)

Note: It may take more than 100 years for the formation of a few centimeters of soil.

10.8 FOREST

The word Forest is derived from the Latin 'foris' meaning outside, the reference being to a village boundary or fence and it must have included all uncultivated and uninhabited land. Today, a forest is any land managed for diverse purposes of forestry, whether covered with trees, shrubs, climbers etc.

Forest can become a raging inferno which in a short time can reduce the growth of centuries to ash and scarred black remains.

A ground fire burns on the forest floor and destroys the organic matter in the live soil, which is necessary to maintain humus. Nitrogen in the decamped leaf mould is also last in fire. Repeated burns expose dead soil when trampled over by grazing animals gets pulverized and the dissolved particles are washed away with the first heavy rain/showers. Great fires known as ground and crown fires can kill standing trees and may even scorch or completely burn them. In the presence of a thick mat of leaf litter on the ground, a fire may prove profitable because it may bring about favorable conditions for the inducement of regeneration of valuable forest species. Thus a light annual fire in the semi-moist Teak forest result forest in a crop of teak seedlings. Because of this fire has been termed a bad master but a good servant of the forester.

Forest fires are mostly caused spontaneously during heat wave conditions in summer, by burning of dry inflammable leaves, by the rubbing of culms of bamboos, from thunderstorm lightning, during fog (st. Elmo's fire). Sparks from coal-fired railway locomotives also trigger forest fires. Villagers sometime deliberately set fire to drive away wild animals, to burn the stubs of perennial grasses. During peak summer thick layer of dead useless pine trees which are highly susceptible to fire.

Causes of major forest fires-smokers (19%), incendiarists (about 30%), Debris burners (about 19%) lightning (about 9%). The potential loss due to forest fire is both material and human life. Fire fighting with minimal equipment poses great risk in all forest fires. Strict 'no-fire' policy is relevant in ecological and societal context rather than trying to control forest fire through technology, which is still ineffective.

Most of forest fires in India occur in the tropical dry forests. About 70% forests, which have scrub savanna, grassland, dry and moist deciduous forests.

Field ecological research indicates that many tree species (distinct to dry forest) have co-evolved with fires and have developed fire-resistance features like thick, spongy bark and re-sprout from rootstock in response to fire. Current fire safety is the blanket implementation of no-fire forest policy. Small forest fires preferable than to infrequent catastrophic fires.

In India, forest dwellers set fire to forest to clear walking paths, to collect non-timber forest products like gooseberry, mahua flowers, and to encourage the fresh growth of grass for their livestock.

Agriculturists set fire to hill forest so that the fertilizing ash from the fire washes down to their fields with monsoon rains. On the other hand fire is prevented to save timber stocks which developed fire lines around timber compartments or coupes. By burning the fire lines, before the on set of summer-fires they could loose a few compartments at best, but saves the forest.

Forest fire protection comprises-prevention of forest area (from fire damage), detection, pre-suppression and suppression. Pre-suppression includes fire protection activities with the organization training instruction and management of a fire control force while suppression of fire includes inspection, maintenance of fire control improvement, equipment and supplies to ensure fire suppression.

Conservation of forests includes-the establishment of check-posts, fire lines, wireless communication, construction of fencing, bridges & culverts, dams, water holes, trench marks, boundary marks, pipelines etc.

Fire lines: For maintenance, moderately, broad, cleared strips all round the forest and criss-cross inside the forest at suitable intervals are called fire lines. Fires lines are kept clean of inflammable materials during fire season because they act as a barriers for advancing fire. Roads also act as fire lines. Creation of public opinion not to undertake incendiarism and not to cause fire by carelessness. When a fire occurs, people should help to put it out.

Land use (1987-88) in India

1. Total geographical area of India is 328.72×10^6 hectares
2. Forest area 64.06×10^6 hectares (19.3% of total area)
3. Agricultural/cultivated area 152.27×10^6 hectares (12.9% of total area)
4. Non-agricultural use area 26.76×10^6 hectares (8.1% of total area)
5. Barron and non-cultivable land 43.45×10^6 hectares (13.3% of total area)

According to statement of forest report 1991, the total area 3287263 Km^2, forest area 770078 Km^2 (23.4% of total)

According to forest survey of India (2011), the total/green cover area = 697888 km^2 (21.23 % of geographical area. Degraded forest is about 8% of countries total cover.

Main forest types Moist tropical forests-

(a) (I) Tropical evergreen forests, (II) Tropical semi evergreen forests, (III) Tropical moist deciduous forests, (IV) Tidal forest
(b) Dry tropical forest-(V) Tropical dry deciduous forests, (VI) Tropical Dry Evergreen
(c) Montane Subtropical Forests- (VII) Sub-tropical moist (point), (VIII) Sub-tropical Dry Evergreen Forests, (IX) Sub tropical Wet Hill Forests

(d) Montane Temperate Forests- (X) Wet Temperature Forests, (XI) Himalayan Moist Temperature Forests, (XII) Himalayan Dry Temperate Forests, (XIII) Alpine Forests

10.9 CORAL REEFS

Total area 2375 Km^2 there are four intensive conservation and management area given below.

1. Gulf of Manner 2. Gulf of Kachchh 3. Lakshadweep and 4. Andaman & Nicobar Islands-they have 80% of global coral diversity, India's National Coral Reef Research Centre is located at Port Blair.

10.10 THE INDIAN FOREST ACT 1927

The British Govt. of India set up a Forest Dept, enacting legislation in the 19th century for making Forest Policy and management. In 1878, the then British govt. in India enacted the India Forest Act 1878 relating Forests in India.

The Indian Forest Bill having been passed by the legislature received its assent on 21stseptember 1927. It came on the statute Book as THE INDIAN FOREST ACT 1927.

The Act emphasized to consolidate the law relating forests, the transit of forests produce and the duty leviable on timber and other forest produce. The Act extends to the whole of India (all states in India).

Explanation of some terms in relation to this Act follows.

Cattle: Includes elephants, camels, buffaloes, horses, mares, geldings (horses), ponies, colts, fillies (young mares), mules, asses, pigs, rams, sheep, lambs, goats and kids.

Forest-officer: Any person appointed by State Govt. empowered to carry out the forest Act (e.g., Collector).

Forest-offence: An offence punishable under this Forest Act 1927.

Forest-Produce: This include mostly timber, charcoal, caoutchouc, catechu, wood-oil, resins, natural varnish, bark, lac, mahua flowers, mahua seeds & flowers (kuth), myrobalans. Also includes the following found in or brought from o forest.

(a) Trees and leaves, flower & fruits and all other parts or produce of trees

(b) Plants (including grass, creepers, seeds and moss) and all parts/produce of such plants

(c) Wild animals and skins, tusks, horns, bones, silk, cocoons, honey & wax and all other parts or produces of animals and

(d) Peat, surface soil, rock and minerals (including lime-stone, laterite, mineral oil and all other products of mines and (quarries)

10.11   RESERVED FOREST

Any forest land or waste land or any other land or in a village property of Government notified in official gazette the as Reserved Forest described by the limits by roads, rivers, ridges or other well-known boundaries. All lands included in a reserved forest are acquired by Forest Settlement-officer under the land acquisition Act 1894. No right acquired over Reserved Forest, and Forest-officer may stop any public or private way or water-course.

10.11.1   ACTS PROHIBITED IN RESERVE FOREST AREA

The following activities/actions area prohibited

(a) Any fresh clearing
(b) Setting fire to a reserved forest or kindle any fire or leaves, any burning that endanger forest
(c) Trespassing or pastures cattle or permits cattle to trespass, cause damage by felling trees or cutting or dragging any timber.
(d) Fells, girdles, lops or burns any tree or strips off the bark or leaves/damages
(e) Quarries stonesburns lime or charcoal, subjects to any manufacturing process or removes any forest produce
(f) Clearing or breaking up any land for cultivation or any other purpose
(g) Hunting, shooting, fishing, poisoning water, setting traps or snares
(h) Killing or catching elephants or any wild-animal

10.11.2   FOREST CONSERVATION ACT 1980

Deforestation causes ecological imbalance and leads to environmental deterioration. In India deforestation had been taking place on a large scale and it caused widespread concern

Chipko andolan-1974

A non-violent resistance for the protection of natural resources was led by Sunderlal Bahuguna, the Chipko Andolan. In 1974, to stop the falling of trees, 27 women in Garhwal hugged the trees. This act of women inspired several environmental movements. The Narmada Bachao Andolan (anti-dam agitation) includes in the above non-violent resistance.

Keeping this in view and to stop further damage to the reserved forests for use of forest land for non-forest purposes an advisory committee constituted to look into the matter and to advise the Central Government in (1975-1979). As a result, The Forest conservation Bill having passed by both the Houses of Parliament received the assent of the president of India. It came on the Statute Book as THE FOREST (CONSERVATION) ACT 1980). Subsequently it was amended in 1988. The Forest Act 1980 extends to the whole of India (except Jammu & Kashmir).

The 1980 Act provides the following general restriction on the use of forest land for non-forest purposes.

No one permitted that any forest land/part of it allowed to be used for non-forest purpose like cultivation, industry etc. However this does not include works relating to conservation, development and management of forests and wild life protection, establishment of check posts, fire lines, wireless communication and construction of fencing, bridges & culverts, dams, water holes, pipelines etc.

Assistance to Forest Authorities in Protection

Every person is linked to forest by way receiving forest produce or by employment. Hence it is the duty of every man to assist. Forest officers & police officers when required for protection-(1) To extinguish forest fire, to pass the message, any leno ledge of information to arrest forest fire, to prevent the commission of offence to forest produce and arresting the offender. Furnish the information about unlawful activity for destroying forest/forest fire to forest officer police officer.

Assist any forest/police officer when required in preventing the commission of any forest offence or assist when there is reason to believe that any forest offence has been committed, assist in discovering and arresting the offender.

The office of forest survey of India (FSI) is located in Dehradun.

Note: Out of 20, 000 medicinal plants species listed by WHO, about 5000 species are in India (of this 581 species in India under the threat of loss recorded by G.R. Divisional Forest Officer India).

10.12 FOREST PRODUCTS

These are classified as Major and Minor products. Major forest products consists of timber, small wood, fuel wood or charcoal. The estimated demand of raw materials of wood in 2000 AD were :

Industrial wood	$50 - 65 \times 10^6$ m^3,
Fuel wood	$320 - 350 \times 10^6$ m^3,
Printing paper	1.78×10^6 tonnes
Writing paper	7.0×10^5 tonnes (7 lakh tonnes)
Industrial paper	7.8×10^5 tonnes (7.8 lakh tonnes)
News print	7.6×10^5 tonnes (7.6 lakh tonnes)

Minor or non-wood forest products consists of Fibres and Flosses, Grasses, Bamboos/canes, Essential oils (includes those from grasses); Tans and Dyes, Gums, Resins and Oleoresins, Drugs, Spices, Poisons and Insecticides; Edible products, leaves and Animal, Mineral and Miscellaneous products.

Uses

Fibres are obtained from tissues of certain woody species, used for rope making.

1. Flosses obtained from certain fruits, used for stuffing pillows, mattresses etc.

2. *Grasses*, bamboos and canes used as fodder, for making paper.

    Tall grass used for making chicks, stools, chairs etc, leaves twisted for making strings. Bamboos used for rafters, roofing basketry etc. Bamboo seed is eaten as grain, its fibre used for making paper. Bamboos are used for making bridges. Canes used as plaiting material, ropes, for furniture, sports goods, walking sticks etc.

    Essential oils are used in soap making, cosmetics, pharmaceutical preparations, confectionary, incense etc. Eucalyptus globulus oils used in medicine, Rusa grass, Khus and Sandalwood are most valuable in commercial field.

3. *Lemon grass* oil contains 80% citral which is used for manufacture of vitamin A. Also used in aromatic chemicals, perfumery, soap, cosmetics, making pain balms and disinfectants.

    Sandal wood used in valuable art carvings, distilling oil which is used in Indian medicine, perfumery, toilet soaps etc.

    Rusa (palmarosa) oil extracted from var motia and certain other grass, which contains 90% geraniol - used in perfurmery, cosmetics, as rose scent in soaps, flavouring tobacco.

    Khus oil distilled from the roots of vetiveria zizaniodes grass - used in perfumery (Itra of Khus).

4. **Tans and Dyes:** Tannins are secretion products of plant tissues. Tanning materials are got from Harra tree, Babul bark, Avaram bark and Wattle bark.

    Dyes are derived from Red sanders which bright red. Khair wood gives chocolate colour (is a preservative) used for dyeing canvas, fishing nets, mail bags, sail cloth etc. The flowers of palash (boiled in water) gives orange colour. Fruits of Mallotus Phillippensis and roots of Morinda tinctoria used as dyes.

5. **Gums, Resins and Oleoresins:** Gums and resins are oozed by plants (due to injury to the bark or wood). Resins when mixed with high percentage of essential oil is called oleoresins. The gum is used in textile trade, cosmetics, dentifices, cigar and food industry.

6. **Drugs, Spices, Poisons and Insecticides:** They are derived from trees, shrubs, climbers, herbs, primitive plants, fruits, flowers, leaves, stems and roots.

Important Drugs are

Root and underground parts: Ipecac   from   Cephaelis   ipecacuanha cultivated in Darjeeling, Shillong and the Niligiris. It gives emetine which is effective for treating amoebic dysentery. Liquorice derived from glycyrrhiza glabra, is a tonic, expectorant, demulcent and mild laxative.

Sarpagandha are the roots Rauwolfia serpentina contains reserpine which is very effective in high blood pressure. Kuth or costus are the roots of saussarea lappa, used as tonic, stomachic, stimulant and spasmodic.

Bark Drugs

Quinine is the most important Indian vegetable drug. Quinine is extracted from the bark of Cinchona ledgeriana cinchona hybrida, effective in treatment of malaria.

Kurchi is the bark of Holarrhena antidysenterica, a shrub. It is a remedy against dysentery.

Wood Drugs

Alkaloids Ephedrine used in the treatment of bronchial diseases.

Leaf Drugs

Hemp is derived from cannabis sativa, grows wild in the foot hills of the Himalayas. It is a source for well-known narcotics - Ganja, Charas and Bhang. The flowering tops are Ganja, the resinous exudation from flowers and leaves is Charas, (the basis for Hashish). Bang is the drink made out of pounded Ganja.

Drugs from Fruits and Seeds

Kala-Jira is a seed cultivated at attitudes 3 to 3.5 km asl. It is stomachic Carminative and Lactagogue, has strong flavour.

Chaulmoogra is a common tree in Western Ghats. Its seed oil is specific for Leprosy treatment.

Spices

Spices are aromatic vegetable products, used in cooking for flavour. The important spices of forest produce are :

(i) *Galangal/Alpine galangal* is found in the Himalayas and Western Ghats. It has aromatic perennial rhizomes which are used for treatment of rheumatism and catarrhal infection.

(ii) *Dalchini or Cinnamon* is a bark of a tree found in Western Ghats south of Konkan. It is stimulant, carminative. Used in candy, gums, incense and perfumes. Recent American research indicates it is very effective in the treatment of blood sugar (diabetes) and high blood pressure.

(iii) *Choti Elaichi and Badi Elaichi* grown in ever green forests of South India and in East Himalayan forests respectively. The seeds are carminative and used as flavouring for sweet dishes and in cooking.

Poisons

Many forest plants such as strychnine, aconite Datura etc., are poisonous but in small regulated doses it acts as good medicine. Barks of various trees are used as fish poisons. Gunja seed paste inserted into the flesh of cattle kills them.

Insecticides

Several of Derris and Tephrosia contain rotenone, which is an insect poison.

Edible Forest Products

Several forest fruits, flowers and leaves, tuberous roots are eaten. A few important species of fruits are given below.

Aam (mango), Bel, Ber, Jamun, Kathal, Khirni, Phalsa, Sitaphal, Tendu etc.

Species of Kernals Eaten

Achar, Akhrot or Walnut, Cashewnut, Chilgoza, Kamal, Singhara etc.

Parts of Species used as Pickles and Vegetables

Amla, Amra, Anar or wild pomegranate, Imli or tamarind, Karonda, Kokum, Munga (drumstick), Kachnar, Kaith (or wood apple)

There are several other tree products such mushrooms, palmyrah, mahuva, Bhilama etc., used for many purposes making liquors etc.

Management of State Forests

In India planning commission, under five year plans, emphasized the need to rehabilitate the depleted forests. It stressed for the creation of plantations of valuable species and conservation of wild life by establishing wild life sanctuaries. Further stressed to preserve natural environment and ecology for scenic beauty, sports, recreation, scientific study and conservation, preservation of biologically diverse flora and fauna. Wild life conservation include the protected areas to be linked with its corridors to maintain genetic continuity between artificially sub-sections of migrant wild life. Tribal people be put in protection of and development of forests, in raising trees and raw materials.

10.13 DANGER DUE TO DEFORESTATION

As said earlier forests are the lungs of the earth. Deforestations leads to emission of CO_2 into atmosphere instead of CO_2 absorption from atmosphere.

Ecosystems that host a CO_2 rich type of soil called peat, known as peat lands, are the most efficient natural carbon sink on the planet. When undisturbed, they store more CO_2 than all other vegetation on earth combined. However, when they are drained and deforested, they can release 6% of global CO_2 emissions each year.

CHAPTER - 11

The Biodiversity

INTRODUCTION

The 11th conference of the Parties to the Convention on Biological Diversity (COP 11 to CBD) held at Hyderabad (India) during 1-19 October 2012. It was discussed to take necessary steps in preserving planet Earth, its millions of plant and animal species and their habitats. More than 193 countries participated in the convention. The Conference of Parties (COP) to the UNFCCC held every year.

Bio-Safety concerns agriculture, health and biotechnology, requiring consultations with these sectors. Nagoya Protocol emphasized access to genetic resources and sharing benefits with communities and the Nagoya-Kuala Lumpur supplementary protocol dealt with liability and redress (due to consequences arising from the use of living modified organisms). There was a need to call attention to the issues that were of interest not only to the environmental community but also to the sectors that utilized genetically modified organisms. At Hyderabad (during 1-5 Oct-2012) bio-safety issues like capacity building, handling transport, packaging and identification of living modified organisms (LMOs), notification requirements unintentional trans-boundary movement and emergency measures, risk management and socio-economic considerations discussed.

India took steps to implement the protocol even before ratifying it. India is one of the very few countries (may be 12 to 15) which already have a full-fledged law in place to regulate the access and benefit sharing of natural resources.

Agrarian crisis by the use of GMOs

It will lead to agrarian crisis if India enters into trade relations with countries that do not adhere to global regimes. These countries may agree to bio-safety protocols with India at first, but there will be agrarian crisis if they (trade partners) dump with poor quality genetically modified organisms (GMOs) being forced to buy processed products that do not meet the safety parameters.

In recent, times, because of modern agricultural technology, Andhra Pradesh stuck to 4 crops having vast land (two crore acres) for agriculture. It grows mainly four crops cotton, paddy, maize and groundnut all through the year. According to govt. statistics these crops cover 70% of the state total cultivable land and the rest 30% area for other crops. Cotton and paddy used are mostly genetically developed breeds and produce 80% of the countries hybrid rice. Experts flame it as death of biodiversity in the state. It was questioned by agricultural experts why the state could not cultivate oilseeds, pulses and millets. A study conducted by progressive formers body revealed Bt. Cotton and Bt. Rice would help in improving yields. Contrary to this, yields across the country dropped following the entry of hybrid crops. Prior to the aggressive propaganda of Bt. Verities shown 70% impressive rise, but it fell to just about 2% subsequently. According to National Crime Bureau records the total member of farmers suicides in the preceding 17 years stood at 2.7lakhs and in Andhra Pradesh 33000. Most of suicides from cotton & paddy growers. It became a question mark with all advanced biotechnology and modern agricultural methods famers driven to commit suicides.

Modern methods aided to kill the Indian biodiversity in rice varieties. There were 50, 000 varieties of rice in India as compared to present just about a dozen. Traditionally India had an array of rice breeds which include drought-tolerant, flood-tolerant and even salt-tolerant. Now it lost all that and replaced by genetically developed few varieties. This highlights the agrarian crisis by the use of Genetically Modified Organisms.

11.1   THE BIO-DIVERSITY ACT 2002

In Greek "BIO" means life and "Logos" meaning science or study of Biology together means the science of life. Biology includes: (1) General biology-relationship of living organisms/things to their surroundings, (2) Botany, study of plants, (3) Zoology, study of animals, (4) Human biology, study of man, all studies related to their surroundings. Thus Bio-diversity means Biological diversity or variation.

United Nations Conference of Environment and Development (UNCED) held in Rio de Janerio is in 1992. It was agreed at the Conference to protect the environment and social and economical development for sustainable development. After 10 years at World Summit, world members of UN Assembly in Johannesburg (South Africa) 2-4 Sept 2002 committed for the sustainable development, a collective effort at local, national, regional and global level which ensures economic development, social development and environmental protection.

For conservation of Biological diversity, sustainable use of its components and fair and equitable sharing of benefits arising out of the use of biological resources and knowledge, the UN Convention on Biological Diversity was held at Rio de Janerio in 1992, wherein the Sovereign rights of the States over their biological resources were reaffirmed. India was signatory to this Convention. Accordingly the Biological Diversity Bill was introduced in Parliament (2002) and was approved by both the Houses of Parliament, received the assent of the President on 5th February 2003. It came on the Statute Book as THE BIOLOGICAL DIVERSITY Act, 2002.

11.2 THE NATIONAL GREEN TRIBUNAL ACT 2010

Amendments to the Biological Diversity Act 2002 by the National Green Tribunal Act 2010.

Biodiversity or Biological diversity means the variability among living organisms from all sources and the ecological complexes of which they are parts and includes diversity within species or between species and of ecosystems.

11.3 BIODIVERSITY

Bio-means life and diversity means variations. The variations of all living organisms at the species level may be termed bio-diversity. This includes plants, animals, fungi and microbes. There are more than 1.5 million named species of living organisms, while unknown species estimated to be 10 to 100 millions.

From very ancient times living organisms have been grouped into plants and animals. These large groups further put into smaller and these into still smaller. Scientific classification of biologist of living things into two very large Kingdoms, plant and animals. These kingdoms next divided each into smaller groups called Phyla. Thus we have plants phylum four (I to IV) and animal phylum I to XI which are given below.

1. **Plant phylum & members**

   Phylum I: Thallophyta (110,000 species)–Algae (Blue, Green, Brown, Red algae, found as pond scums and sea weeds possessing chlorophyll. Fungi (Bacteria, yeasts, molds, mushrooms, lichens, which are non-green parasitic)

   Phylum II: Bryophyta (23000 species)–Multicellular green plants, living on land

   Eg: Mosses, Liverworts

   Phylum III: Pteridophyta (10, 000 species)–Terrestrial plants. Eg: Ferns, Club Masses, horsetails

   Phylum IV: Spermatophyta (250, 700 species)

   Eg: seed plants (pines and their relatives and flowering plants.

2. Animal Phylums

Phylum I - Protozoa (20, 000 species), unicellular animals or colonies of cells

Phylum II - Porifera (2500 species). Marine and fresh water sponges

Phylum III - Coelenterate (10, 000 species). Free swimming metazoans, coral, Jellyfish

Phylum IV – Platyhelminthes (6500 species). Flatworms with ribbonlike body

Phylum V – Nemathelminthes (3, 500 species). Roundworms with un-segmented bodies

Phylum VI – Rotifera (1500 species). Wheel animals, Rotifers

Phylum VII – Annelida (5000 species). Segmented worms with the body cavity. Leach, earth worm

Phylum VIII – Arthropoda (674, 500 species). Animals with segmented bodies crayfish, lobster

Phylum IX – Mollusca (90000 species). Soft-bodied invertebrates, octopus, snail

Phylum X – Echinodermata (5000 species). Marine forms with tube feet for locomotion starfish, sea urchins

Phylum XI – Chordata (40, 000 species). Notochord not more than four legs or paired limbs. All the vertebrates.

Vertebrata sharks, reptiles, birds, mammals.

11.4   COASTAL AND MARINE BIODIVERSITY

[Ref. National Biodiversity Authority, Chennai. Prakash Nelliyat & Bala Krishna Pisupati]

The oceans cover about 71% of surface area of the earth and constitutes 90% of the habitable space. The size of life in oceans very from the largest Blue whale to tiny micro-organisms. Coast lines support fragile ecosystems-mangroves, coral reefs, sea grass and sea weeds. Mangrove forest host fish crabs & birds, monkeys deer and even tigers. Coral reefs occupy about 0.1% of the sea but has about 35% all known marine species live in it. Sea grass supports variety of aquatic life like marine turtles, juvenile prawns. Sea grass absorbs more CO_2 as compared to tropical forest ($50:1/Km^2$). Seaweeds have great commercial value and an important marine living resources. At present about 51% of the worlds populations living in 10 km of the coast. The following are the main human dependence on the marine and coastal ecosystems.

1. Sea life produces one-third of the O_2 that we breath

2. It absorbs about 30% CO_2 emitted by human activity since the Industrial Revolution.

3. Fisheries employ about 200 million people and provide more than 15% of the dietary intake of animal protein

4. Marine byproducts are raw materials for manufacturing industries like paints, fertilizers, skin lotions, toothpastes and medicine

5. The divergent chemical deposit in the marine environment are an asset and might yield anti-cancer drugs

6. The shore provides for marine transportation, recreation, tourism and common salt production

7. Mangroves help of protect the coastal aquifers from sea water intrusion and safeguard the coastal communities from natural hazards like cyclones, tsunamis, hurricanes, hazardous tides

8. Coastal wetlands regulate water quality by capturing and filtering sediments and organic waste transits from land.

According to Law of Sea Convention, States have to adopt laws and regulation to prevent and control pollution of the marine environment from or through the atmosphere. Indian peninsula was about 5500 Km and Andaman, Nicobar & Lakshadweep Islands about 2000 km. There are about 63 million people living in low elevation coastal areas which is approximately 11% of the global population living in coastal areas. Man has been benefited from marine and coastal ecosystem from time immemorial. Recent human activities harming the environment. Large scale burning of fossil fuels causing ocean acidic and warning. Between 1950-2012, the average sea surface water temperature rose by $0.4^{\circ}C$, which is harming marine ecosystem and coral bleaching is a symptom.

Interception of river water flows into sea by way construction of dams and diversion depriving the coasts of freshwater, sediments and nutrients. This is causing imbalance in ecosystems. Coastal urbanization adding toxic pollutants, non-biodegradable wastes adversely affecting reproduction, growth and damaging marine wildlife Over-hunting reducing stock of many marine species and great whales have been decimated hunted to a fraction of their original population. At present more than four million fishing vessels, including industrial trawlers are engaged in fishing about 140 million tonnes per year. It is estimated that about 13% of global fishers are collapsed. 30-35% of critical marine habitats like sea grass, mangroves & coral reefs destroyed due to anthropogenic interventions.

Management strategies: Utilize the services of marine and coastal ecosystems in the following ways.

1. For the extraction of coastal resources, adequate knowledge of bio-physiological functions of the ecosystems and their requirement capacity at the regional level is required

2. Coastal and marine ecosystem/products are 'public goods' with an open access character. Biodiversity is an asset and the "Right" of future generation too. Consequently strict enforcement of the law required

3. By establishing Marine protected area/region, it is possible to recovery of coastal and marine biodiversity

4. Introduce pollution tax, user charges & fee. This will act as disincentives to resource degradation and depletion

   Simultaneously introduce gifts and payments for preserving ecosystem

5. Establish control on activities which have adverse effect on coastal marine ecosystem

6. To achieve coastal biodiversity conservation it is essential the participation of stake holders and peoples participation

11.5   WILD LIFE

The animals and the birds that live in the wild is called wildlife. Wildlife includes, any animal, aquatic or land vegetation which forms part of any habitat. Habitat includes land, water or vegetation which is the natural home of any wild animal.

From time immemorial from protophyte to the developed tree forms and from the protozoa to the largest mammals existed in nature in a dynamic harmony-involving complexity, intermingled and, inter dependent floral and faunal chains, which we call eco-complex. The eco-complex changes in composition and in mass, both in space and in time. Thus wildlife with respect to a locality or region presents the entire animal and plant community therein covering all forms. The conservation of wildlife directly influenced by human environment and conservation of forests.

India is endowed with a variety of physiognomic characters viz regional climatic and edaphic types with wide range of habitat-types and accounts a large variety of wildlife.

The fossil beds (the remains of extinct fauna) indicate the existence of Mastodons, great herds of elephants, hippopotamuses, rhinoceroses and four horned ruminants, the Sivatherium. There were giraffes, large and pigmy horses, camels, herds of wild oxen, buffaloes, bison's, deer, many kinds of antelope, wild pigs. Fossil beds also reveal the existence of chimpanzees, orangutans, baboons, langurs and macaques. The carnivores include a type of cheetah, sabre-toothed tigers, wolves, jackals, and foxes etc. The bears similar to our present sloth bear, the rodents by various genera including mole-rats, porcupines and hares. The present day fauna in India has more than 500 species of mammals, of which elephants, the one-horned rhinoceros, the arna (the wild buffalo), the gaur (Indian bison).

The distribution of major mammals in India to three zones-the Himalayas, the sub-Himalayan Terai and the Peninsula.

The survival of wildlife depends on environmental factors that influence its habitat viz the forest, the scrub, the cultivated fields, the marginal wastelands and pastures and marshes, the lakes and the rivers– the availability of sufficient food and conditions that are favorable for breeding.

11.5.1 THE VARIETY OF FAUNA IN INDIA

Biological Diversity in India

India is the home for about one third of known life forms in the world. According to survey of India, Department of science and Technology, there are more than 500 species of mammals and 2060 forms of species of birds, which are truly Indian. The mammals include mighty elephants to smallest of deer in the world (the mouse deer). The largest of the wild OX (the Indian gaur) and colourful birds from peacock to the small Ticket's Flower Pecker are the heritage India. The recent twentieth century rapid human growth and livestock population led to the pressure on land development and on country wilderness.

The 20th century witnessed the loss of a few magnificent wildlife, like Hunting Leopard, while Winged Wood Duck. Indian National wildlife sanctuaries created to protect the wilderness, since 1960 to 1989, 65 to 445 National Parks and sanctuaries increased in the struggle to save wildlife.

Some of the major National Parks and sanctuaries developed are the Kaziranga National Park in Assam which housed unique fauna of one-horned Rhinoceros, the Himalayan sanctuaries homed snow Leopard and tigers homed in 16 Tiger Projects in India. Indian birds, the Grey Pelican homed in Neela pattu of Andhra Pradesh, Golden Langoor, Cap Langoor in Manes sanctuary in Northeast part of India. Keoladeo National Park accounts 331 winter visitors from different parts of Asia Middle East, Siberia, Europe etc. Due to such efforts several species saved from extinction such as great Indian Bustard, the Jerdon's conserved.

11.5.2 HUMAN POPULATION GROWTH

During Holocene epoch (about 8000 BC) the population of the world was about 8 million. Farming, towns, cities came into existence at about 6000 BC. By 100 AD population grew to 300 millions, and by 1800 AD it was 1000 million. Man is a highest form in natural spices in the ecosystem. He has developed brain and evolved language for communication, developed science and technology, agriculture, reservoirs, cities etc. for his better living and thus creating new environment and ecosystems. However his inadvertent application of science & technology becoming bane than a boon to mankind and in general to all life an environment.

Life on earth expanding with favorable environment while unfavorable environments causing life dwindling. After industrialization the human population is growing very fastly. According to British economist Thomas R. Malthus, population is growing in geometrical progression while the means of its subsistence growing only arithmetical progression. This creates imbalance between growth and in the means of subsistence and that would lead to huger, epidemics, war, terrorism etc. Density of population is not uniform and depends on fertile soil, water availability, favourable climate avoiding extremes of hazardous weather like cyclones, floods, droughts, earthquakes, snowstorms, volcanoes, Tsunamis etc. Depending on these factors ancient civilizations flourished in river valleys which were fertile and forest products for food, habitat and other domestic necessities were easily available.

Population density may be defined as number of individuals of species per unit area. A country may have large area but may have very poor habitation because it may have large forest area, maintains, fallow land or desert. As compared to developing countries, developed countries have relatively large land area in relation to population density. The population land ratio of America, Russia are equal and have equal population densities while in Europe this ratio is higher as compared to Asia. For example the Netherland is much smaller than India but has higher population density (319 persons/km^2) than India (168 persons/km^2). All parts of Netherland are habitable while it is not so in India. Population in general depend on two basic concepts of ecology-biotic potential and carrying capacity. Biotic potential means natural increase of population under optimum conditions (abiotic/non-living) while carrying capacity means the maximum availability of available resources.

As the population density of a country increases the per capita income, the land availability, wood, minerals, fuels and other resources decrease. This will lead to problems of economic socio-political and environment. Population distribution largely depends on (1) geophysical factors (like climate, water availability, soil, energy, mineral resources, and transportation), (2) socioeconomic factors, (3) demography.

The urbanization growth causing over-crowding sanitation sewage disposal problem, transportation, traffic jams, environmental pollution due to vehicles, noise pollution, multiplex buildings, water problems, petty thefts and crimes. The cause of urbanization growth is industrial growth, job opportunities, educational facilities, medical facilities (schools, colleges, technical institutions, super specialty hospitals, cinemas and other entertainments).

According to UN population Division, world's population, (2016) present 7.6 billions will rise to 8.6 billions by 2030, 9.9 billions by 2050 and 16.2 billions by 2100. India's current population (2017) estimated at 1.34 billions is projected to rise to 1.51 billions by 2030 and further rise to 1.66 billions by 2050 and then decline to 1.52 billions by 2100. China's current (2017) population 1.41 billions (the most populous country) become 1.44 billions by 2024. According to the projections nine countries India, Nigeria, Democratic republic of the Congo, Pakistan, Ethiopia, Tanzania, United States, Uganda and Indonesia will account for more than half the growth in the global population between 2017 and 2050. Fertility rates are falling globally. In Africa decline from 5.1 to 4.7. In Asia from 2.4 to 2.2, in Europe increase infertility from 1.4 births per 100 man in 2000-2005 to 1.6 in 2010-2015.

Birth rate (or Natality rate) = No of new births of individuals per unit of time

Death rate (or mortality rate) = No of individual deaths per unit of time

Natural increase = Birth rate–death rate

Birth-death rate per 1000 persons 46 :44 in 1901 and fell to 37:14 in 1973, indicate population growth rate variation 0.2% in 1901 to 2.3% in 1973.

According to 2011 census, the population density of twin cities Hyderabad-Secunderabad was 18480 persons/km^2 as against 17648 persons/km^2 in 2001. Total population 40.1 lakh in 2011 as against 36.9 lakh in 2001. This shows decadal growth of 4.71%. Similarly the population density of Ranga Reddy district 707 people/km^2 in 2011 against 477 people/km^2 in 2001 and decadal growth 48.15%, population 52.7 lakh from 35.8 lakh.

Present population density of china 142 persons/km^2,

India 396 persons/km^2, America 34 persons/km^2,

Indonesia 125 persons/km^2, Russia 8.3 persons/km^2.

Causes of human population growth mainly due to development of science and technology and its applications to mankind.

Construction of houses to protect himself and domesticated from wild animals and adverse weather. This lead to community life in villages, towns and cities. Education, agriculture, domestication of animals and their service to man. Industrialization resulted in improvement of transportation, roads, food storage and realized the man improve living, health, sanitation etc, over population resulted in food, land and energy crises. Further industrial revolution resulted in environmental pollution.

The UN conference on Human Environment (1972) stressed the need for protecting & improvement of environment. According to UN report, at present mankind is persisting with thoughtless extravagant consumption of natural resources and damaging the natural resources in an unprecedented manner. At the present consumption rate it requires 21.9 hectares land per person, while earth's capacity is about 15.7 hectares. Because of this 116550 sq km of forest area being lost across the world each year. 60% of world's major rivers have been damaged and 10% of the world's large rivers are running dry every year. The US consumptions of energy increased by 20% during last two decades.

11.5.3 THE PRESENT ENVIRONMENTAL ISSUES

During last 150-200 years, all species of life on earth affected by air, water and soil pollution. The main issues are-global warming, climate change. Deforestation weakening the lungs (green plants) of the earth. Energy shortage crises Harmful toxic and radioactive waste, oil spill, depletion of natural resources. World populations explosion 74 million people per year and Nuclear weapons.

During 2015-2050 the number of people living in cities projected to grow from 3.9 billions to 6.3 billions. The population of urban dwellers will increase from 54% to 67% of the world's population. According to Population Reference Bureau (PRB), the world population will reach 9.9 billion in 2050, increasing by 33% from an estimated 7.7 billions now (2016). The PRB projections show Africa's population will reach 2.5 billions by 2050, while the number of people in the Americas will rise by only 223 millions to 1.2 billion. The population projection by 2050 indicate that the combined population of the world's least developed countries in the world will double by 2050 to 1.9 billion.

According to UN report 2011 (Revision of the world urbanization prospects) India will witness the largest increase in urban population during 2010-2050, followed by China. India will add 497 million to its urban population and China 341 million population into cities, followed by Nigeria

200 million, the United States 103 million and Indonesia 92 million. By 2025, Tokyo (Japan is projected to remain the world's most populated urban agglomeration with almost 39 million inhabitants, though the population increase insignificant. Shanghai in China would be 28.4 million inhabitants, Mumbai in India would come third with 27 million inhabitants. The megacities in India–Delhi, Kolkata & Mumbai are expected to grow faster than those in Egypt & Turkey. The urban population is highly concentrated in a few cities. In 2011, about three quarters of the 3.6 billion urban dwellers lived in 25 countries whose population ranged from 31 million in Ukraine to 682 millions in China. China, India & US accounted for 37% of the world urban population.

11.5.4 BIOLOGICAL DIVERSITY UNDER STRESS

The stress on the biological diversity is being brought by Man's dominance and not Nature. This is of great concern. Look at the world human population growth trend.

World Human Population

Year	In millions
4000 BC	90
AD-1	300
AD-1750	800
AD-1850	1300
AD-1900	2500
AD-1950	3900
AD-2000	6400
AD-2017	7600 (7.6 billions)
Projected AD-2030	8.6 billions
AD-2050	9.8 billions
AD-2100	11.2 billions

According to National Academy of sciences USA study, plants outweigh people by 7500 times and make up more than 80% of the world's biomass, Bacteria about 13% of the worlds biomass. Fungi, mold & mushrooms make up about 2% which exceeds all animals biomass. Since the start of civilization man cut/destroyed plants to its half, and wild animals by 85%. Domesticated animals outweighed all wild animals by 14 to 1. World chicken are triple the weight of all wild birds. About one-sixth the weight is dry carbon. Thus humans form a miniscule part of the earth

During last five decades population growth together with global economy and consumption rise is called "The Great Acceleration". Since 1971, the global food production of wheat grown 116% and rice grown 133% to feed the growing population. 30% of the earth is covered by forests. To feed the growing population, 40% of forests destroyed during 2000-2010 and converted into agricultural land. Man is the highest form of life on earth but he is not independent of other forms of life. It is known that all forms of life is made up of protoplasm and carry identical process, but live in totally different environment. There exist a critical relationship between living things and its environment. Throughout the history of the world man has been harming the environment unknowingly for his comfort. He developed agriculture to meet the needs of food, constructed houses, towns, cities to protect from adverse weather & comfort. In this process he has harmed the environment and other species of life unknowingly. This act pushed the ever expanding species to backwards. During the last few centuries human beings wholly dominated over the global other species of life. As a result of this domination other wild life particularly vertebrates declined because of their habitat loss, environmental pollution and diseases. Equally responsible are the climate changes, weather pattern changes with rise in extreme weather events. Global warming causing decline of polar bears along with polar ice melt. During 1990-2015, 177 mammal species lost, at least 30% of the geographical area (habitat) that they inhabited lost. The National Academy of Sciences mapped 27600 species loss by population and range, about half of the terrestrial vertebrate species. Scientist named this loss biological annihilation or sixth mass extinction.

The living planet index (LPI) compiled by WWF (World Wildlife Foundation) explanation is given below.

Humans are stretching nature too thin. Nature is providing man about $ 125 trillion in terms of natural resources, like fossil fuels, land for agriculture etc.

Bio-capacity is the ability of an ecosystem to renew itself, (measured in global hectares – gha) and Ecological foot print is a measure of human consumption of natural resources. During (1970-2018) last 50 years bio-capacity has increased by 27% with new technology, solar energy, land management practices, while ecological foot print increased by 190% (which is nearly 10 times the growth rate of bio-capacity). Since 1970, bio-diversity showing down trend.

LPI: It gives the state of global bio-diversity and plants health, taking into account the populations of thousands mammals, birds, fish, reptile and amphibian species.

LPI considers five types of biological threats, of which Habitat loss is common and over exploitation. These two together account two-thirds of all threats. According to LPI (1970-2015), (a) there is globally decline of 60% bio-diversity and (b) 89% decline in south & central America. (c) There is

75% decline of all plant, amphibian, reptile, bird and mammal species and that have gone extinct since 1500 AD were either harmed by over exploitation or agriculture. It is observed that was struggle for existence in all species.

11.5.5 Vareity of Fauna in India

The zoological Survey of India (HQ located in Kolkata) is responsible for surveying the faunal resources. With vast climate variation, India has a large variety of fauna, about 90, 000 species. Protista (about 2600), mollusca (about 5000), Anthropoida (7000), amphibian (240), mammalia (400), reptilia (460), members of protochorodates (120), pisces (2550), aves (1240) and other invertebrate (8330).

India is a treasure house of biodiversity and a perfect host. The Olive Ridley turtles endure thousands of kilometers of journey for nesting in the welcoming sounds of odisha's coast. The long distance migratory birds from freezing Siberia, seeking warmth is Chilka Lake, the second largest brackish water in the world.

Andaman & Nicobar Islands is the home of to a tenth of India's fauna species. The area of the islands is only 0.25% area of India's geographical area but has 11009 species, according to Zoological Survey of India (ZSI). Publication titled faunal diversity of Biogeographic zones. The population of islands is about 4 lakhs include vulnerable tribal groups- Great Andamanese, onge, Jarwa, Sentinelse and shompens. The Narcondam hornbill, its habitat restricted to a lone island, the Nicobar Megapode, a bird that builds nests on the ground; the long-tailed Nicobar Macaque, and the Andaman day gecko, are among the 1067 endemic faunal species found only on the Andaman and Nicobar Islands and nowhere else. Similarly eight species of amphibians and 23 species of reptiles are endemic to the islands. Another unique feature of the islands ecosystem is its marine faunal diversity, which includes coral reefs and its associated fauna. In all, 555 species of scleractinian coral (hard or stony corals) are found in the island ecosystem.

Wildlife

Wildlife connotes the plant and animal communities coexist in dynamic equilibrium or simply means the animals and birds of the wild or the entire animal and plant kingdom of a locality. They are jointly called the eco-complex, which varies in mass and composition both in time and space. The variations depend on climatic and edaphic factors.

India has a rich heritage of wildlife because of traditional compassion for all life. Emperor Ashoka's edicts prohibited killing of animals and preservation of forests. Now the national animal of India is 'Tiger' and the national bird 'Peacock'. We have briefly learnt about the Indian flora and now we shall learn about the Indian fauna.

Based on climate and physiography, India provides a wide range of habitat which accounts for large variety of wildlife. The fossils of extinct animals in Sivalika range reveals that wildlife belongs to Tertiary period of geological times. There were elephants, hippopotamuses, rhinoceroses, fourhorned ruminants (the Siva thereums), giraffies, large and pigmy horses, camels, wild oxen, buffaloes, bisons, deer, antelope, wild pigs. Fossil beds indicate the existence of chimpanzees, orang-utans baboons, langurs and macaques. The carnivores include a type of cheetah, sabre-toothed tigers, wolves, jackals, foxes, also civets, martens, ratels and otters porcupines, hares, bears etc.

The present day Indian fauna has about 500 species of mammals. The important is the elephant, one-horned rhinoceros, the Arna (the wild buffalo), the Gaur (the Indian bison). Important mammals in India are distributed in three zones – the Himalayas, the Sub-Himalayan Terai, and the peninsula.

The animals found in the first two zones are :

The Himalayan yak (the snow Ox) found in Ladakh. The Himalayan sheep are shapu (or Urial), bhoral (the blue sheep), nayan (huge sheep with curved horns); Goat-antelopes are Serow and Goral found in the eastern Himalayas. The Himalayan wild goats found in Kashmir are markhor, the ibex, the Himalayan thar, the beautiful Kashmir stag, (counter part of European red stag) and Kastura (or the musk deer). Thamin is a pretty deer found in Manipur. The tigers, panthers, snow leopards, the brown and black bears, the red panda, the Indian otters, the beech marten, the yellow-throated marten, the Himalayan Weasel, Marmots, a few species of squirrels, bats, the elephant and the tiger. The one-horned rhinoceros, the hog deer, the clouded leopard are all the typical animals that are found in the Himalayan foot hills. The other main carnivores found are sloth bear, the fox, the wild dog, feline cats. The chinese pangolin is found in Assam besides elephants and Arna, deer-barasinga, sambar, the chital (spotted deer), the muntiac (barking deer), giant and flying squirrels, Indian porcupine and wild pigs.

Animals Found in Peninsular India

The elephants, gaur, chinkara (or the Indian gazelle), the black buck (Indian antelope), the Nilgai (blue bull), the mouse deer, the chousingha (four horned antelope), the Nilgiri thar, the Indian lion, cheetah (hunting leopard). All deer species of the Himalayan foothills except hog deer are also found.

Carnivores

Tiger, Panther, Sloth bear, Cats, Wild dogs, the fox, the jackal, the hyena, the mongoose, civet cat, rattel. A few arna found in Bastar district of Chhattisgarh, giant and flying squirrel, hares, the pangolins are also found.

Avifauna

The Indian bird life is rich. There are hundreds of species of aquatic, gallinaceous and arboreal birds. The chief aquatic birds are : storks, herons, egrets, cormorants and flamingoes, ducks. The waders and shore birds are : the snipes, ibises, cranes and lapwings. The main ground birds are : the great Indian Bustard, pea fowl, jungle fowl, partridge and quail.

The species of perchers and song birds are : Babblers, barbits, bulbuls, cuckoos, mynas, doves, parakeets, pigeons, rollers, beaters, flycatchers, hoopoes, drongos finches, finchlarks, wagtails, wagtails, warblers, orioles are important. Pied Malabar hornbill is a rare arboreal bird. A variety of owls, eagles, kites, falcons, kestrel, the shikra are main birds of prey, the scavenging species of vultures and crows. The famous Siberian crane visits Bharatpur sanctuary every year.

Reptiles

India has a large variety of snakes – poisonous ones are Cobra, Dhaman large snake. Pythons of large size are found amongst rocks along streams. Crocodile (Mugger) and gharial are famous for their skins. Lizards include chameleon, monitor lizard or varanus.

Fishes

A variety of fishes are found in ponds, marshes, streams, inland rivers, estuaries and the sea. The important river species are : Mahseer, Chilwa, Olive Carp, Lesser Barils, Seetul, Batchwa, Murral, Eel, Carnatic carp.

Tanks fish: Rohu

Estuaries fish: Bahmin, Surmai, Horse mackerel, Queenfish, Morwasa, Gobra, Rock Perch, Banasia and Woli Herring.

Sea fish: Shark

The rainbow trout common in Himalayan rivers. During twentieth century and earlier there has been human and cattle population explosion. To meet their needs extreme pressures have been put on natural resources particularly on forests, pasture lands, forest lands to agriculture, grazing cattle. This resulted in encroachment of forests and wildlife habitat and their destruction. Many species have become locally extinct, and our wild life slowly decimated. In order to improve the wildlife Government of India constituted Indian Board for Wildlife in 1952. Under the advisory of Board conservation and protection of wildlife, constitution of national parks, sanctuaries, zoological gardens and promoting public interest and awareness in conservation of wildlife in harmony with natural environment was taken up by the Central and State Governments on priority. The wildlife (protection) act 1972, was enacted by parliament in September 1972. In April 1973, Project Tiger was launched. Under this project Tiger reserve (area 400- 4000 sq. km) created under different types of habitat which are given below.

1. Manas (Assam), 2. Palamuru (Bihar), 3. Simlipal (Orissa), 4. Corbett Park (UP), 5.   Ranthambor (Rajasthan), 6.   Kanha(MP) 7. Melghat (Maharashtra), 8. Bandipur (Karnataka), 9. Sundarbans (West Bengal), 10. Sariska (Rajasthan), 11. Periyar (Kerala), 12. Dudhwa (UP)

In each tiger reserve sufficient area has been set apart as the core, where no forestry operations will be done, rigid protection against poaching, stopping grazing cattle, creating water bodies are important. This helped not only in wildlife protection but also wildlife tourism and recreation value. The important National Parks and Sanctuaries are given below.

Manas (Assam/north Kamrup) – for wild buffalo and rhinoceros.

Kaziranga (West Bengal) – for rhinoceros, hog, deer, barasinga

Jaldapara (Jalpaiguri district) – for rhinoceros

Hazaribagh National Park (Bihar)

Palamau National Park (Bihar) – for gaur

Gir National Park (Gujarat) – for Indian Lion

Bandipur National Park (Karnataka) – for elephant, gaur

Ranganthittu – for bird sanctuary

Dachigam National Park (Kashmir) – for Kashmir stag

Periyar National Park (Kerala) – for elephants and gaur

Kane National Park (M.P) – for Tiger, Barasinga, gaur, blackbuck etc.

Shivpuri National Park (M.P) – for Chinkara

Bandhavgarh National Park (M.P) – for tiger, sambar, neelgai

Sanjay National Park (M.P) – for tiger, deer etc.

Taroba National Park (Maharashtra) – for tiger, bear

Bharatpur (Keolaghana) (Rajasthan) – for bird sanctuary

Sariska National Park (Rajasthan) – for tiger, panther

Ranthambor National Park (Rajasthan) – for sambar

Corbett National Park (UP) – for tiger, elephant, hog, deer, gharial

Rajaji National Park (UP) –

Dudhwa National Park (UP) – for tiger, elephant

Explanation of some terms

Animals includes mammals, birds, reptiles, amphibians, fish and chordates and invertebrates.

Animals articles-articles made of captive animals or wild animal which may be any part or whole of such animal and ivory imported into India and article made of that.

Habitat includes land, water or vegetation which are the natural homes of any wild animals.

Hunting includes killing or poisoning of any wild animal or captive animal and every attempt to do so; capturing, coursing, snaring, trapping, during or baiting any wild animal; injuring or destroying whole or part of body of any animal or in the case of wild birds or reptiles damaging the eggs or disturbing the eggs or nests of such birds or reptiles.

Livestock means farm animals like buffaloes, bulls, camels, cows, donkeys, goats, sheep, horses, mules, yaks, pigs, ducks, geese, poultry etc.

"National Park" means an area declared as a National Park.

Captive animal means any animal, specified in schedule (I-IV), captured or kept or bred in captivity

Protected area means a National Park, a sanctuary, a conservation reserve or community reserve notified.

Reserve forest means the forest declared to be reserved by the State Govt.

Sanctuary means an area declared as a sanctuary by notification.

Land includes canals, creeks and other water channels, reservoirs, rivers, streams and lakes.

Biodiversity: This includes genetic, species and ecosystem diversity. It is the variability among living organisms from all sources (terrestrial, marine and other aquatic ecosystems) and the ecological complexes of which they are part. It includes diversity within species, between species and ecosystems.

Biological resources: It means plants, animals and micro-organisms or parts thereof, their genetic material and by-products with actual or potential use or value. This does not include human genetic material.

Value added products: The products which may contain portions or extracts of plants and animals in unrecognizable and physically insuperable form.

Bio-survey and bio-utilization: It means survey or collection of species, sub-species, genes, components and extracts of biological resource for any purpose and includes characterization, inventorization and bioassay.

Fair and equitable benefit sharing: Sharing of benefits as determined by the National Bio-diversity authority.

Sustainable use: Use of components of biological diversity in such manner and at such state that does not lead to the long-term decline of biological diversity (thereby maintaining its potential) to meet the needs and aspirations of present and future generations.

11.5.6 COASTAL AND MARINE BIO-DIVERSITY

The law of the sea convention states that states shall adopt laws and regulation to prevent and control pollution of the marine environment from or through the atmosphere.

From time immemorable people have lived on the coasts and fishing. About 40% of the world's population living within 10 km of the coast. The main human dependence on the marine and coastal ecosystems are:

1. Life in the sea produces one-third of the oxygen that we breath.

2. The ocean absorbs about 30% of CO_2 emitted by human activity (ever since industrial revolution).

3. Fisheries (globally) employ 200 million people and provide about 15% of dietary intake of animal proteins.

4. Marine by-products are raw materials for manufacturing industries like paints, fertilizers, skin lotions, tooth pastes and medicines.

5. The divergent chemical deposits in the marine environment are an asset and might yield to new anti-cancer drugs.

6. The coastal shore provides for marine transportation, recreation, tourism and salt production.

7. Mangroves help to protect the coastal aquifers from sea water intrusion and safeguard the coastal communities from natural hazards like cyclones, tsunamis, hurricanes.

8. Coastal wetlands play an important role in water quality regulation by capturing and filtering sediments and organic waste transits from inland.

India has a coastline of about 7500 km, of which about 5400 km belong to peninsular India and the remaining to the Andaman, Nicobar, Lakshadweep Islands. With less than 0.25% of the worlds coastline India accommodates 63 million people (about 11% of the global population living in low elevation coastline areas).

Humanity has been benefited from marine and coastal ecosystem. However, human activities both on land and sea have made adverse impacts on the marine and coastal quality and renewability. Large scale burning of fossil fuels causing both Ocean acidic and warmer. The average sea surface water temperature rose by 0.4°c between 1950-2012. This warming is threatening marine ecosystem and coral bleaching. The sea level rose about 2.2 cm on an average since 1980. Adding toxic pollutants, non-biodegradable wastes adversely affecting marine wildlife. Noise pollution generated by shipping and industrial activity contributing to the prevention of species (like whales) from communicating with each other across kilometers of the Ocean.

Over hunting has reduced the stock of many species. The auk and the sea mink have become extinct and great species like whales declined.

At present more than four million fishing vessels (including industrial trawlers) are engaged in fishing about 140 million tonnes per year. It is estimated that about 13% of the global fisheries are collapsed and 30 to 35 % marine habitats (like seagrasses, mangroves, coral reefs) have been destroyed, due to mainly anthropogenic interventions. It is the duty of man to preserve environment and biodiversity, otherwise it will be wiped out. Old saying "Paryavarno Rakshati, Srusti Rakshita" [i.e., if you preserve environment, it will save life on earth].

According to IPCC latest draft report, with the present rate of GHGs emission, there is a high probability (90-100%) that the Arctic ice cover will shrink through the 21st century. The complete Arctic ice will melt if the global temperature rises more than 2 °C by 2100. The global mean temperature estimated for the period 2016-2035 would rise 0.4 to 1.0 °C as compared to 1986-2005. The combined land and ocean temperature rose by about 0.8 °C over the period 1901-2010 and 0.5 °C over the period 1979-2010. It reaffirmed that man-made climate change heating the atmosphere that would cause dramatic change in global ecosystem. Rainfall across the globe would increase in frequency and intensity. An increase in temperature more than 2 °C will cause sea level rise. On an average 50 countries exposed to 11% of the world's natural disasters and suffer 53% deaths due to these calamities each year. The developed countries would be exposed to 15% of all hazards but they account 1.5% of deaths. Nature alone cannot be blamed for the disaster because poverty, inequality, political decisions and prompt action also contribute to the disasters.

11.5.7   THE WILD LIFE (PROTECTION) ACT, 1972

India had been one of the richest and most varied wildlife in the world. The rapid decline of India's wild animals and birds became the cause of grave concern. Some wild animals and birds had already been extinct and some others were in the danger of being lost forever. Areas which were once teeming with wild life become devoid of it. As a result the Wildlife (Protection) Bill was passed by both the Houses of Parliament and approved by the President on 9th September 1972, and is called "The Wildlife (Protection) Act, 1972". The Act provides the protection of wild animals, birds and plants and ensures the security of countries ecology and environment.

The future of global flora and fauna in the changed climate

A warming of earth would create a variety of hazards for millions of people who never drove a car or flew in aircraft. Many animal and plant species will suffer because of man's unlimited fossil fuel burning. More droughts, heat waves and more extreme weather events during next 100 years would reflect on crops sown and grown in various climatic zones. Agricultural practices would have to be modified with the use of latest technologies. More adverse impacts would be on agriculture in poorest countries, and they will be more vulnerable to droughts and famine. In the changed state the effects on biodiversity is unimaginable. It was predicted by the University of Leeds, UK that 15-37% of plant and animal species over the world would face extinction by the year 2050. In Britain and California, during 1970 dry spell 5 to 21 species were decimated. Several species moved northward to cooler areas.

Destructions of forests directly effects the green lungs of the Earth (oxygen, carbon, sink). According to one study deforestation puts four times more carbon into atmosphere than the Nations fossil-fuels burning does. In the event of global warming, climate change, more droughts, extreme weather events likely rise. As regards the effect of climate on forestry towards the poles is uncertain. Permafrost melt is high-altitude forest stress. Both forest fire and insects are the part of natural ecosystems of North America. In summary climate change will have both positive and negative effects on crops.

Explanation of some terms

Biotechnology: The technological application that uses biological systems, living organisms, to make or modify products or processes for specific use.

Domesticated or cultivated species: The species in which the evolutionary process influenced by humans in order to meet their needs.

Ecosystem: It means a dynamic complex of plant, animal and micro-organism communities and their non-living environment interacting as a functional unit.

Ex-situ conservation: The conservation of components of biological diversity outside their natural habitat.

Genetic material: Any material of plant animal, microbial or other origin containing functional units of heredity.

Genetic resources: The genetic material of actual or potential value.

Habitat: The place or type of site where an organism or population naturally occurs.

In-situ conservation: The conservation of ecosystems and natural habitats and the maintenance and recovery of viable populations of species in their natural surroundings and in the case of domesticated or cultivated species, in the surroundings where they have developed their distinctive properties.

Protected area: A geographically defined area which is designated or regulated and managed to achieve specific conservation objectives.

11.5.8   THE NATIONAL BOARD FOR WILD LIFE

The board consists:

1. Chairperson-the Prime Minister
2. Vice chair person-the Minister in charge of Forests and Wildlife
3. Members
   (a) Three Members of Parliament,
   (b) Member of Planning Commission in-charge of Forests and Wildlife (now Neeti Ayog, EOF, W, CC) [Environment of Forest, Wildlife and Climate Change]
   (c) Five persons of NGOs, nominated by Central Govt. from among of eminent conservationists, ecologists, and environmentalists
   (d) The Chief of Army
   (e) Govt. officials-Secretaries, Director etc.

11.5.9 STATE BOARDS FOR WILD LIFE

1. Chair person-The Chief Minister of the State
2. Vice chairman-The Minister in-charge of Forests and Wildlife

3. Three MLAs

4. Three members of NGOs (nominated). [Line department specialists]

5. Ten members-Eminent persons, two Scheduled Cast/Tribes members and other line, Dept. Govt. officials.

11.5.9.1 PROTECTED AREA/SANCTUARY

By notification, the State Govt. declare any area (other than any reserve forest or traditional waters) as a Sanctuary which has ecological, faunal, floral, geomorphological natural or geological significance, for the purpose of protecting, propagating or developing wild life or its environment.

11.5.9.2 PUNISHABLE OFFENCES

According to the wild life protection Act 1972, hunting of any wild animal specified in schedules I-IV Prohibited.

A few examples are given below

Mammals like, Andaman Wild pig, Himalayan Brown bear, Capped Langur, Chinkara (or Indian gazelle), Gangetic Dolphin, Indian Elephant (or India, Indian Lion, Wild Ass, Musk Deer, Nyan (or Great Tibetan Sheep), Tibetan Antelope, tiger, Wild Yak etc.

Amphibians & Reptiles like Audithia Turtle, Gharial ganges soft-shelled Turtle, Hawksbill turtle, Pythons, water lizard etc.

Critically Endangered Birds

White-Bellied Heron, Ardea, Insignias, Spoon-billed Sandpiper.

11.6 BIOGEOGRAPHIC ZONES IN INDIA

There are about 25 biogeography provinces in India. Trans Himalayan, Himalayan, Indian desert, Semi-arid, Western Ghats, Deccan Peninsula, Gangetic plains North-east India and Islands. The aim is to designate one representative site as Biosphere Reserve in each biogeography province for long term conservation. This programme was started in 1986 and only 16 sites have bean designated. The following are the 16 designated Biosphere Reserves.

1. Nilgiri (Tamil Nadu, Kerala and Karnataka)

2. Nanda Devi (Uttarakhand)

3. Nokrek (Meghalaya),

4. Manas (Assam),

5. Dibru-saikhowa (Assam)

6. Sundarban (west Bengal)

7. Gulf of Mannar (Tamil Nadu)

8. Great Nicobar (Andaman and Nicobar Islands)

9. Similipal (Odisha)

10. Dehang-Debang (Arunachal Pradesh)

11. Khangchand zonga (Sikkim)

12. Pachmarhi (Madhya Pradesh)

13. Achanakmar-Amarkantak (Chhattisgarh & MP)

14. Agasthyamalai (Tamil Nadu, Kerala)

15. Katchchh (Gujarat)

16. Cold Desert (Himachal Pradesh)

11.7 THE PUBLIC LIABILITY INSURANCE ACT, 1991

The growth of industries with hazardous processes operations lead to the increase in risks from accidents. The people involved in such accidents, workmen employed and innocent people residing in the vicinity. Such accidents may cause deaths, injury to the people and cause damage to public and private properties. Economically weaker sections suffer great hardships because of delayed relief and compensation. In order to ameliorate the suffering of people due to accidents in such hazardous installations, it is essential to provide for mandatory public liability Insurance. To achieve this objections the Public liability Insurance Bill was promulgated by the President of the India on 31st January 1992 as the industrial owners handling hazardous substances had to take Insurance policies by March end of 1992.

Act 6 of 1991: The public liability Insurance Bill having been passed by both the Houses of Parliament, received the assent of the President on 22nd January 1991. It came on the Statute Book as the Public Liability Insurance Act 1991. This act was further amended in 1992.

The objective of this act is to provide immediate relief to the persons affected by accidents occurring while handling any hazardous substance and all other matters connected with accident.

Explanation of some terms

Accidents means an accident involving a sudden or unintended occurrence while handling any hazardous substance resulting in continuous or intermittent or repeated exposure to injury or death to any person or damage to any property (except war or radioactivity).

11.8   SUSTAINABLE DEVELOPMENT

Johannesburg Declaration on Sustainable Development 2002

World summit on sustainable development held in Johannesburg South Africa 2-4 September 2002 to reaffirm the commitment sustainable development. The salient features are:

1. To build a humane, equitable and caring global society and need for human dignity for all.

2. The children of the world would inherit free from indecency, poverty, environmental degradation and unsustainable development.

3. To create a new and brighter world of hope.

4. A collective responsibility to strengthen the pillars of sustainable development- (i) Economic development, (ii) social development and environmental protection-at all levels.

5. Humankind is at a crossroad. There is need to eradicate poverty and human development.

6. The conference held at Rio de Janeiro agreed to protect environment, social & economic sustainable development.

7. The UN conferences held at Monterrey and Doha defined a comprehensive vision for the future humanity of the world.

8. It is a challenging task for (i) eradicating poverty, (ii) changing consumption and production patterns and (iii) protection of natural resources with the objectives of (essential requirements for) sustainable development.

9. The deep fault line that divides human society between the rich and the poor and the ever increasing gap between the developed and developing worlds pose a major threat to global prosperity, security and stability.

10. The adverse effects of climate change observed in (i) loss of biodiversity, depletion in fish stock, desertification in the global environment and also in the rise of intensity of natural disasters. These effects are more evident in developing countries which are more vulnerable to air, water and marine pollution that rob their a decent life.

11. Globalization added rapid integration of markets, mobility of capital but put new challenges for the pursuit of sustainable development. The benefit and costs of globalization are unevenly distributed because developing countries facing difficulties.

12. The global disparities between the developed and developing countries likely to increase both economically & in vulnerability.

The ways to global sustainable development

1. The biodiversity is the collective strength for the global sustainable development.

2. Human solidarity, cooperation among world's civilizations required irrespective of race, religion, language, culture and tradition.

3. Human dignity achieved through clean water, sanitation, adequate shelter, energy, health care, food security and biodiversity protection. It requires capacity building, use of modern technology for development, technology transfer human resource development, education and training.

4. Human pledge required sustainable development to fight against conditions with particular focus on chronic hunger, malnutrition, foreign occupation, armed conflicts, illicit drug problems, organized crime, corruption, natural disasters, illicit arms trafficking, trafficking in persons, terrorism, intolerance and incitement to racial, ethnic, religious and other hatreds, xenophobia and endemic, communicable and chronic diseases, in particular HIV/AIDS, malaria and Tuberculosis.

5. The millennium development goals (i) women's empowerment & emancipation and (ii) gender equality

6. The global society challenges: (i) poverty eradication and (ii) sustainable development to ensure available resources used for the benefit of humanity.

7. Developed countries are urged to contribute towards internationally agreed levels of official development assistance.

8. Support regional alliance [like the New Partnership for Africa's development (NEPAD)] to promote regional cooperation, improved international co-operation and sustainable development.

9. Special attention be given for the development of small Island developing states and the least developed countries.

10. Affirm the vital role of the indigenous people in sustainable development.

11. Sustainable development requires a long-term perspective, decision making and to work with all major groups respecting the independent roles of each of them.

12. The private sector, both large and small companies have a duty to contribute for the equitable and sustainable communities and societies.

13. Provide assistance to increase income generating employment opportunities in a accordance with the international labor organizations (ILO) declaration of fundamental principles and rights of work.

14. Private sector corporations enforce corporate accountability.

15. Undertake and strengthen and improve governance for implementation of the millennium development goals.

The Biological Diversity the guidelines for international collaboration

1. The guidelines for international collaboration research projects transfer or exchange of Biological Resource between institutions in other countries.

2. National Biodiversity Authority require transfer and exchange of biological resource with international collaborations research projects. All research collaborations have to be approved by the concerned Ministry or Department of Government of India.

3. The collaborations shall abide by the provisions of existing national laws

4. The projects to be used only for research, and approved by central Government.

5. Specified proposals and material transfer guidelines be developed National Biodiversity Authority

6. Sharing the benefits of the results of research be in accordance with the provisions of Intellectual Property Rights.

7. The voucher specimen of the biological resources occurring in India transferred or exchanged under the project be sent to the designated repository in accordance with the Act.

8. Publication of research papers, books, bulletin(s) etc based on the results of the research shall not be done without the approval of the Indian collaborator.

9. All research work & results shall be reported to the National Biodiversity authority.

10. Any enquiries/results details about biological diversity etc related issues can be obtained on www.nbaindia.org

International Biodiversity Day-(May 22, 2014 India)

The aim of celebrating IBD is to focus the attention of people to awareness on Biological Diversity, the importance of preserving it in villages and Islands.

State Biodiversity Board, opened the Biodiversity Register, which empowers villagers to document their own flora and fauna. Biodiversity laws make it difficult for outsider to export natural resources of village without consent of villagers. Plants and Animal species endemic to the state are on the verge of extinction, proactive people's participation was necessary to protect them. It is the urgency that local people inculcate proactive environments, activities in their daily life to save the biodiversity.

Importance of Biodiversity: Man and diverse species are inter-dependent and hence the wellbeing of the planet is dependent on biodiversity. Man has been dependent on plant and animal species for all his needs food, shelter & health. All these needs are derived from the other species in the environment like air, water, soil and plant species, animal species which all include in natural resources. The Example of present day world over is Broiler chicken. This type chicken may be called man-made, because its existence depends mainly on man. As this species is the largest of all the other existing birds and is widely used by man all over the world. Similar by man's dependence on domesticated animals is well known from a very very long time.

The earth's biodiversity is an essential asset to mankind. Genetic diversity has its importance in evolution, LPI also gives a clear biodiversity importance and its threats.

Importance of Biodiversity conservation

Biological diversity tends to conservation of ecological diversity which tends to preservation of food chain. Conservation of genetic diversity leads to preservation of animals, which in turn preserves or helps sustainability. Sustainability is key to human development in all fields.

In summary, under sustainable development, India hopes to address the needs of poor, vulnerable and marginalized communities. UN would help for sustainable development with regard to health, nutrition, to eliminate poverty.

India aims to address: (i) Education & employment, (ii) Nutrition and food security by increasing agricultural productivity, (iii) Health, water and sanitation, (iv) climate change, disaster resistance and (v) over all national youth development.

National Biodiversity Authority

National Biodiversity Authority consists of:

1. Chair person: Appointed by the central Govt. having adequate knowledge & experience in conservation and sustainable use of biological diversity

2. Three ex-officio members appointed by central Govt.

   One from the Ministry of Tribal Affairs, two from the Ministry of Environment and Forests (one Additional Director general of forests or the Director General of Forests)

3. Seven ex-officio members the ministries of the central govt.

   (i) Agricultural Research Education

   (ii) Biotechnology

   (iii) Ocean Development

   (iv) Agriculture and Co-operation

    (v) Indian System of Medicine

    (vi) Science and Technology

    (vii) Scientific and Industrial Research

4. Five non-official members (specialists & scientist) relating to conservation of biological diversity, sustainable use of biological resources, representatives of industry conservers, creators and knowledge-holders of biological resources.

Functions and Powers of the National biodiversity Authority

    (i) To regulate activities & issue guidelines

    (ii) Grant approval for undertaking any activity

    (iii) Advise the Central Govt. on relating matters of conservation of biodiversity

    (iv) Advise the State Govt. on importance of biodiversity heritage

    (v) Perform functions necessary for Biodiversity

    (vi) Take any measure to oppose the grant of intellectual property rights derived from India

Duties and Strategies of Government for Conservation of Biological Diversity

1. Develop national strategies, plans, programmes for conservation and sustainable use of biological diversity

2. Identification and monitoring of areas rich in biological resources, promotion of in-situ and ex-situ conservation of biological resources, incentives for research, training, public education, create awareness to biological diversity

3. Issue directions to concerned officials to arrest overuse, abuse or neglect biological resources, habitats and provide technical assistance

4. Integrate the conservation, promotion and sustainable use of biological diversity

5. Take measures to assess environmental impact on biological diversity and include public participation

6. Protect the knowledge, generic of local people relating to biological diversity

7. Notify conservation of all heritage sites of biodiversity importance

8. Notify, prohibit or regulate collection, appropriate steps to rehabilitate and preserve those species under threat of extinction

9. Designate institutions as repositories of biological resources, newtaxon discovered

Estimated species on the Earth

The estimated plant and animal species on the earth is about 9 million of which 1.2 million have been identified (till August 2011). Scientists are yet to discover and classify about 90% of the plant and animal species. The number of species figure out varied widely from 3 million to 100 million. The quest is not mere scientific fancy. Humans derive enormous benefits from the richness of life on the planet earth from food medicines, to clean air and water. There appears as surge in extinction rates.

Notable UN studies point that the world is facing the worst losses since the time of dinosaurs vanished 65 million years ago. A team studied existing species data bases and taxonomic data. They examined well known groups and found the relative number of species assigned to phylum, class, order, family and genus follow consistent patterns. This study indicate that 6.5 million species on land and 2.2 million in ocean depths. The study had an error margin of 1.3 million in total.

The scientist's results conclude that 86% of existing species on land and 91% of species in the ocean still won't description.

Biodiversity in simple terms is the variety of living organisms (flora and fauna) found in a specified geographical area/region on earth.

Sustainable Development: The environmental protection Act (1986) is aimed to guarantee 'right to life', to ensure sustainable development by increasing greenery (by plantation of trees/afforestation) that protects ecology create awareness about hazardous effects of pollution air, water, noise pollution and rain-water harvesting or take steps to store water underground.

For regulating pollution, lay down standards for air, water & soil, procedures for handling of hazardous substances and restricted substances like potassium/ammonium nitrates and explosives. Further procedures and safeguards for prevention of accidents which are resistive to environmental degradation and provide remedial measures for such accidents.

Pollution control: Industry operation/process should not discharge or emit pollutants in excess of standard prescribed. For example pollution of Ganga water by dairy farms, severs, public latrines, throwing of corpses/animal dead bodies, industrial effluents. It is the liability of construction workers and public not to commit nuisance by letting out effluents from their drainage system. The management for drainage effluents should follow stringent conditions only after treatment. A tannery cannot be set up without primary treatment plant otherwise it will cause adverse effect to public at large.

Indiscriminate mining operations not permitted that damage natural wealth, environment (forests, Lakes, rivers, wildlife). Government should lease mining after scientific evolving long-term plan so that it will not exploit mineral wealth, damage environment, ecology and deprive local people (Girijans) of their winning bread. It is the liability of local residents to provide information to the concerned authorities if there is any accident or apprehension of unforeseen act or event that may cause damage to environment. It will be bounden duty of authorities/agencies to act promptly for remedial measures to prevent or mitigate the environmental damage.

CHAPTER - 12

The Climate – Role of Ocean in Climate Fluctuations

INTRODUCTION

According to IPCC, the heat content of the entire global system has increased since 1950s. The oceans have absorbed 90% of the heat increase. The heat content of upper ocean (0 – 700 m slab) increased from 1961 to 2003 was about 16×10^{22} J. This heat caused average increase in sea temperature by about 0.1 °C in the upper 700 m slab. The SST changes (1961-2003) by about 0.4 °C at the surface and 0.1 °C rise in temperature in the upper ocean depth 700 m. This heat amount would be equivalent to an almost 100 °C change (rise) in atmospheric temperature. The oceans have much greater capacity to store heat than the atmosphere. The ocean is one thousand times more dense than air and the specific heat of ocean water is about four times that of air. The thermal adjustment of the atmosphere alone is about a month. The thermal adjustment time scales in the ocean are very much longer. The oceans have a much greater capacity to store heat than the atmosphere. This explains the buffering of atmospheric temperature by the oceans enormous heat capacity. (Buffer meaning protection from damaging impact).

The time scale for adjustment of the deep ocean is 1000 years, because the slow circulation of the abyssal ocean limits (obstructs) rate at which its heat can be brought to the surface. These long time scales buffer atmospheric temperature changes, consequently ocean would play a very important role in climate. Oceans reduce the amplitude of seasonal extremes of temperature and buffering atmospheric climate changes.

The world's oceans occupy about 71% of the surface area while the land occupies only about 29% area. The following properties about atmosphere and oceans must be noted.

The mass of the atmosphere; 5.26×10^{18} kg

Global mean surface temperature; $288\,^{\circ}$k

Global mean atmospheric pressure; $1.013 \times 10^{5}\,P_a = 1013$ hp$_a$

Global mean atmospheric density $\simeq 1.225$ kg/m^3

Global ocean surface area 3.61×10^{14}m^2

Global ocean mean depth 3.7 km

Global ocean volume $3.2 \times 10^{17} m^3$

Global ocean mean density $1.035 \times 10^3 \, kg/m^3$

Global ocean mass 1.3×10^{21} kg

Specific heat of water $4.18 \times 10^3 \, J/kg/^{\circ}K$

Latent heat of fusion (of water) $3.33 \times 10^5 J/kg$

Latent heat of evaporation (of water) $2.25 \times 10^6 \, J/kg$

Density of fresh water $0.999 \times 10^3 \, kg/m^3$

Viscosity of sea water 10^{-3} kg/m/s

Specific heat of ocean is about 4 times that of (atmospheric) air.

Process	Time Scales
Weather	hrs/days to weeks.
Land surface	hrs/days to months.
Ocean mixed layer	hrs/days to months.
Sea ice	weeks/months to years
Volcanoes	weeks/months to years
Vegetation	hrs/days to millions of years.
Thermocline	yrs/decades to centuries
Mountain glaciers	decades to centuries
Deep ocean processes	centuries (10^2) to 10^4 years
Ice sheets	10^2 to 10^5 years
Orbital forcing	10^3 to 10^5 years
Tectonics	10^6 to 10^9 years
Weathering	10^5 to 10^9 years
Solar constant	$10 - 10^2 - 10^3$ to 10^9 yrs $1-10^9$ years
Natural CO_2 cycle	1k to 10k y (k = thousand)
Anthropogenic CO_2	decades to centuries $(10 - 10k$ years$)$

12.1   WORLD'S LAND UTILIZATION (1990)

Arable land area about 10% worlds land surface (WLS).

Meadows and pasture area about 20% of WLS

Forest area is about 27% of WLS (4035 million hectares)

Deserts area is about 20% of WLS. They occur mostly in high air pressure belts between lat 10° and 35° N/S. Cryosphere or Ice cover area is about 10% of WLS. It was occupied about 30% WLS during the peak of Pleistocene ice-age. If all ice melted the sea level would rise about 60 – 90 meters (200 – 300 ft) flooding many densely populated low lands, and great cities.

12.1.1 CLIMATE

The ever changing physical state of atmosphere at any location constitutes the weather. Weather is described in terms of instantaneous values of meteorological elements on surface of the atmosphere and in atmosphere, like temperature, wind (direction and speed), pressure, humidity, state of sky (clouds), precipitation, visibility etc. The average conditions of weather over a long period at a place together with its extremes and their probability of occurrence constitutes the climate of the place.

Climate normals are average values of weather elements over a place for about 10 days or one month (say 1-10 Jan, January month etc) averaged over a long period 30 to 100 years of record. These normals are used as measuring standard yardstick for describing behavior or variation of weather or climate.

Weather and climate have been changing from the origin of atmosphere and the life on earth. It reached a state that is suitable for life on earth. Climate is a complex phenomena and never fits into rigid demarcations. It changes one place to another, one generation to the next, from one century to the next and one ice age to the next.

The phenomena of weather and climate have universal jurisdiction and they do not owe fealty to any man or manmade institution. Weather is a subject of interest to everyone, whether rich or poor, learned or illiterate, young or old and all through the year. At present global warming and climate change is the hot topic of discussion all over the world.

For billions of years earth has supported life, which indicate climate remained within a narrow limits with all variations, like warm epochs to cold epochs of ice ages. During past 4.5 billion years, the incoming solar radiation intensity increased by about 30%. Climate models suggest that the input rate of CO_2 (carbon-dioxide) from volcanic activity into the atmosphere is balanced by the rate of removal of it by the photosynthesis and chemical weathering (sink). Carbon from the earth's interior is pumped into the atmosphere (fossil fuel burning), chemical weathering of continental rocks, which finally washed into the oceans and deposited as sediments.

Present global warming is attributed to the use of fossil fuels, like diesel, natural gas, wood, coke etc. As long as we use these fuels, we like it or not global warming and climate change will continue.

Human intervention: Before industrial revolution world human population size, was about 1000 millions. By 1800 AD, which rose to more than 2.5 billions, by 1900 AD. It rose to more than 7.5 billions in 2005 and at present (2017) the population was 7.6 billion and likely to exceed 8.6 billions by 2030 and to 11.2 billions by 2100 AD. India's population in 1950 was about

360 millions, in 2000 it was 1028 millions and at present (2017) it is about 1.3 billions and projected to 1.5 billions by 2030.

During last 150 years Industrial activity grown 25 to 40 folds, fossil fuel consumption rose 30 to 50 times and this added CO_2, SO_2, N_2O emissions to the atmosphere together with toxic chemicals like As, Cd, Zn, & Hg and particulate matter. Burning of fossil fuels in motor vehicles, trucks, furnaces, factories and in production of electricity etc., increased CO_2 in the atmosphere from about 275 ppm (parts per million) in pre - industrial state to 354 ppm by 1990 and recently (2017) exceeded 405 ppm. Population growth compelled forest land to bring in agricultural cultivation which surpassed all the earlier human history. Water consumption increased more than fivefold during this period. Deforestation reduced one-third of forest land (about 6 billion hectares) which sniffed carbon sink. According to WMO No. 735 publication, in totality man succeeded in adding more than 30 gigatonnes of carbon to the atmosphere annually.

World size carbon reservoirs are:

World vegetation 500 billion metric tones (bmt),

World soils 1500 bmt,

Atmosphere 735 bmt

The life time of CO_2 in atmosphere 50-200 years

CH_4 (methane) 7-10 years

N_2O (nitrous oxide) 150 years

CFCs (chloroflourocarbons) 75-110 years

According to IPCC (Inter government Panel on Climate Change) 2007, global heat content between 1960-2003 increased by 16×10^{22} J, of which 99% is absorbed by Oceans. This heat increased the sea surface water temperature (SST) by 0.4°C and sea top surface depth (0-700m) water by 0.1°C. This amount of addition of heat to the atmosphere could have increased the temperature of atmosphere by about 100 to 150°C in the absence of ocean (absorption). Oceans huge heat capacity buffering the atmospheric temperature.

Since 1750 human activities caused increase in radioactive forcing 1.6 w/m² and rise in global temperature by 0.74°C between 1906-2005 (100 years). The mean annual temperature over India rose by 0.52°C during 1901-2000 (100 years) and continued to be above normal since 1990. The mean annual temperature of India during 2009 was 25.5°C as against the normal 24.64°C, which was 0.914°C above normal. During 1901-2009 (108 years), there were 12 hottest years, of which 8 were during 1999-2009. The statistics of hot years given in the following table in order of decreasing magnitude.

S.No	Year	Temperature departure from LPA ($^{\circ}$C)	S.No	Year	Temperature departure from LPA ($^{\circ}$C)
1	2009	0.913	7	1998	0.514
2	2002	0.708	8	1941	0.448
3	2006	0.60	9	1999	0.445
4	2003	0.560	10	2001	0.429
5	2007	0.553	11	1987	0.413
6	2004	0.515	12	2005	0.410

The year 2015 is the earth's hottest year, recorded 0.908 $^{\circ}$C above 20th century average temperature. Subsequently two years 2016 and 2017 in some respects were equally hot. Though there are some indications of climate, but India as a whole does not show significant impacts of climate change on regional basis in terms of moisture and on the agriculture on the global scale global atmospheric moisture appear to rose by about 2%. This is likely to increase the severity of tropical storms in equatorial belt, but it may not increase the tropical disturbances.

Consequences of global warming: Globally extreme weather events on rise, like long dry spells, which will have significant negative impact on fodder and grain production and finally affect livestock. Other extreme events include high intensity of few hours or single day rainfall, heat wave conditions, severe thunderstorms with lightening, flash floods, urban flooding, Himalayan and Arctic, Antarctic glacier melts, avalanches, severe snow storms, super cyclones/hurricanes accompanied with strong winds and precipitation etc.

Vulnerability: All human societies are not equally vulnerable to climate change. Poor people undeveloped countries are more likely to be affected as compared to rich developed countries. The 2017 year Atlantic hurricanes, Harvey (August), Irma (Sept), Arctic snow storm "beast from the east" (Feb 2018) are examples. Harvey caused catastrophic flooding in Houstan, turned streets into flooded rivers, trapped residents to shelter in higher floors. Wind speed exceeded 240 kmph (category-4). It poured 75 cm rain in 24 hrs in Texas, blasted chemical plants causing 30 to 40 foot flames & damaged oil industry 114000 liters of Crude oil. In this event most of the sufferers economically poor. The same is the case of Irma (Category-5 hurricane). Wind speed exceeded 295 Kmph, destroyed 95% of homes in Caribbean Barbuda, Puerto Rico.

Beast from the east "Emma" shivered Europe, Rome, London etc. In France temperature dropped to -10 °C, in Germany -27 °C, Moscow -20 °C & Poland -26 °C.

Climatic disasters: The word climatology is derived from the Greek word Klima meaning inclination, which refers to the angle of incidence of sun's rays on earth. The hazardous events like droughts, floods, tornadoes, hurricanes etc, cause lot of hardships. Extreme cold winters in parts of North America in 1977 and unusually hot summers in the same region in 1900 were also hazardous. The environmental abnormal conditions cause great human stress and suffering, particularly in aged people, children and infirm.

The past civilizations show large climatic changes caused major impacts on human society. Only a few thousand years ago, the present subtropical deserts, such as the Sahara & Rajasthan were much wetter than today and were able to support much larger populations. Some 18000 years ago, at the climax of the most recent ice-age, much of North America, northwest Europe were covered by sheets of Ice. If such large changes that occurred in the past, they can surely be repeated in future. During such events, some places may be covered by deserts and thick forests.

The natural causes of past climate changes, are attributed to: Variation in earth's orbit around the sun, fluctuations in solar energy output, dust in the upper atmosphere from volcanic eruptions, changes in the precision of earth's rotational axis.

Mitigation: In recent time man made activities, industrialization, burning fossil fuels etc. contributing to climate change. Adverse effects of climate change can be mitigated by creation of awareness and people's participation help in balancing the weather from extremes to moderation. Environmental changes will have grave impacts on mankind and also on living creatures and vegetation. These have been fore warnings of every adverse weather like depletion of natural resources. As yet little has been achieved in applications of improved technologies, shearing of knowledge, experience and resources to mitigate the adverse effects globally.

12.2 CLIMATOLOGY

Weather and Climate: The physical state of atmosphere at any particular location rarely exhibits a steady state even during short intervals of time. The ever changing physical state of atmosphere constitutes the weather. It is described in terms of instantaneous values or short period mean values of meteorological elements on surface of the earth and in the atmosphere, such as temperature, pressure, humidity, wind, state of sky (cloud), precipitation etc. The average conditions (not only numerical values but available information) of weather at a place over a long period (more than 30 years), together with its extremes and its probabilities constitutes the climate of the place. Weather and climate has been changing right from the origin of atmosphere and life on earth and reached a balanced state that is suitable for life on earth. Climatology may be broadly described as statistical meteorology in relation to geography or natural environment. Climate never fits into rigid demarcation. It changes from one type to another, one generation to the next, from one century to the next and from one ice age to the next. Climatic normals are average values of weather elements over a place for about 10 days or one month (say 1-10 January or January month), averaged over a long period about 30 to 100 years of records. These normals are used as yardstick for describing its behaviour or variation of weather / climate.

Climatic Controls: The factors that control the climate are: (i) Latitude or sun's inclination (ii) Altitude, (iii) Topography, (iv) Land and sea distribution and (v) Ocean currents.

(i) *Latitude*: This determines the amount of heat received from the sun. The sun is the main driving force of the earth-atmospheric systems.

(ii) *Altitude*: The height above sea level generally determines precipitation. With increasing heights pressure and temperature falls and increases precipitation.

(iii) *Topography*: Large mountain barriers stand as climate boundaries. The orientation of the Himalayas is an outstanding example that acts as a climate barrier. The Himalayas prevent the northern cold wind entering over Indian sub-continent during winter. It also acts as a barriers in southwest monsoon that entraps the moist maritime air over India.

(iv) *Land and Sea Distribution*: The amount of heat absorbed or radiated by a surface is a function of nature of surface and its thermal capacity. Because of this there is lot of difference in temperature and pressure over land and sea / ocean at the same latitude. Secondary circulations like land and sea breeze, monsoons are developed due to land-sea distribution.

(v) *Ocean currents*: Ocean currents act as transporters of heat from equator to pole wards. They also greatly alter the local coastal climates.

In addition to the above factors, climate is affected by semi-permanent Highs and Lows.

Climate Classification: Climate classification is a complex subject. Most of the classifications are based on the relationship between climate and vegetation or soil. Natural vegetation is the best basis for classification of the world's climate. The following is the simplest and the climate zones are given below.

1. **Equatorial Climate:** It is also called humid tropical or rainforest. This zone extends from latitude 10 °N to 10 °S. Over the oceans it is called

doldrums. The main features of this zone: light winds, high humidity and temperature. Diurnal range of temperature exceeds the annual range. There is no dry season. It has two rainy seasons with the movement of thermal equator. It has most unstable atmosphere which is associated with convective clouds and thunderstorms. Annual rainfall is a function of topography.

2. **Savannah or Trade Wind Littoral Climate:** It extends from latitude 10° to about 25° in both hemispheres. These zones have alternate dry, wet seasons. Wet season is associated with the sun overhead. The rainfall amount and duration decreases as one moves from lower to higher latitudes. Similarly annual and diurnal temperature ranges increase with increasing latitudes. During winter dry trade winds prevail. Fairly high uniform temperatures prevail all through the year. In this region flying conditions on land are hot, dusty and turbulent.

3. **Steppe or Arid Sub-tropical Climate:** This zone extends from Savannah to about latitude 30° in both hemispheres. The world's great hot deserts - the Sahara, Arizona, Arabia, Kalahari, South America and Australia deserts are all located in this zone. The main features of this zone is subsidence of air (surface level divergence), very hot summers with large diurnal range of temperature, meagre rain with very short rainy season bordering areas. Bordering Savannah rainfall confined to summer while bordering pole wards winter rainfall. Annual temperature range small. In summer dust/sand storms observed.

4. **Warm temperature zone or Mediterranean type:** This zone extends between latitude $30 - 45^{\circ}$ and includes parts of California, central Chile, extreme South Africa, parts of southern Australia. In summer it comes under the influence of arid-subtropical zone but in winter it experiences unsettled weather of cool temperate zone. The later is due to the shifts of Sub-tropical High towards the equator. Summers are dry and hot but winters may be wet and cool.

5. **Cool temperature zone:** This zone extends between latitude 35-50°. The main features of these zones mobile westerlies, frequent passages of frontal depressions and cold anticyclones or ridges. No dry season and winters can be cold and very clod in areas which are away from oceans. Winds are predominantly westerlies with frequent gales.

6. **Boreal Zone:** This zone roughly extends between latitude 40-60 $^{\circ}$N and observed only in northern hemisphere. It is well developed and covers North America, Scandinavia and former USSR. The chief feature of this zone are cool and moist summer, very cold winters and

large annual range of temperature. It is dominated by tropical continental air mass in summer and polar continental air mass in winter.

7. **Polar and Tundra Zones:** This zone extends between latitude 55°N to north pole and latitude 50°S to south pole. The chief features of these zones is that it has 20 hours of day sunshine for three months in summer and 20 hours per day of darkness in winter. Mean temperature in warmest months is little higher than freezing point (0°C). It has some tundra vegetation like mosses, lichens and grasses otherwise the surface is continuously covered by snow or ice. The ground is permanently frozen. There is no summer and climate is humid inspite of being very cold. Precipitation is mainly snow or snow grains.

The above 7 zones of climate are as shown in Fig. 12.1.

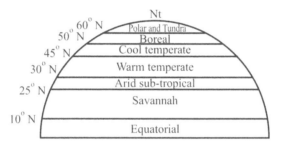

Fig. 12.1 Climatic zones of Northern Hemisphere

8. **Monsoon Climate:** We have already studied about monsoons. The seasonal reversals of wind direction and temperature cooling is the main feature. The cause of changes is the differential heating of large land mass and sea area altering with the season (like land and sea breeze). Monsoon climate is best explained by summer monsoon over the Indian sub-continent. The onset of southwest monsoon heralds the rainy season over India and the northeast monsoon of winter returns to the dry weather in most of India.

12.3 CLIMATE OF INDIA

Physiography of Indian Sub-continent

In the north of the Indian sub-continent the mighty Himalayan ranges are spread roughly in east-west direction. The ranges are emanating from Pamir knot in the northwest of India, one range Hindukush spread towards west-south west, another range Sulaiman passes southwest wards to Balochistan from Indus plain. In the northeast India, from Assam to Myanmar Khasi,

Jaintia and Garo Hill ranges are spread. South of the peninsular India is occupied by the Indian Ocean. West of the peninsula, the Arabian Sea and east of it the Bay of Bengal. In south India, Western Ghats are spread in a north-south direction along the west coast, while Eastern Ghats are spread along east coast.

The Indian sub-continent is spread over tropical, sub-tropical and temperate zones. The coastal area of peninsular India is influenced by maritime air mass. India has widely contrasting climatic conditions. The annual rainfall varies from less than 22 cm (217 mm Jaisalmer, Rajasthan) to more than 1100 cm (1142 cm Cherrapunji, Assam). Temperatures vary between − 45 °C (Dras, December 1910) in winter to more than 50 °C in summer (Alwar, east Rajasthan 1956).

Seasons

The dominant climate over India is monsoon circulation. In one half of the year wind blows from cooler maritime area to dry land (which is summer monsoon or southwest monsoon). The other half season cold dry wind from Asian anticyclone to warm Indian Ocean (which is winter monsoon or northeast monsoon). Thus there is a reversal of pressure and wind pattern between the halves. The change in pattern takes place gradually. The transition between these two predominant systems lies two intermediate seasons. The seasons in India are thus classified into four seasons. Which are given below.

1. Winter (December-February)

2. Pre-monsoon season (summer season) (March-May)

3. SW-monsoon season or Rainy season (June-September)

4. Post-monsoon season or cool season (October-November)

1. **Winter (cold) season: (December-January-February)**: During January NE-trades are prominent with Siberian High (msl) or Anticyclone centred at about latitude 45 °N, longitude 105 °E and India lies at its periphery. Continental cold air blows over central parts of Asia and a part of which comes over north India, while a major part of it is prevented by the mighty Himalayas. Intense Siberian high persists during December, January and February. Sub-tropical jet stream core shifts to south of the Himalayas and is located at about latitude 22 °N.

   During this period the STH in southern hemisphere is found between equator and latitude 10 °S. It extends from north Australia to eastern parts of Africa. A marked trough on sea level chart runs from Kerala to Gujarat and another Tannasserin coast to northern Myanmar and Assam. A weak ridge runs from NW-India to central Bihar (see Fig. 12.2).

This is the driest season for the country, except Kashmir, most part of India receives an insignificant percentage of rainfall. However this winter rainfall is important for crops in northern India. The highest rainfall of 20 cm is received in Kashmir and at some places in Nicobar islands. About 5 cm of rain received in most of the states in north India including Assam. In south India south of lat 10 °N and east coastal strip south of lat 15 °N and island groups receive rainfall. The chief synoptic features during this season are about 4-5 western disturbances pass over in north India.

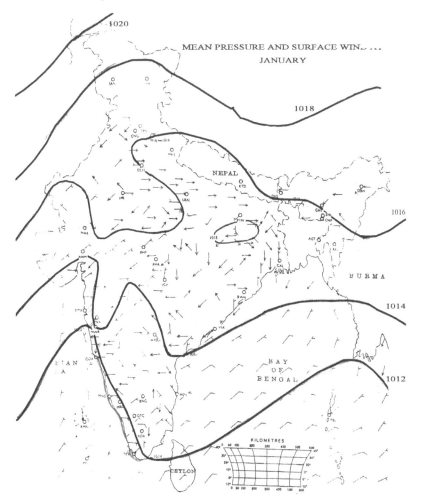

Fig. 12.2 Mean Pressure (hPa) and surface wind (January)

In this season cold waves occur in various parts of the country, particularly north of lat 20 °N. Cold dry air blows from Northwesterly/northerly direction and lowers the day and night temperatures significantly (below normal by 5 to 8 °C). Cold waves do not occur over Bay and Arabian sea Islands, Tamil nadu, Coastal Andhra Pradesh, coastal and south interior Karnataka and Kerala. Jammu and Kashmir is haunted by severe cold waves. Ladakh recorded maximum duration of severe cold wave for 30 days.

Fog and mist are very common in most part of the country affecting visibility particularly in early mornings. Conditions in the rear of the passage of the Western Disturbance in north India thick fog occurs. The most susceptible areas are northwest India, Uttar Pradesh, Bihar, North Bengal and Assam. Advection fog occurs in the coastal belts. The Brahmaputra valley, Sundarbans, the coastal belts and the hills and valleys of the Peninsula experience this type of fog/mist. The smoke near the cities and towns is being changed into smog and reduces the visibility to about 2 km in the inversion layer. Poor visibility very hazardous for landing of aircraft. Very low thunderstorm activity.

2. **Summer Season (March to May):** Pressure gradient becomes slack and begins to readjust as the sun moves north (see Fig. 12.3). A feeble High lies over Arabian sea and another over Bay of Bengal. Feeble lows develop over western parts of Rajasthan, Bihar and Myanmar. In this season the Siberian High weakens and shifts to north, while the sub-tropical High in the southern hemisphere shifts to lat 30 $^{\circ}$S. By May the summer heat low extends from north Africa to Asia.

Mainly convective type of rain occurs in various parts of the country. Assam (60–80 cm), Jammu and Kashmir (20 – 40 cm), Kerala and neighborhood (30-40 cm) receives considerable rain due to thunderstorm activity. Hail is also associated with thunderstorms, which decreases with the advance of the season. Cherrapunji receives more than 200 cm of rain. The second important area of rainfall activity extends from west Bengal to North Coastal Andhra Pradesh. Western parts of India, except extreme north and south are driest. Western Rajasthan and adjoining Saurashtra and Kutch are practically dry (about 0.5 cm). Norwesters, Kalbaisakhi and Andhi (local names of severe thunderstorm dealt separately) are confined to this season only. Tropical cyclones and heat waves are the other synoptic features. During hot weather period (March-June) surface temperatures over many parts of India abnormally shoot up practically over north India where temperatures will be above normal by 5 to 8 $^{\circ}$C or more. The incidence of severe heat waves occur mostly in Uttar Pradesh, but there is no region where recurs successively every year. Bay Islands Lakshadweep, Tamil Nadu, Kerala, Coastal and south interior Karnataka are not affected by heat waves. Rest of the country is prone to the incidence of heat waves. Maximum, period of 15 days heat wave lasted over Jammu and Kashmir.

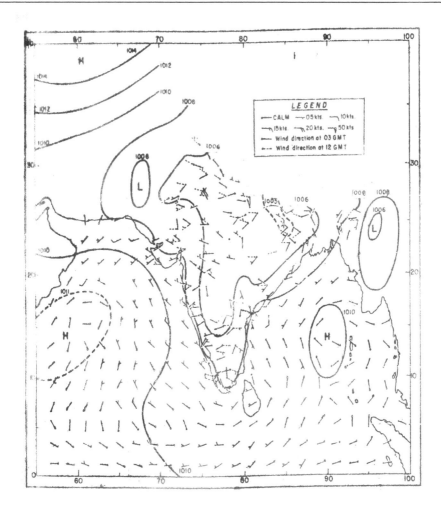

Fig. 12.3 Mean pressure (hPa) and surface winds (April)

Thunderstorm activity is maximum is this season, exceeding 16 thunderstorm days in May over Bengal/Assam and Kerala. Squalls generally accompany thunderstorms in this season. Violent thunder / squalls with heavy showers and occasional hail are associated with Norwesters over Bengal, Assam, Orissa and East Bihar. Duststorms (Andhies) are common in north India in this season.   sometimes visibility will be less than 2 km and on few occasions dust remains suspended in the atmosphere for a few days.

In coastal areas land and sea breeze affects are pronounced in this season.

3. **Monsoon season (June-September)**: In this reason low pressure area extends from north Africa to Siberia via Baluchistan and becomes

intense. Another trough of low extends from Rajasthan (heat low) to Head Bay of Bengal (which is called monsoon trough in India) which is of great importance to India (see Fig. 12.4). The sub-tropical high in southern hemisphere become more marked and centered around lat 30°S and long 60°E. A weak ridge extends over Arabian sea off west coast of India and another ridge in the Bay of Bengal off Tannasarin coast to Myanmar. By August Afro-Asian low weakens and by September it becomes an east-west trough.

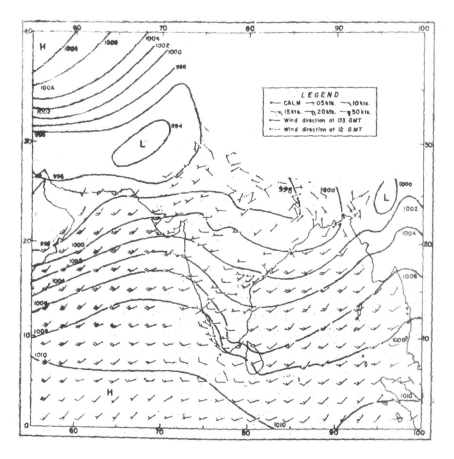

Fig. 12.4 Mean pressure (hPa) and surface winds (July)

Summer monsoon or southwest monsoon is the main rainy season for India. Except Kashmir and neighborhood and Kerala, the remaining area of the country receives more than 75% of the annual rainfall in this season. Orography and wind direction (wind ward side or lee ward side) plays an important role in distribution of seasonal rainfall as observed in Western ghats and Khasi-jaintia hills, where wind blows at right angles to the ghats/hills. In north Indian plains the minimum rainfall belt runs from northwest Rajasthan to the central parts of west Bengal, roughly along the axis of monsoon trough. Rainfall rapidly decreases along east coast south of latitude 9.5 °N to Kanyakumari. Pamban records only about 4 cm rain during monsoon. In general, July is the rainiest month. Seasons heaviest rainfall zones are Assam (200-400 cm), west coast and adjoining ghats north of latitude 10 °N (200-300 cm). Ladakh valley in the extreme north, western parts of Rajasthan and southernmost part of Tamilnadu gets very small amount (5-10 cm) of rainfall.

The chief synoptic features of southwest monsoon are lows, depressions and their movement along monsoon trough, mid-tropospheric circulations off Gujarat-Konkan coasts and off Andhra, Orissa coasts.

The sub-tropical westerly jet stream migrates to north of Himalayas and tropical easterly jet stream appears over peninsula.

During monsoon low cloud base comes down to 450 ft to 600 ft agl and during depressions the cloud base may be lower to 100 ft to 200 ft agl. Surface winds of the order 25-30kt are common and during depressions areas it may be 40-50kt. Low level jet also appears in low level westerlies. Visibility is good but deteriorates in heavy rain to a few hundred meters.

4. **Post monsoon season (October, November):** In this season pressure field becomes slack, by September end monsoon trough shifts to latitude 13°N (see Fig. 12.5). Low pressure belt extends from Africa to west Pacific through Arabian Sea and Bay of Bengal between equator and latitude 20°N, with centers over Africa, Bay of Bengal and western Pacific Ocean. The Asian high runs roughly along latitude 50°N and centered about longitude 9o°E. The Indian Ocean high centered near 30°S shifts to east and by November it is centered at longitude 80°E. The trough in the south Bay of Bengal persists till November end but disappears in December. During this season rainfall activity is mainly confined to south peninsula, the east coast, Assam and parts of Kashmir. It is the wettest season for Tamil Nadu (40-80 cm average rain) where NE- monsoon prevails. (In fact the period Oct-Dec is called Northeast Monsoon). Rest of south peninsula gets 5-10 cm of average rain. Western India including Jammu and Kashmir gets rainfall less than 5 cm.

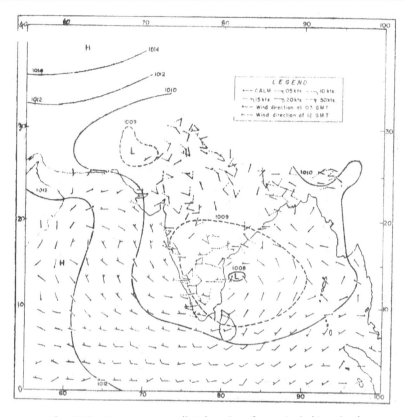

Fig. 12.5 Mean pressure (hPa) and surface winds (October)

The chief synoptic features are tropical cyclones in Bay of Bengal and Arabian Sea. In the north western disturbance start making their appearance and affecting the weather over northwest India. With gradual setting of cold weather, poor visibility due to fog / mist is experienced in the early mornings. Snowfall may be common in the Himalayas in association with Western disturbances. The sub-tropical jet stream core shifts to south and located between latitude 30-35°N at 200hPa level with the average speed 60-70 KT.

12.4   THE ROLE OF OCEAN IN CLIMATE FLUCTUATION

According to IPCC, the heat content of the entire global system has increased since 1950s. The oceans have absorbed 90% of the heat increase. The heat content of upper ocean (0 – 700 m slab) increased from 1961 to 2003 was about 16×10^{22} J. This heat caused average increase in sea temperature by about 0.1°C in the upper 700 m slab. The SST changes (1961-2003) by about 0.4°C at the surface and 0.1°C rise in temperature in the upper ocean depth 700 m. This heat amount would be equivalent to an almost 100°C change (rise) in atmospheric temperature. The oceans have much greater capacity to store heat than the atmosphere. The ocean is one thousand times more dense than air and the specific heat of ocean water is about four times that of air. The thermal adjustment of the atmosphere alone is about a month. The thermal adjustment time scales in the ocean are very much longer. The oceans have a much greater capacity to store heat than the atmosphere. This explains the buffering of atmospheric temperature by the oceans enormous heat capacity. (Buffer meaning protection from damaging impact).

The time scale for adjustment of the deep ocean is 1000 years, because the slow circulation of the abyssal ocean limits (obstructs) rate at which its heat can be brought to the surface. These long time scales buffer atmospheric temperature changes, consequently ocean would play a very important role in climate. Oceans reduce the amplitude of seasonal extremes of temperature and buffering atmospheric climate changes.

The world's oceans occupy about 71% of the surface area while the land occupies only about 29% area. The following properties about atmosphere and oceans must be noted.

The mass of the atmosphere $\square$ 5.26×10^{18} kg

Global mean surface temperature $\square$ $288\,^{\circ}$k

Global mean atmospheric pressure $\square$ 1.013×10^5 $P_a = 1013 \mathrm{hp_a}$

Global mean atmospheric density $\simeq 1.225$ kg/m^3

Global ocean surface area $3.61 \times 10^{14} \mathrm{m}^2$

Global ocean mean depth 3.7 km

Global ocean volume $3.2 \times 10^{17} \mathrm{m}^3$

Global ocean mean density 1.035×10^3 kg/m^3

Global ocean mass 1.3×10^{21} kg

Specific heat of water 4.18×10^3 J/kg/$^{\circ}$K

Latent heat of fusion (of water) 3.33×10^5 J/kg

Latent heat of evaporation (of water) 2.25×10^6 J/kg

Density of fresh water 0.999×10^3 kg/m^3

Viscosity of sea water 10^{-3} kg/m/s

Specific heat of ocean is about 4 times that of (atmospheric) air.

Process	Time Scales
Weather	hrs/days to weeks.
Land surface	hrs/days to months.
Ocean mixed layer	hrs/days to months.
Sea ice	weeks/months to years

Volcanoes	weeks/months to years
Vegetation	hrs/days to millions of years.
Thermocline	yrs/decades to centuries
Mountain glaciers	decades to centuries
Deep ocean processes	centuries (10^2) to 10^4 years
Ice sheets	10^2 to 10^5 years
Orbital forcing	10^3 to 10^5 years
Tectonics	10^6 to 10^9 years
Weathering	10^5 to 10^9 years
Solar constant	$10 - 10^2 - 10^3$ to 10^9 yrs 1-10^9 years
Natural CO_2 cycle	1k to 10k y (k = thousand)
Anthropogenic CO_2	decades to centuries ($10 - 10k$ years)

12.4.1   ROLE FACTORS (IMPORTANCE) OF OCEAN IN CLIMATE VARIABILITY

Ocean is a pivotal component of the climate system because heat, water, momentum, GHG and many other substances cross the sea surface. Ocean releases heat and water vapour to the atmosphere. More than 80% of water vapour of the atmosphere comes from the oceans. Ocean currents transport heat and salt around the globe. Buffering of atmospheric temperature changes is by oceans huge heat capacity. An Ocean current has immense effect on climate. Example, the confluence of two main ocean currents, the Warm Gulf Stream and the cold Labrador Current over north & west of New found land. The main fluxes or forces that determine ocean currents are – prevailing wind, earth's rotation and buoyancy (or salinity differences). Ocean currents generally called drifts.

Large – scale circulation is setup whereby the warm sea affects western coasts and cold seas affect eastern coasts.

Cold water coasts have low rate of evaporation hence feeds low water vapour content. This factor aids desertification, aridity.

Warm water coasts have high humidity which favours thunderstorm activity and high precipitation.

Important Ocean Currents: Cold water current Oyashio current (that affects Japan), Warm Kuroshio current, cold Labrador current, warm Gulf Stream and Cold Falkland current and Antarctic circumpolar (or west wind drift).

Almost all dynamical process in the ocean are driven by the sun and the atmospheric winds drive the ocean surface circulation up to a depth of top one kilometer, while winds and tidal mixing drive the deeper currents in the ocean.

The principal ocean's sources and sinks of energy are insolation, evaporation; I R emission from sea surface and sensible heating of the sea by warm and cold winds blowing over it.

Similarly, the heat sources and sinks of energy for atmosphere – insulation, uneven heat loss & gain by ocean causes winds, evaporation at sea surface and then condensation.

Regional climate is strongly influenced by thermal properties of the earth's surface and neighbouring ocean. In general ocean moderates extreme conditions. Similar influence exerted on global climate, because of ocean's large heat capacity and its capacity to hold substances in solution. The ocean surves as a reservoir of energy and carbon. As a consequence it (ocean) gives thermal inertia to the climate system mainly in exchanges with the atmospheric heat and CO_2.

Ocean circulations are driven by atmospheric wind stress, which transfers momentum to the ocean. Ocean circulations are also driven by transfer of heat and moisture (by way of density which influence buoyancy of water). The poleward transfer of heat achieved by the general circulation of earth – atmosphere system, of this 60% is contributed by the atmospheric circulation and 40% is contributed ocean circulation.

The spatial distribution of the ocean's heat content and SST (sea surface temperature) trend over the last three decades is not uniform. Climate change models predict non-uniform changes when projected over the next century (2100). It predict greatest warming in Arctic, little change in the sub-polar North Atlantic and Antarctic circumpolar current

12.5 PRESENT AND PAST RADIOACTIVITY OF THE EARTH

Radioactivity is the most important property of our earth. Modern Radioactivity of the earth is mainly associated with radioactive isotopes ^{238}U, ^{235}U, ^{232}Th and ^{40}K which decays as below

$$^{238}U \rightarrow p_b^{206} + 8\alpha, \qquad ^{235}U \rightarrow P_b^{207} + 7\alpha, \qquad ^{232}Th \rightarrow P_b^{207} + 6\alpha,$$

$$^{40}K + e^- \rightarrow \begin{matrix} ^{40}C + \beta \\ a \\ A_r^{40} \end{matrix}$$

During radioactive decay heat energy is liberated. Models of the radioactive earth generates (2.3 to 10) × 10^{20} calories of Radiogenic heat per year. Modern geothermal evidence indicate that by thermal conductivity the earth loses (1.9 ± 0.1) 10^{20} cal of heat annually, which is less than the amount of heat produced by the radioactive model of the earth.

4.5 billion years ago there was twice more ^{238}U on earth than now, which released twice more energy. In addition, some radioactive elements which were present initially are now extinct (some transuranium elements).

The radioactivity of earth is an important source of its internal heat and can cause melting of material in the earth's interior. Relatively high radioactivity of the young Earth contributed to the rise of its temperature and melting of material. The isotopes of the most long lived radio-active elements including the transuranium ones $\left(^{244}P_u, ^{247}C_m, z = 112 - 116 \right)$ existed for sometime in the early history of the earth.

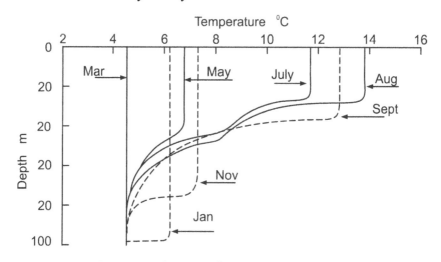

Fig. 12.6 A schematic of seasonal of thermocline

Temperature records of past few centuries show that the average temperature of the earth remained nearly constant at about 15 °C. This implies that the earth is radiating back (emitting) the same amount of energy as it is receiving from the sun. Hence the earth is in radiative balance with its surroundings.

The Paleo (fossil) records show that the state of radioactive balance as at present was not existing in earlier eras. In the geological history of the earth, there had been warm and cold epochs or ice-ages. This indicate that there were periods in which solar radiation received was above normal during warm epochs and below normal during cold epochs. These warm and cold epochs tell us the important of radioactive balance of the earth for life on earth. The most recent major glaciation, called the Great Ice Age, lasted for a period of between 6×10^5 and 10^6 years until about 2×10^4 years ago.

12.6   CLIMATIC CONTROLS

The factors that control the climate broadly are:

1. Latitude or sun's inclination
2. Altitude
3. Topography
4. Land and sea distribution and
5. Ocean currents

In addition to the above factors, climate is affected by semi – permanent pressure Highs and Lows.

Climate is thus the average condition of the atmosphere, ocean, land surfaces and the ecosystems that dwell in them.

Recently the importance of climate has increased due to rapid changes after industrialization. It is realized that climate change is not restricted to past-eons but it is occurring on time scales that affect human activities.

The surface area of oceans covered over the globe is about 71% and about 80% of the atmospheric water vapour is pumped from the ocean surface. Regional climate is highly influenced by thermal properties of the earth surface, particularly by neighbouring ocean. In general ocean moderates extreme conditions. Similar influence is exerted on global mean climate because of oceans large heat capacity and oceans capacity to hold substances in solution. The ocean serves as a reservoir of energy and carbon. As a result ocean gives thermal inertia to the climate system, mainly in exchanges with atmospheric heat and CO_2.

According to IPCC studies the heat content of entire global system has increased since 1950s. The oceans have absorbed 90% of the heat increased, because there is large heat storage capacity in sea water as compared to the atmosphere, land-ice or the continents.

Major oceans are interconnected by currents. Circulation system exchange mass between ocean basins. Ocean circulations are driven by atmospheric wind stress which transfers momentum to the oceans. Ocean circulations are also driven by transfer of heat and moisture (by way of density which influence buoyancy of sea water). The poleward transfer of heat achieved by the general circulation of earth-atmosphere system; of this

60% is contributed by the atmospheric circulation and 40% by ocean circulation.

The special distribution of ocean's heat content and SST trend over the past 50 years is not uniform. Climate change models predict non–uniform changes when projected over the next century (2100). The model predict greatest warming in Arctic, little change in the sub - polar North Atlantic and Antarctic circumpolar current.

Table 12.1 Global monthly surface temperatures for the period 1901-2000
(*Source:* NCDC website)

Month	Mean temperature °C		
	Land (area 29%)	Sea (area 71%)	Global (land-sea combined)
Jan	2.8	15.8	12.0
Feb	3.2	15.9	12.1
March	5.0	15.9	12.7
April	8.1	16.0	13.7
May	11.1	16.3	14.8
June	13.3	16.4	15.5
July	14.3	16.4	15.8
Aug	13.8	16.4	15.6
Sept	12.0	16.2	15.0
Oct	9.3	15.9	14.0
Nov	5.9	15.8	12.9
Dec	3.7	15.7	12.2
Annual	8.5	16.1	13.9
Range	11.5	0.7	3.8

The above data shows that global land area (29%) average temperature 8.5 °C, range 11.5 °C, while global ocean area (71%) average temperature 16.1 °C, range 0.7 °C. Combined global land and ocean average temperature 13.9 °C and range 3.8 °C.

The following points are important in respect of the role of ocean on climate variability

1. Heat, water, momentum, GHGs and many other substances cross the sea surface. This shows that oceans play a central component role in climate system.

2. Oceans/seas release heat and water vapour into the atmosphere.

3. Ocean currents transport heat and salt around the globe.

4. In global hydrological cycle, oceans maintain pole to equator temperature gradient and fresh water transport.

5. Buffering of atmospheric temperature changes by oceans enormous heat capacity.

6. In middle latitudes, atmospheric changes tend to precede oceanic changes through air-sea interactions.

7. In tropical latitudes, variation in SST and tropical air temperature and winds are in phase with one another that is they are interdependent.

8. In tropical oceans (particularly in Pacific) El Nino – Southern Oscillation phenomena are closely related.

9. According to Houghton and Woodwell (WMO-No. 735), the physical and chemical processes at sea surface release carbon fluxes about 100 Gt into atmosphere, while it absorb about 104 Gt. (1 Gt = 10^9 metric tons)

10. The global ocean carbon reservoir 36000 metric tons while the atmospheric carbon reservoir 735 mt.

11. The specific heat of ocean waters is about 4 times that of atmospheric air.

12. According to Lvovich MI, over periphery land area average annual precipitation depth 910 mm, evaporation 560 mm, over inland area precipitation depth 238 mm, evaporation 238 mm while on world ocean precipitation depth 1140 mm and evaporation 1251 mm.

13. Global land surface annual mean temperature 8.5 $^\circ$C, range 11.5 $^\circ$C while global sea surface annual mean temperature 16.1 $^\circ$C and range 0.7 $^\circ$C.

12.7 COMPONENTS OF CLIMATE SYSTEM

Climate variation and climate change: The changes that take place in climate over a period due to natural causes is called climate variation. However the changes that take place in climate over a period of time due to man-made activities is called anthropogenic climate change or simply climate change.

Climate variation includes ice-ages, long term warm climate enjoyed by dinosaurs and the event of prolonged drought over Sahel region in Africa. Climate change includes ozone hole over Antarctica and present global warming.

Climate models are mathematical representations of climate system by equation containing elements like temperature, wind, ocean currents, precipitation, evaporation, snow melt and other climate variables. Climate modeling includes interaction of fields or interlocking systems of atmosphere, ocean, land surface, sea-ice, land-ice, part of biosphere, hydrosphere. Earth system models also include physical, chemical and biological aspects over the same period.

Global warming is associated with changes in climate which are attributed to increased GHGs pumped into atmosphere by human activities.

Environmental changes include air pollution, water pollution, deforestation, soil erosion, endangerment of species or ecosystem (by loss or pollution of habitat). Climate prediction models largely depend on statistical theories (time series, probability occurrence of extreme natural events).

In order to arrive at a decisive prediction of future climate the study and clear understanding of climate components is essential. Regional weather, ELNinos, North Atlantic oscillation, Asian monsoon variations, North American monsoon variations, droughts, floods, ocean circulation processes, ice-ages, must be understood. In addition to these environmental chemistry, biosphere evolution and linkages are required. The fig.12.7 shows schematic climate system components.

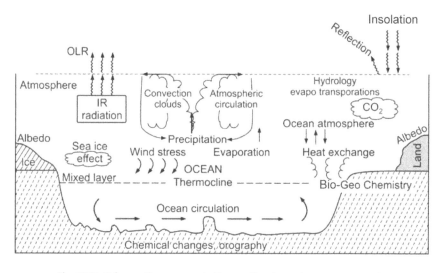

Fig. 12.7 Schematic representation – climate system components

The Atmosphere, the ocean, Land surface, cryosphere, the Biosphere, and Lithosphere.

Cryosphere consists of land-ice (ice shelves & glacier) snow and sea-ice.

The Biosphere is all living beings & vegetation on the earth & oceans.

The Lithosphere-solid earth (volcanoes, earthquakes).

12.8 CLIMATE SYSTEM OR EARTH SYSTEM

The global interlocking system of atmosphere, ocean, land surfaces, sea ice and land ice and parts of biosphere and also solid earth. Thus earth system emphasizes the simultaneous study of all parts in this system, that contribute due to chemical reaction and biological contribution. Thus in a nut shell earth system models include physical, chemical, biological aspects.

Observed Impact of Human activities during last 150 years on climate change and variation.

Overall, man has achieved to add green house gases and reduced the sinks. Of the greenhouse gases increased by human activity, 70% account to energy sector.

12.8.1 EXTINCTION OF DINOSAURS ON THE EARTH

During Triassic period the great reptiles Tyrannosaurs and Dinosaurs rose and ruled the world in Triassic period. The giant reptiles became extinct by the end of cretaceous period due to extreme heating of the environment.

It was originally thought that an asteroid that smashed the earth created a giant crater chicxulub, off the Caribbean coast of Mexico was responsible for the extinction of Dinosaurs on earth. The impact of asteroid created earthquakes of magnitude scale 11 and waves of volcanic eruptions. The vast lava flows known as the Deccan Traps in India. The expulsion of huge lava over a million years left the Deccan trap 1200 meters thick in places today. A very large volume of hot lava which cover an area of France to a depth of several hundred meters. The flow of hot lava for a long period resulted in hampering photosynthesis in plants, and coincided with mass extinction event lend to extinction of mass of life together with the extinction of giant reptiles dinosaurs on the earth.

12.8.2 CLIMATE CHANGE

Climate change is restricted to remote or recent past, but it is continuing on all time scales that affect human activities. All atmospheric, geological, environmental changes that are observed at present on various scales all being attributed to global warming. It is echoed by IPCC. World over governments 48% main problem is how to combat climate change and next 41% problem is how to combat/prevent human terrorism. All over world the scientific talk is how to mitigate problem that arose from climate change.

Climate models are mathematical representation of the climate system which depends on the climatic elements and its changes. Earlier this was handled by climatology. At present it is dealt by fluid dynamics using equations of dynamic meteorology which involved physics, mathematics, chemistry, engineering or biological environmental sciences. Thus the climate system became Earth system, which is coupled with atmosphere, ocean, land surfaces, ice both on land & sea, biosphere. In other words it involves atmosphere, hydrosphere, (cryosphere, lithosphere and biosphere). In addition to these the introduction of man-made fluxes called anthropogenic effect on climate change.

Atmosphere, ocean, land surface and ecosystem are all interlinked and the activities of man that contribute to global warming manifesting in climate change. Oceans occupy 71 percent of the earth's surface and 80% of the atmospheric water vapour comes from the ocean's surface. More than 90% of the earth's CO_2 is dissolved in sea waters. The total mass of CO_2 in atmosphere is less than one-sixth mass of CO_2 in Oceans.

Ocean is a central component of the earth's climate system. Heat, water, moment, GHGs, all cross the sea surface. In this process water vapour, heat is transported from Ocean to the atmosphere. Ocean currents transport heat and salt around the globe. Oceans huge heat capacity buffers atmospheric temperature changes. Climate and weather had been changing from the origin of the earth's atmosphere and life on earth. And no doubt it will continue to change in future. There is no accepted theory of climate that one-one relates to earth's system.

According to recent UN report, at present mankind is persisting with thoughtless and extravagant consumption of earth's natural resources and damaging resources in an unprecedented manner. As a result about 30% amphifians, 20% of mammals and about 12% of bird species under the threat of extinction. Human activities are pumping GHGs into atmosphere and 70% of this accounts to energy sector.

IPCC noted the global warming would melt polar ice, Himalayan glaciers retreat, sea level rise, drop in production of wheat, rice, maize and more than one billion people may face fresh-water shortage by 2020. In relation to last two and half centuries climate change, there will be no effect or negligible effect on western boundary currents, little effect on major wind systems like trade winds, zonal winds, no effect on Hadley, Ferrel and Polar cells of wind circulation and little or no effect on Gulf stream and Ocean Gyres.

Regional climate is highly influenced by thermal properties of the earth's surface particularly neighboring ocean. Ocean moderates extreme conditions by virtue of its large heat capacity as compared to atmosphere, land-ice and continents.

Ocean water both in time and space always acted by same forces gravity (momentum), Coriolis force, pressure, temperature, wind and salinity, ocean flow is mainly guided by the sun, moon and tectonic process.

12.8.3 DISASTER (MEASUREMENT) MANAGEMENT SYSTEM-INSTRUMENTS

1. **Plume RAE:** It provides first responders the information they require along with decision support so that they can answer critical questions and take immediate, effective action.

    It provides, where toxic plume heading, and its concentration. It indicates the area to be evacuated, and the place convenient for shelter. It shows hazardous road that is to be closed. It provides the sequence of actions with available limited resources.

2. **Chemical Warfare Agent Detector:** It is based open loop Ion mobility spectrometry technology. Designed to detect chemical warfare agents and toxic industrial chemicals. It can be used as stand-alone portable monitor or integrated into an Area RAE network. It can be used as a personal detector, a monitor for surveying contaminated areas or a fixed installation detector and provides continuous operation.

3. **Weather Monitoring station:** It is a self contained digital meteorological monitoring system. it provides instant access to weather information. It collects data-(i) Wind speed & direction, (ii) Air Temperature, (iii) Relative Humidity, (iv) Solar radiation /Sunshine duration, (v) Barometric pressure, (vi) precipitations, (vii) Data logger with 200 days storage capacity and provision for pass-word protection.

4. **Ionising Radiation meter (gamma rays/Neutrons):** It is used for (i) Rapid detection of gamma ray sources (even at low levels), (ii) Personal warming device and to locate sources, (iii) To provide rapid detection of both gamma rays and neutron sources (even in potentially flammable environment).

5. **Radiation Measurement:** Dosimeter for gamma rays. Self reading electronic radiation dosimeter offers personal dose and dose-rate monitoring for gamma rays.

6. **CEMS (continuous Emission Monitoring System):** (i) CEMS employs a micro computerized, multi component, Ratio-NOIR gas analyzer for measurements of NO_x, SO_2, CO, CO_2 and O_2 (ii) Measures three (NO_x, SO_2, O_2), four (NO_x, SO_2, CO, O_2) or five (NO_x, SO_2, CO, CO_2, O_2) components simultaneously. Teaching institutions, NGOs can make use Air pollution monitoring instruments:

    (i)   Sorbent tubes, sample bags, impinges, coated filters.

    (ii) For particulate matter-cyclone, 10M samples parallel particulate impacts, real time dust monitor-Haz-Dust I, Haz-Dust IV

(iii)   For Bio-aerosol monitoring, sampling pumps-Quick Take 15 and Quick Take 30, Bio-stage, Bio-sample, spirometer, Audiometer

(iv)   Gas Monitors-Multi gas (4 gas), multi gas with VOC (5 gas) continuous gas monitor, Area gas Monitor

(v)   For water and waste water analysis TOC (total organic carbon) Analyser on line TOC analyzer, BOD equipment

(vi)   Automatic kjeldahl apparatus-used for determination of total nitrogen, ammonium nitrogen, proteic nitrogen, nitric nitrogen, phenols, Hydrocyanic acid, cyanides alcohol content etc. in water and waste water, soil and chemicals.

(vii)   Toxic monitor (to measure the toxicity is effluents)

(viii)   Trimeter (to measure, Turbidity, chlorine, colour)

(ix)   Rapid total coliform and E-coli system

(x)   Continuous Emission monitoring systems

(xi)   Continuous stack particle emissions monitor

(xii)   Land fill gas monitors

(xiii)   Aerosol sampling impactor (MOUDI)

(xiv)   Continuous Ambient particulate monitor

(xv)   Real-time Aerosol monitoring

(xvi)   Aerosol particulate profiler

(xvii)   Air quality monitoring system

(xviii)   Urban, road side pollution monitor

(xix)   Micro environmental pollution monitoring

(xx)   Sound level meters

12.8.4 ENVIRON AUTOMATIC WEATHER STATION

It is compact, portable and easy to use. The system is solar powered and can be sited (established) almost everywhere. The assembly can be mounted on the mast switch it on, set wind direction north.

The weather station is applicable in-

(i) Drought monitoring, (ii) Disaster Management, (iii) Climate studies, (iv) Environmental monitoring, (v) Flood monitoring (vi) Industrial monitoring, (vii) Agricultural monitoring, (viii) Early warming, (ix) Scientific research.

Elements that it can measure are – (i) Air temperature, (ii) Atmospheric pressure, (iii) Wind (speed & direction), (iv) Relative humidity, (v) soil moisture, (vi) Solar radiation, (vii) leaf witness and (viii) river gauge.

Bubbler: OTT compact Bubbler sensor (a) Range 0-50 ft / 0-100ft (b) resolution 0.003 ft (c) Accuracy ± 0.02 ft.

Sensors and specification

(i) Air temperature. (a) Range - 40 °C to + 123 °C (b) Accuracy ± 0.3 °C (c) Resolution 0.01 °C

(ii) Atmospheric pressure (a) Range 150 to 1100 hPa (b) Accuracy ± 1.5% (c) Resolution 1 hPa

(iii) Wind speed (a) Rang 0-200 kmph (b) Accuracy ± 1.0 kmph (c) Resolution 0.1 kmph

(iv) Wind direction (a) Range 0-360° (b) Accuracy ± 2° (c) Resolution 1°

(v) Relative humidity (a) Range 0-100 % (b) Accuracy ± 2% (c) Resolution 0.1%

(vi) Soil moisture (a) Measuring range 0 to 2.39 centibar (b) output 0 to 3 volts linear (c) Depth of observation 10 to 100 cm (d) soil type: To be used with all types of soils

(vii) Solar radiation (a) Global in coming radiation with cosine correction (b) range 0 to –1750 W/m^2 (c) Resolution 1W/m^2 (d) Accuray ± 5% (e) Cosine corrected accuracy ± 3%

(viii) Leaf witness (a) operating temperature –10 to + 60 °C (b) Accuracy ± 0.5 hr or better (c) unit dry/witness duration in hours (d) Measurement 10 m seconds (e) power 2 SVDC @ 2mA to 5 SVDC @ 7mA

Summary of IPCC studies 1996, 2001 and 2007

1. In higher latitudes, surface warming is greater than global average in winter and smaller in summer, comparatively smaller ocean warming on global scale.

2. IPCC 2007 estimates. As compared to 1980-1999, by 2090-2099 temperature rise (depending on emissions) vary 1.8 to 3.4 °C and 2.8 to 4.0 °C likely.

3. Earth surface and lower atmospheric (troposphere) warms while stratosphere cools.

4. Precipitation and evaporation would increase depending on warming. For double CO_2 emission the precipitation average increase is 3 to 15%. There is shift in tropical rain bands from one model to another.

5. Increase in soil moisture in higher latitude winter but decrease in northern latitude summer.

6. Overall snow cover and sea ice decrease. There would be decrease in mountain glacier mass and area.

7. As compared to 1980-1999, sea level rise due to ice melt by 2090-2099 would be 0.23 to 0.51 meters and 0.18 to 0.38m and eventual rise because of the thermal expansion 1 to 4 meters.

8. Extreme weather events, heat waves, tropical cyclones and flooding would rise. Rise in number of hot days, maximum temperature days.

9. Rise in minimum temperature over land and decrease in frost days.

10. Rise in heavy rain/precipitation events.

11. Rise in tropical cyclone intensity.

12. According NASA studies, rise in sea level by global warming would be 65 cm in next 80 years (2100 AD)

13. Green land ice sheet melt would rise sea level by 7 meters.

14. If global warming melts all ice at the poles (north & south) and mountain glaciers would raise sea level to 65.8 m (216 feet).

15. The eventual value of GHG range from category radiation I forcing 2.5-3.0 W/m^2 to category IV radiation forcing 6.0-7.5 w/m^2

16. Climate change is the net complex interaction of solar radiation, atmosphere, hydrosphere, Biosphere cryosphere, soils (lithosphere), vegetation, urbanization, nuclear explosions, use of mineral oils/natural gas, transportation, ecosystem and activity of human being.

According to UN Environment programme (UNEP), Global Environment outlook-4, Report (2007), "Our living way beyond our means". UN report says that at present humankind persists with thoughtless and extravagant consumption of natural resources, planet Earth is hurting towards unprecedented resources crunch.

At the present consumption rate it requires 21.9 hectares per person, while earth's capacity is on average is only 15.7 hectares per person. As a result of this 45000 sq miles of forest area being lost across the world each year. 60% of the world's major rivers have been dammed or diverted and fish population declined by 50% in the last 20 years. About 30% of amphibians, 23% of mammals and 12% of birds are under threat of extinction due to human activity, while, 10% of the worlds large river's is running dry every year before reaching the natural end (into sea). The US has increased its energy consumption by 20% over the last two decades.

Climate change lead to decline of underwater kelp forests by impacting their microbes. Ocean warming can change microbes on the kelp surface which cause disease. This is potentially put fisheries at risk. Blistering and bleaching of the kelp's surface effect their ability to photosynthesis and this warming threatens underwater forests.

Global Environment Outlook – 6 (UN meet at Nairobi 2018)

According to GEO-6, the world will exhaust all its fossil fuels (energy related carbon budget) resources in less than 20 years, which will keep the global temperature rise to well below 2 °C. It will take even less time to exhaust the carbon budget to keep the global temperature rise below 1.5 °C.

There will be demand for renewable resources, particularly solar energy and to better (improve) solar batteries. Indian Universities should develop to technologies to substitute the fossil fuel energies. Developing countries like Costa Rica, have pledged carbon neutrality by 2021. It is important, investigating in education for sustainable development technology like compact cities, public transport, renewable energy, drastic reduction in plastic use, reduction of food waste. All these contribute to reduce global warming.

Climate Vulnerability Index (CVI)

Poland conference on climate change found that Assam, Arunachal Pradesh and Uttarakhand vulnerable to climate change. The risks are: climate, socioeconomic. An index was developed based on the vagaries of climate change. Vulnerability is a measure of the inherent risks that a district faces viz- (i) by virtue of its geography and (ii) Socio-economic situation.

Eight parameters of vulnerability score-

1. Percentages of area in districts under forest
2. Yield variability of food grain
3. Population density
4. Male literacy rate
5. Infant mortality rate
6. Percentage of population below poverty line
7. Average man-days under MGNREGA (Mahatma Gandhi Rural Employment Guarantee Act)
8. Area under slope more than 30%

Scale: 0-1 (1 is the highest vulnerability)

1. Assam score 0.72
2. Mizoram 0.71
3. Jammu and Kashmir 0.61
4. Manipur 0.6
5. Meghalaya
6. West Bengal
7. Naga land
8. Himachal
9. Tripura

10. Arunachal

11. Uttarakhand

12. Sikkim 0.42

Impact of Climate Change on Agriculture

Increase in GHG emissions impact on agriculture and water resources. There appears to be both positive and negative effects due to increase of CO_2 in the atmosphere. With increased pressure on the limited cultivated area, a faster depletion of nutrients and groundwater appear to be a definite positivity (CRIDA 2007). Extreme events like daily rainfall amount and intensity of rainfall cause more erosive rains. The effects of climate change can be mitigated gated through proper rain water management, crop planning, alternate crops, cropping systems, integrated farming systems, enhancement of water productivity, precision farming and agro-advisories.

The specific interventions are:

1. Agronomic manipulation

2. Evolving crop varieties to withstand warming

3. Varieties to face possible newer pest-disease complexes

4. Crop and varieties that fit into new cropping systems and seasons.

5. With skewed distribution of rainfall, the specifications of water harvesting, structures in wet semiarid and humid regions.

According to CRIDA (2009), under national action plan for climate change under "Dry land Agriculture" the following are included:

1. Development of drought, temperature and pest resistant crop varieties.

2. Improved methods of soil and water conservation that help crops to adopt extreme weather condition.

3. Stakeholders consultation and awareness generation on climate change and sharing and dissemination of information.

4. Devise financial support mechanisms to enable formers invest in risk mitigation.

Impacts of increased CO_2 in atmosphere on Vegetation (crop yield)

In plants, photosynthesis process absorbs CO_2 converts into food, fiber and other forms of biomass in the presence of solar radiation. Plants use sunlight to convert CO_2 and water into carbohydrate, that supply almost all animal and human needs for food, oxygen but some water released as byproduct. Increased CO_2 levels in atmosphere in the presence of superior efficiency of photosynthesis results in reduced loss of water per unit area of leaf. Elevated CO_2 in fields increase in plant growth, grain yield and canopy photosynthesis simultaneously reducing evaporation and negative impact of temperature stress. In totality, food crops with C_4 metabolism, including corn, sorghum, mullet and sugarcane, show yield increases ranging 10 to 55%. In case of

Tuber and root crops (including potatoes sweet potatoes) shows increase in output ranging 18-75%. Legumes including peas, beans and soyabeans, show yield increases 28 to 46%.

In general rise in mean seasonal temperature by 2-4°C reduces the yield of annual crops of determinate growth habit, such as wheat, grown in well-watered conditions. This fall in yield is due to shorter crop durations at warmer temperatures. In summary impact of warmer mean seasonal temperature by 2-4°C and elevated (increases) CO_2 concentrations to about 700 ppm at the end of 21st century on the current yield of annual crops grown in environments with adequate water may not be great. [Ref: Advances in plant-atmospheric interactions-CRIDS-2009]

12.9 ROLE FACTORS (IMPORTANCE) OF OCEAN IN CLIMATE VARIABILITY

Ocean is a pivotal component of the climate system because heat, water, momentum, GHG and many other substances cross the sea surface. Ocean releases heat and water vapour to the atmosphere. More than 80% of water vapour of the atmosphere comes from the oceans. Ocean currents transport heat and salt around the globe. Buffering of atmospheric temperature changes is by oceans huge heat capacity. An Ocean current has immense effect on climate. Example, the confluence of two main ocean currents, the Warm Gulf Stream and the cold Labrador Current over north & west of New-found land. The main fluxes or forces that determine ocean currents are – prevailing wind, earth's rotation and buoyancy (or salinity differences). Ocean currents generally called drifts.

Large – scale circulation is setup whereby the warm sea affects western coasts and cold seas affect eastern coasts.

Cold water coasts have low rate of evaporation hence feeds low water vapour content. This factor aids desertification, aridity.

Warm water coasts have high humidity which favours thunderstorm activity and high precipitation.

Important Ocean Currents: Cold water current Oyashio current (that affects Japan), Warm Kuroshio current, cold Labrador current, warm Gulf Stream and Cold Falkland current and Antarctic circumpolar (or west wind drift).

Almost all dynamical process in the ocean are driven by the sun and the atmospheric winds drive the ocean surface circulation up to a depth of top one kilometer, while winds and tidal mixing drive the deeper currents in the ocean.

The principal ocean's sources and sinks of energy are insolation, evaporation; I R emission from sea surface and sensible heating of the sea by warm and cold winds blowing over it.

Similarly, the heat sources and sinks of energy for atmosphere – insolation, uneven heat loss & gain by ocean causes winds, evaporation at sea surface and then condensation.

Regional climate is strongly influenced by thermal properties of the earth's surface and neighbouring ocean. In general ocean moderates extreme conditions. Similar influence exerted on global climate, because of ocean's large heat capacity and its capacity to hold substances in solution. The ocean surves as a reservoir of energy and carbon. As a consequence it (ocean) gives thermal inertia to the climate system mainly in exchanges with the atmospheric heat and CO_2.

Ocean circulations are driven by atmospheric wind stress, which transfers momentum to the ocean. Ocean circulations are also driven by transfer of heat and moisture (by way of density which influence buoyancy of water). The poleward transfer of heat achieved by the general circulation of earth – atmosphere system, of this 60% is contributed by the atmospheric circulation and 40% is contributed ocean circulation.

The spatial distribution of the ocean's heat content and SST (sea surface temperature) trend over the last three decades is not uniform. Climate change models predict non-uniform changes when projected over the next century (2100). It predict greatest warming in Arctic, little change in the sub-polar North Atlantic and Antarctic circumpolar current

12.10 THE EARTH-ATMOSPHERIC SYSTEM

Annual Carbon Fluxes

According WMO No. 735 (1990) the atmosphere of the living planet earth the annual carbon fluxes were in 20th Century end.

  (i)   Photosynthesis on land removes about 100 Gt of carbon (in the form of CO_2) from the atmosphere annually.

  (ii) Plant and soil respiration each returns (C) about 50 Gt (total 100 Gt).

  (iii) Fossil fuels burning (5 Gt) and deforestation (2 Gt) release (C) into atmosphere 7 Gt.

  (iv) Physico-chemical process at sea surface release (C) into atmosphere about 100 Gt.

  (v)   Physicochemical process at sea absorb (C) about 104 Gt.

Net carbon added to the atmosphere 3 Gt annually $(-100 + 100 + 7 + 100 - 104 = 3)$ in the form of CO_2 $1 Gt = 10^{12} kg$.

Evolution of Atmosphere and Life on Earth

The primordial substance of the solar system consisted mainly water and carbon dioxide (CO_2).

The mass of the earth is about 6×10^{24} kg,

The mass of the atmosphere is about 5.6×10^{18} kg.

The mass of ocean water is about 1.4×10^{21} kg.

That is the mass of ocean is more than 250 times the mass of atmosphere.

The mass of the hydrosphere is about 0.023% of the mass of the earth, while the mass of the atmosphere is about 0.00009% of the mass of the earth. The mass of the O_2 (oxygen) in the atmosphere is about 10^{18} kg, that is one-fifth of the mass of atmosphere.

The plant–kingdom over the globe produces O_2 about 3×10^6 kg and it provides about 10^{17} kg of biomass annually. An average size of tree supplies about 3500 kg of O_2 per year which is sufficient for three people.

The present atmosphere of the earth resulted from the evolution of life on earth. All living organisms are primarily composed of C, O_2, H and N. These are also the basic chemical elements of water and air shells of the earth. The biosphere constitutes about 1440×10^{15} tons of water, 233×10^{10} tons of CO_2 and 11.8×10^{14} tons of O_2.

In summary, at the beginning 4500 million years ago, the surface of the earth was depleted in free oxygen, and UV- solar radiation could have penetrated into the top ocean surface to a depth of about 10 m. However the illuminated top layers of sea water were favourable for development of living matter. This living matter gave rise to uni-cellular photosynthesizing Blue- green algae or their ancestors might have evolved in the zone depth (below 10 m) in ocean where they were protected from the lethal UV- solar radiation, but visible radiation reached into the sea. These ancient life developed at constant density zone, which was responsible for water decomposition to form free oxygen. As a result the biosphere became oxidizing in nature. In this state all CO_2 from the atmosphere used by photosynthesis and carbonatization. Free oxygen released into atmosphere. The free oxygen in the atmosphere formed an ozone layer, which absorbed the UV-solar radiation. This helped algae spread over the land & sea. Subsequently animals developed as superstructure relative to the photosynthesis and then developed species of animals which consumed plants for food. These species breathed in oxygen and breathed out CO_2. With oxygen fixation in the pigments blood and thus animals spread over the land side by side plants. It must be noted, initially animals have long evolved in marine water in the zone of sea saturated with oxygen. Thus hydrosphere proved to be the home for ancestral plants and animals. The top layer (surficial layer) of the ocean water for long period was the principal zone of intense cycle of C, H, O_2 and other biophile elements. Life will cease when the energy resources of the planet earth are exhausted. It is now beyond doubt the dynamic ocean had played a tremendous role in moderating the climate, that is suitable for life on earth. It is now essential to study ocean dynamics along with studies of Atmosphere, Environment, Biosphere and other basic sciences to understand the climate change in the near future and in the long run.

In order to understand the intricacies of climate change it is required to know the past history of the earth's climate. For billions of years earth has supported life which implies climate remained within narrow limits with all variations in the climate from warm epochs to cold epochs of ice ages.

During past 4.5 billion years the insolation increased by about 30%. Climate models suggest that the input rate of CO_2 from volcanic activity into atmosphere is balanced by the rate of removal by chemical weathering (sink).

A continuous record of atmospheric conditions (climate) over Greenland and Antarctica dating back to about 4 – lakh years is traced with the help of several ice-cores through the Greenland ice sheet and three through the Antarctic ice sheets. Annual layers in the ice core (count) gives the age. The fallout of volcanic ash also provide common markers in ice cores. Oxygen isotope ratios of the ice give temperature over the ocean upwind of the glaciers. Bubbles in the ice give atmospheric CO_2 and methane concentration. Pollen, chemical composition and lava particles provide information of volcanic eruptions, wind speed and direction. Thickness of annual layers gives rate of snow accumulation. Isotopes of some elements give solar activity and cosmic ray activity. Deep-sea sediment cores in the North Atlantic (made by ocean drilling program) give information about:

  (i) Sea surface temperature and salinity above the core.

  (ii) Ice volume in glaciers,

  (iii) Production of ice bergs.

Findings

  1. The oxygen-isotope record in the ice cores indicate that there were abrupt temperature changes during the past one lakh years (10^5 years).

      During the last ice age (on many occasions) Greenland temperature warmed over the periods 1-100 years. Subsequently, followed by gradual cooling over long periods.

      About 11500 years ago Greenland temperature warmed by about 8 °C in 40 years in three steps of 5 years. Such abrupt warming is called Dansgaard or Oeschger event.

      The ice core studies show that much of northern hemisphere warmed and cooled in phase (with temperatures).

  2. During the past 8000 years, the climate was roughly constant.

      During all of recorded history, our perception of climate change is speculative (as it was during warm and stable climate).

  3. Henrich events: North Atlantic sediment studies show coarse material was deposited on the bottom and mid–ocean (Atlantic). Only ice-

bergs carry such material out to sea. This shows times during which large numbers of ice-bergs were moved into the north Atlantic. These are called Henrich events.

4. The correlation of Greenland temperature with ice-berg production is related to the deep circulation.

12.11 CLIMATE OVER EARTH HISTORY

The primordial atmosphere was similar in composition to volcanic and meteorites gas. The largest composition of volcanic gas is water vapour. The second component CO_2 which is rapidly consumed by green plants by way photosynthesis. The chemical evolution of atmosphere and ocean involved living organisms with the role of photosynthesis of green plants.

The origin of our planet earth is still obscure. It is believed that the earth matter before the formation was in the form of plasma, a fourth state matter. Plasma is an electrically neutral mixture of negatively charged electrons and positively charged nuclei of atom. Plasma can be reliably considered as a mixture of ideal gases.

For billions of years earth has supported life indicating climate remained within narrow limits with all variations in the climate from warm epochs to cold epochs of ice ages. During past 4.5 billion years, the insulation increased by about 30%. Climatic models suggest that the input rate of CO_2 from volcanic actively into the atmosphere is balanced by the rate of removed by chemical weathering (sink). Carbon from the earth's interior pumped into the atmosphere (fossil fuel burning), chemical weathering of continental rocks which finally washed into the Ocean and deposited as sediments.

According to prof. George Voitkevich, the mobile gas components of the atmosphere and water experience the fastest cycle, while a much slower cycle in continental matter. For atmosphere CO_2, the time required for a full cycle of matter is 7 year, 4000 years through photosynthesis of atmospheric oxygen, nearly one million years for ocean water by way of evaporations and 80-100 million for continental matter by way of weathering and removal from land surface.

According WHO, the life time of GHGs in the atmosphere vary from hour/weeks to more than more than 100 years. Tropospheric Ozone has a life periods a few hours/days, CFCs (chlorofluorocarbons) have about 75 to 110 years, N_2O (nitrous oxide 150 years, CH_4 (methane) 7-10 years CO_2 (Carbon-dioxide) 50-200 years.

12.11.1 ANNUAL CARBON FLUXES IN THE EARTH ATMOSPHERE SYSTEM

(Gt = gigatonne = one billion metric tons)

(i)   Photosynthesis on land removes about 100 Gt of C from the atmosphere annually (in the form of CO_2)

(ii) Plant and soil respiration each returns about 50 Gt (total 100 Gt)

(iii) Fossil fuel burning (5 Gt) and deforestation (2 Gt) release CO_2 into atmosphere 7 Gt

(iv) Physicochemical process at sea surface release CO_2 into atmosphere about 100 Gt

(v)   Physicochemical process at sea absorb about 104 Gt CO_2

(vi) Net 3 Gt [– 100 + 100 + 7 + 100 – 104] CO_2 is added to the atmosphere annually

(vii) Global warming is mostly is attributed to use of naturally available fossil fuels coke, petrol, diesel, wood etc. As long as we use these fuels, wc like it or not global warning and climate change will continue. Thus the adverse climate effect can be mitigated by curtailing the use of these fuels together with pollution control.

The major carbon reservoirs of world given below (10^9 metric tons)

World vegetation   560

World soils   1500

Atmosphere   736

Oceans   36000

Fossil fuel reserves 5000 to 10000

Atmospheric carbon can be removed one to two billion (10^9) tones year by planting trees (a forestation) each of 10^8 hectares of forest (a sink)

WMO (World Meteorological Organization)

Main functions-WMO facilitate worldwide cooperation in establishment of network of meteorological observatories and promote development of centers for providing meteorological services to promote rapid exchange of weather information and the standardization of meteorological observation and their publication to help forward/further the application of meteorology for human activates, encourage research and training in meteorology.

12.12   THE EVOLUTION OF THE EARTH'S ATMOSPHERE

The atmosphere, wreathing the world with white clouds has gradually evolved with the blue seas and lakes, the green vegetation and brown soils of the land. The surface of the earth is teaming with life, which would be impossible without the atmosphere, and the atmosphere would not be the same without life on Earth.

Air is a mechanical mixture of gases. The various gases, solid and liquid particles in the air, that envelope the earth are bound to it by the gravitational attraction of the earth. This envelope of gases, solid and liquid particles is called atmosphere. The atmosphere extends above the surface of the earth to outer space. However the height or the atmosphere is taken as 1000 km. More than 50% of the mass of atmosphere lies below an altitude of 5.5 km and 98% of the mass lies below an altitude of 30 km. Above 700 km of altitude a very thin atmosphere of Hydrogen and Helium atoms exist. The lowest one kilometer is called planetary boundary layer, contains 10% of the mass of atmosphere. All of biological, human activities are confined to this planetary boundary layer baring aircraft flights. The mass of the atmosphere about 5.6×10^{18} kg, while the mass of the earth is about 6×10^{20} kg, and the mass of oceans (1.4×10^{21} kg) is more than 250 times the mass of the atmosphere. The density of the dry air of the surface is 1.225 kg/m^3 and at an altitude of 5.5 km it drops to 0.66 kg/m^3 and at 30 km altitude it is about 0.013 kg/m^3. The gases are well mixed in the troposphere (the lowest 8-12km) and contain nitrogen 78%, oxygen 21%, argon 0.94% CO_2 (0.03%) which is essential plant growth. Other trace gases present include methane, hydrogen, ozone, helium and eon. The mass of O_2 in the atmosphere is 10^{18} kg, that is one fifth of the mass of the atmosphere. The plant kingdom over the globe produces oxygen about 3×10^6 kg per second which is consumed by the living beings over the world. About one kg of hydrogen/per second escapes to outer space from the top of the atmosphere and 12 kg/second cosmic dust falls from the space on the earth. The volume of water vapour varies from place to place but it is about 14×10^{12} m^3 in the whole atmosphere.

The atmosphere was not always like this. About 4500 million years ago when earth was formed and before primitive forms life appeared, the atmosphere was without oxygen. Being radioactive, so heated internally as well as by the sun, the gases poured from fissures, cracks and volcanoes and created an early atmosphere of methane, ammonia water vapour and CO_2 (greenhouse gases). The resulting GHGs effect ensured mean temperature at the earth's surface of 15-30°C, which is suitable for the development of primitive life forms.

Geological evidence suggests that the first forms of life began about 600 to 1000 million years after the earth was formed. The primitive life were bacteria and have been called "our ultimate grand parents". As the bacteria evolved into cyano-bacteria forms blue-green in colour, they began to use insulation to produce oxygen and organic materials. While bacteria colonized the planet, they also influenced the composition of the atmosphere which had first nourished them. With the passage of another 1500 million years (or about 2000 million years ago) the evolution of green plants and photosynthesis set the stage for the development of animal forms that breathed in O_2 and breathed out CO_2 which lead to the present day atmospheric composition.

Water vapour regulates the weather systems and produces precipitation. Atmosphere also transports many air borne particle. Throughout the ages, atmospheric winds transported heat poleward from equatorial region and created climate which is hospitable for living of some form of life almost everywhere of the globe.

The fluctuations in composition of the atmosphere and changes in temperature over the past 200000 years can be determined by analysis of tiny air bubbles trapped in ice that has accreted year by year in the Antarctic and Arctic. The cores drilled in ice caps is a record of changes, which can be measured and deducted. The Vostok core in Antarctica (in Antarctic station Vostok) by USSR, the CO_2 concentration and temperature were estimated for the past 160000 years. The analysis shows CO_2 fluctuated from about 180 to 220 ppm in the atmosphere and temperature rose and fell over the range of 9 to $10°C$ in a roughly perfect Rhythern with the rise and fall of CO_2 concentrations (or broadly by GHGs concentration). In the past 100-150 years, human activities become dominant force in changing the composition of the atmosphere. The current CO_2 levels 350 to 400 ppm demonstrate the rapid increase in CO_2 concentration as a result of human activities, mostly burning of fossil fuels. In 2013, the world passed the landmark of 400 ppm at Mauna Loa.

A continuous record of atmospheric conditions over Greenland and Antarctica dating back to about to about 4 lakh (4×10^5) years traced with the help of several ice-cores through the Greenland ice-sheet and three through the Antarctic ice sheets. To get the age, annual layers in the ice core are counted. (Note : The interpretation of a particular proxy record in terms of climate is sometimes uncertain).

The fallout of volcanic ash (occasional, world over) also provide common markers in cores:

Oxygen-isotope ratios ($^{18}O/^{16}O$) of the ice give temperature over ocean upwind of the glaciers. Bubbles in the ice give atmospheric CO_2 and methane concentration. Pollen, chemical composition and lava particles provide information of volcanic eruptions, wind speed and direction.

Thickness of annual layers give rate of snow accumulation. Isotopes of some elements give solar activity and cosmic ray activity .

Deep-sea sediment cores in the north Atlantic (made by ocean drilling program) give information about sea-surface temperature and salinity above the core, ice volume in glaciers, production of ice bergs.

Findings

The oxygen-isotope record in the ice-cores indicate abrupt temperature charges during the past one lakh years (10^5 years).

During the last ice age (on many occasions) Greenland temperature warmed over the periods of 1 to 100 years. Subsequently followed by gradual cooling over longer periods.

About 11500 years ago Greenland temperature warmed by about $8°C$ in 40 years in three steps each spanning 5 years. Such abrupt warming is called Dansgaard/Oeschger event.

Much of the northern hemisphere warmed and cooled in phase with temperatures calculated from the ice-core.

During the past 8000 years the climate was more or less constant.

Note: It may be noted that our perception of climate change is highly speculative, because all of recorded history has been during a period warm and stable climate.

Henrich events: North Atlantic sediment studies indicate coarse material was deposited on the bottom and mid-ocean (Atlantic). Only ice-bergs can carry such material out to sea. This indicates time when large number of ice-bergs were released into he north Atlantic. These are called Henrich events.

The correlation of Greenland temperature with ice-berg production is related to the deep-circulation.

12.13 CLIMATE DURING GEOLOGICAL CRETACEOUS PERIOD OF EARTH WARM EPOCH 135-180 MY AGO

During the cretaceous period the earth was bereft of ice sea level was 100-200 m higher than the present (20th century). There were no ice caps over the entire world. The super continent of Pangea began to break apart and only by 100 MY ago the present day continental structure was seen. Cretaceous meaning plenty of Chalk (limestone). During this period the land area had broad leaved plants, dinosaurs, tussles and crocodiles existed (north of Arctic circle) in polar region. During this Cretaceous period CO_2 levels were much higher than the present about five times the CO_2 level of pre-industrial concentration and temperature about the poles $25°C$ warmer than the present. This was because of pole ward transfer of heat from equational region in the Tethys Sea. Paleo records of Cretaceous period indicate the presence of palm trees and reptiles in the interior of continents, which shows that temperatures were above $0°C$ even in peak winter in the polar area north of lat $60°N$.

Cold epochs: [Holocene epoch $10^7 - 10^6$ MY ago]

Between 18 to 23 KY ago (about 21KY ago) the earth was covered by ice sheet and called the recent glacial cycle-the last glacial maximum (LGM).

Reconstruction of climate at LGM carried out in the project CLIMAP using proxy data from ocean sediments. Sea level was 120-130 m lower than present. Sea ice was more extensive covering most of the Greenland and Norwegian seas and persisted all through the summer.

Canada, northern parts of United States, northern parts of Europe (including large parts of Scandinavia. British isles and Wales) and parts of Eurasia was covered with thick ice. Present changes, Glasgow & Stockholm were covered with 1 km thick ice. In southern hemisphere Argentina, Chile and New Zealand and parts of Australia and South America were covered with ice. The average sea surface temperature (SST) was $4°C$ colder than the present and North Atlantic SSTs were colder than the present by $8°C$ and equatorial latitude SSTs were $2°C$ colder than the present. At the last glacial maximum cold and windy raid conditions prevailed equator ward of the ice. At the LGM deserts expanded into Asia including North Africa & Arabian deserts.

The above historical warm & cold epochs awaken us about climate.

12.14   GEOLOGICAL TIME SCALE

According to James Hutton (1726-1797), a Scottish scientist, the following modern geological time scale came into use since late eighteenth century. Based on sediment deposits of sequence of layers, he propounded principle of super–position and includes principles of uniformity of process.

William Smith (1764-1839) used rock layers identified by fossils they contained or possessed. This classification is called fossil correlation. Fossils are the remains of animals and plants found preserved in rocks.

Geological time scale divided into four main Eras, built on rock strata. Periods are based on fossils discovery.

Eras derived from Greek words "palaes" (old) "Mesos" (middle) and "Kainos" (recent), combined with "zoe" (life).

Discovery of radioactivity helped to determine actual ages of rocks. Isotopes of some elements enables scientists to know about solar activity and cosmic activity. Radiocarbon dating used to fix the age of wood, bones, shells, peats etc.

The Table 12.2 gives geological time scale.

Table 12.2 Geological Time Scale

Era	Period	Epoch	Time scale in year
Cenozoic	Quaternary	Recent	10^4-10^6
		Pleistocene	1-10
	Tertiary	Pliocene	10^7
		Miocene	2.5×10^7
		Oligocene	4.0×10^7
		Eocene	6.0×10^7
		Paleocene	7.0×10^7
Mesozoic	Cretaceous		1.35×10^8
	Jurassic		1.80×10^8
	Triassic		2.25×10^8
Palaeozoic	Permian		2.70×10^8
	Carboniferous		3.50×10^8
	Devonian		4.00×10^8
	Silurian		4.40×10^8
	Ordovician		5.00×10^8
	Cambrian		6.00×10^8
Pre –			7.00×10^8-
Cambrian			4.50×10^9

MY = millions years.

1. **The pre-Cambrian Era (700 – 4500 MY ago):** It was the oldest and longest of the main divisions. During this period mountain–building took place and oldest rocks were formed.

2. **Cambrian Period (600 – 700 MY ago):** In this period shallow seas covered over large part of surface of the earth. Life existed in sea as sea weeds, sponges, marine invertebrates, however no life existed on land.

3. **Ordovician Period (500 – 600 MY ago):** Volcanic eruptions. Seas continue to expand. All life confined to sea (no life on land). Fishes developed, invertebrates.

4. **Silurian Period (440 – 500 MY ago):** Periodic rise and fall of sea levels. Continuous change on land. Plants began to adopt on land. New species of vertebrate animals formed. First air–breathing animals evolved.

5. **Devonian Period (400 – 440 MY ago):** In this period volcanic activity increased. Land areas expand. Mountains began to form. Vertebrate animals develop. A variety of fish appears in sea water. Primitive amphibians. Invertebrates slowly move over to land.

6. **Carboniferous Period (350 – 400 MY ago):** Seas spread. Most of present Europe and Russia lie under water, which slowly emerge as swampy areas.

    Coal began to form in swamp vegetation. Large trees over tropical swamps. Amphibian creatures develop. Various marine life develop and spread. Reptiles breed on land.

7. **Permian Period (270 – 350 MY ago):** Warping of earth crust. Northern Hemisphere covered with ice. Deciduous plants appear. End of marine creature domination, creatures increase on land. Insects emerge. Spiders and primitive reptiles.

8. **Triassic Period (225 – 270 MY ago):** Shrubs cover mountains and deserts. More development of mountain ranges (in NH). Fish shaped reptiles, flying fishes and Lobster like creatures formed. Rise of dinosaurs. Primitive mammals appear.

9. **Jurassic Period (180 – 225 MY ago):** Rockies, Andes and Panama develop. High mountains of previous period suffer from erosion. Lime stone forms. Coniferous forms develop, flowers begin to bloom. Aquatic reptiles dominate in sea. On land birds evolve. Giant dinosaurs lived in swamps Turtles, egg laying mammals appear.

10. **Cretaceous Period (135 – 180 MY ago):** Swamps deltas appear. Rivers flow, chalk deposits appear. Major mountains build up. Deciduous trees and flowering plants spread. Flying reptiles dominate in sea. On land birds evolves. Giant dinosaurs become extinct.

11. **Epochs: 11. Paleocene (70 – 135 MY ago):**

12. **Eocene (60 – 70 MY ago):** Severe volcanic activity, warming of climate, flowering plants dominate. Present species of fishes developed in sea. Big whales and sea-cows appear. On land modern animals and giant reptiles and primitive monkeys appear.

13. **Epoch of Oligocene (40 – 60 MY ago):** Alps formed. In seas crabs, snails evolve. On land animals and plants found in abundant. Primitive anthropoids (ape like man) and Mesohippus appeared.

14. **Epoch of Miocene (25 – 40 MY ago):** Earth's crust completely formed. Alps and Himalayas formed. Boney fish, sharks, protohippus, whales develop in sea. On land mammals, water birds, penguins in Antarctica appeared.

    Equatorial regions cooled by 3 to 4°C, while Europe, America warmed up by 7 to 10°C.

15. **Epoch of Pliocene (10 – 25 MY ago):** Continents and oceans develop into present form. Europe and Asia land masses join. Vegetation on land limited. Rise of Man and Pliohippus.

16. **Epoch of Pleistocene (1- 10 MY ago):** A great ice-age, ice sheets and glaciers cover most of Europe and America. On land rise modern horse. Emergence of Homosapiens (man).

17. **Recent epoch or Holocene (10^4 to 10^6 years ago):** Ice sheets retreat. Sea level rises. Man's dominance grows. Domestication of animals.

KY = Thousand years.

LGM (Last Glacial Maximum) (18 – 23 KY ago).

CLIMAP (Climate Long range Investigation, Mapping and Prediction).

This is based on proxy data from ocean sediments. Thick ice covered over Canada, Northern United States of America, Northern Europe and parts of Eurasia.

During this period 1km thickness ice layer covered over Chicago, Glasgow & Stockholm. Antarctica sea level was about 120 – 130 meters lower than the present.

The average SST was 4 °C colder than the present.

North Atlantic SST was colder by more than 8 °C. Low latitude temperatures were about 2 °C lower than the present.

12.15 MILANKOVITCH CYCLES

Climate variation occurs due to earth's position (elliptical orbit) relative to the sun, orientation of earth's axis with respect to the orbital plane and precession of axis. These variation of climate is on the scales of 10 KY to 100 KY.

(i) Eccentricity (e) of elliptical orbit, time scale 100 KY to 400 KY.

If e = 0 (circle) to e = 0.07.

(ii) Obliquity-spin axis of earth varies 22.1° to 24.5° on time scale 41 KY.

Present obliquity 23.5°. At position 22.1° it will be ice age.

(iii) Precession-the earth's spin vector precesses with period 23 KY.

The earth is 3% closer to the sun in January perihelion than in July – aphelion. At perihelion earth receives 7% more solar energy than when the earth is at aphelion.

1. When the 'e' is small there is little change in earth-sun distance.

2. When the 'e' large, earth will receive about 20% more solar energy at perihelion than when at aphelion.

When the earth's axis tilt is more, earth experience stronger summer sun and weaker winter sun (particularly at high latitudes).

When the tilt is small ice-ages set in. Cooler summer will not melt winter ice/snow. The precession scale 23 KY. Earth will be closest to the sun in July instead of January. This causes stronger summer.

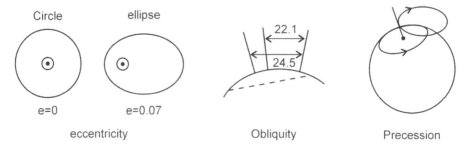

Circle	ellipse		
e=0	e=0.07		
eccentricity		Obliquity	Precession

Paleoclimate

The study of ancient climate of earth using records of ocean and lake sediments (deposits), glaciers and ice sheets and continental deposits is called Paleoclimate. Proxies of post climate are very large, which include measurements of isotopic ratios of: (1) shells using in ocean sediments, (2) thickness and density of tree rings, (3) chemical composition of ice and (4) radioactivity of corals.

Paleoclimates, however based on sparse data yet they provide a lot of history of past climate. Paleoclimate records provide is climate variations on time scales of decades and longer. Theory and modeling of paleo climate variations is still rudimentary (or elementary).

Climate models suggest that the input rate of CO_2 from volcanic activity into the atmosphere is balanced by the rate of removed by chemical weathering (sink).

Carbon from the earth's interior pumped into the atmosphere. Chemical weathering of continental rocks which finally wash into the ocean and deposited as sediments.

Based on isotopic measurements of oxygen, sediments at the bottom of the ocean give us a proxy record of climate conditions in the water column.

Key proxy is $\delta\ ^{16}O$ –a measure of ratio of isolopes of oxygen $^{18}O = \dfrac{^{18}O}{^{16}O}$, which is recorded in seabed sediments by the fossilized calcite shells of foraminifera (organisms that live near the surface or bottom of sea).

$\delta\ ^{18}O$ in the shells is a function of $\delta\ ^{18}O$ of the ocean and the temperature of the ocean.

The vapour pressure of $H_2O^{16} > H_2O^{18.}$

During evaporation of sea water to $H_2^{16}O$ in atmosphere, vapour pressure the residual sea water becomes rich in ^{18}O.

^{18}O = in shells of surface planktonic

^{16}O = in bottom dwelling (benthic) foraminifera made of Ca Co$_3$.

$$\delta^{18}O = \frac{\left(\dfrac{^{18}O}{O^{16}}\right)_{sample} - \left(\dfrac{^{18}O}{O^{16}}\right)_{stand}}{\left(^{18}O/O^{16}\right)_{stand}} \times 1000\%$$

O^{18} is enriched with calcification temperature decreases.

$1°C$ decrease in temperature result in 0-2 % increase in $\delta\ ^{18}O$

The record of $\delta\ ^{18}O$ over the last 55 million years indicate a cooling of the deep ocean by a massive 14 °C. This implies deep ocean temperature were close to 16 °C compared to 2 °C as observed at present.

12.16 IMPACT OF CLIMATE CHANGE ON AGRICULTURE

Rain fed Agriculture in India

Introduction: According to CRIDA, Fifty years of National Resource Management Research: [NBSSLUP, 2001]

1. Net cultivated land (NCL) in India is about 143 million hectares
2. Dry land/rainfed area 97mha is about (68 % of NCL) which produces 44% of the India's food requirements, supports 40% of human population & 60% livestock population.
3. (a) About 15m ha of dryland lies in arid region which receives rainfall < 500 mm per annum (PA)
   (b) another 15 mha receives rainfall 500-750 mm PA
   (c) 42 mha receives rainfall 750-1150 mm PA and the remaining
   (d) 25 mha receives rainfall > 1150 mm PA
4. About 74% of annual rainfall occurs during monsoon period (June-September) which has coefficient of variation 0.3 to 0.6
5. Global rainfed cropland 73%
6. In India rainfed cropland 60%, which provides about 40% of total production (mainly coarse cereals, oilseeds, pulses and fruits etc.)

7. Cultivated area under sorghum about 93%; 94% under pearlmillet; 79%under corn; 87% under pulses; 76% under allseeds; 64% under cotton and 59% under tobacco.

CWR (Crop Weather Relationship)

According to Theophrastus, Greek Philosopher.

"It is the year that bears the fruit and not the field".

The principal weather parameter of agriculture are:
(i) Precipitation
(ii) Air temperature (Maximum and Minimum)
(iii) Relative humidity (moisture content of the air)
(iv) Solar radiation
(v) Wind (speed)

Crops are generally grown for their storage organs like tubers, pods or grain and not for the total biomass. Plant growth or various plant organs development is represented by phonological stages through which it passes. Factors that influence crop production are: (i) Yield defining factors like radiation (ii) yield limiting factors like water availability and nutrients and (iii) yield reducing factor like weeds, pests and diseases.

Crop pass through three phases:

1. The growth of increase in first phase depends on assimilation which depends on leaf area, which is proportional to the energy (insolation) intercepted

2. Dry matter accumulation in the second phase. Dry matter accumulation depends on growth rate and duration of the phase.

3. Crop maturing in the third phase and the growth rate is decreasing.

Insolation intercepted is given by the formula

$$F_n \frac{I_b}{I_a} = \left[1 - e^{K * CLAI} \right]$$

Where F_n = Fraction of the radiation intercepted by the crop (i.e. crop canopy)

I_a = Radiation received above the crop canopy

I_b = Radiation received below the crop canopy

K_C = Extinction coefficient

LAI = Leaf area index of the crop i.e. the leaf area of the crop per unit area of the ground. Generally, the fraction of radiation intercepted by the crop will be unity when LAI is more than 3.5 for most crops.

Harvest Index is the ratio of grain yield to the total dry matter production

$$\text{Harvest Index} = \frac{\text{Grain yield}}{\text{Total dry matter production}}$$

Harvest index ranges from 0.3 to 0.4

Radiation use efficiency (RUE) is defined as below

$$\text{RUE} = \frac{\text{Total dry matter production per unit area}}{\text{Radiation used by the crop}}$$

Water use efficiency (WUE): It is defined as the "Amount of dry matter produced by a crop per unit of water transpired".

Yield = Radiations absorbed × WUE × HI

Yield = Total water use × WUE × HI

Where HI = Harvest Index

12.16.1 AGROCLIMATOLOGY

Agroclimatology relates the climate and weather to the crop production. The crop production has several operations which are dependent on weather and climate. The main crop production operations are: field preparation for sowing, weeding, irrigation, harvesting threshing transportation and marketing which are all weather dependent. Agroclimatic analysis vary with location, crop and time specific. Growing degree days (GDD) is used for crop phenology study. The thermal responses differ crop to crop. Climate water balance is used for delineation of different climates, for assessing the droughts and agriculture drought.

Growing degree days: The growth and development of any crop depends on temperature, moisture, sunshine or light. Irrigation helps to nullify the moisture problem. Thermal regime influences developmental activities of a crop and light regime effects growth. Thermal stress plays role in all stages of crop, particularly during reproductive and mature stages of crop. Thermal stress reduces the duration of each phenophase consequently reduces productivity. It must be noted that some plants are photo-sensitive & also thermosensitive. Because of this GDD varies from season to season for same crop and same variety. Hence GDD computed as follows.

Method-1 : $\text{GDD} = \sum \left[\frac{(T_x + T_n)}{2} - T_b \right]$

Where $\quad T_x = $ Maximum temperature (°C)

$\quad\quad\quad\quad T_n = $ Minimum temperature (°C)

$\quad\quad\quad\quad T_b = $ Base temperature

Method-2 : $\text{GDD} = \sum (T_x - T_b)$

Method-3 : $GDD = \sum [T_c - (T_x - T_c) - T_b]$

Where T_c = Ceiling temperature, varies from crop to crop. For ratio crop it is $35°C$

Method-4 : $GDD = \left[\dfrac{T_x + T_n}{2} - (T_x - T_c) \right] - T_b$

Method-5 : $GDD = \sum \dfrac{DL}{12} \left[\dfrac{T_x + T_n}{2} - T_b \right]$

Where DL = Day length hours

Method 6, 7 & 8 are same as methods 2, 3 and 4 but it is multiplied by $\dfrac{DL}{12}$

Thermal sensitivity index (TSI)

TSI defined as below

$TSI = \dfrac{\text{Range of duration in different thermal environments}}{\text{average duration}} \times 100$

TSI of crop or genotype can be assessed as follows

TSI	Thermal sensitivity
< 5	Tolerant
5 – 10	Moderately tolerant
10 -5	Moderately susceptible
> 15	Susceptible

12.16.2 TOPOCLIMATOLOGY

In hilly terrain, the effective rainfall varies from top to bottom of the topography depending upon the slope, vegetative cover, soil type etc. The total run-off water would be much higher than the actual computed values using total rainfall.

There are several factors that influence the effective rainfall.

Factor	Relative Characteristics
Rainfall	Amount, intensity, distribution and time
Other Met facture	temperature, radiation, RH, wind speed
Land	Topography, slope, types of use
Soil	Depth, structure, texture, bulk density, organic matter
Ground water	Depth, quality
Management	Tillage, leveling, bunded or not etc
Drainage channel	Size, shape, roughness
Crops	Nature, rooting, depth, ground cover, stage, rotation

CHAPTER - 13

International Geosphere-Biosphere Programme IGBP-1989

The environmental conditions on the surface of the earth more or less stable over long geological periods. However there had been continental drifts and major and minor ice ages and also warm epochs during this period. These changes took place gradually due to natural processes from inside and outside the earth. In contrast to this there are changes after industrialization during 150 years, 1840-1990 due to human intervention. The changes that observed-sea level rise, major changes in climate patterns and increase in frequency of extreme natural events/disasters. IGBP-1989 aimed to mitigate the adverse effects by developing global predicting models incorporating the interactions between Geosphere and Biosphere. Interactive processes (geochemical) linking the life forms on land and sea. The trend of international programmes related to environment, climate, ecology and human impacts.

The objectives of IGBP-"to describe and understand the interactive physical, chemical and biological processes that regulate the total Earth System, the unique environment that it provides for life, the changes that are occurring in this system and the manner in which they are influenced by human actions".

The IGBP envisaged-

1. Documenting and predicting global change
2. Observing and improving our understanding of dominant forcing functions.
3. Improving our understanding of interactive phenomena in the total earth system
4. Assessing the effects of global change that would cause large scale and important modifications that affect the availability of renewable and non-renewable resources.

13.1 THE LINKAGES OF GENERAL CIRCULATION MODELS (GCM) WITH ECOSYSTEM MODELS

In GCM grid cells used $1/2^\circ \times 1/2^\circ$ and ecosystem dynamics are largely controlled by this scale or smaller, hence it should be scaledup in space.

Atmospheric boundary layer model provides adequate and appropriate interface with the Global Scale Circulation Model.

Global geosphere-biosphere models are extensions of global climate models which incorporate transfer of chemical compounds and key chemical and biological reactions and transformation. For this purpose the data includes solar flux, stratosphere and above, clouds and earth radiation budget, tropospheric, chemistry, role of vegetation in hydrological cycle, land surface, oceans, sea level and snow & ice.

13.2   GLOBAL CHANGES OF THE PAST

Studies of the physical, chemical and biological parameters found in natural archives such as ice cores, tree rings and in ocean, lakes and terrestrial sediments (or lacustrine sediments) have revealed a wealth of information on both the "natural" and perturbed behavior on the coupled earth system.

The general nature of climate change from glacier to interglacial in the Northern and Southern hemispheres including the records of surface temperature and precipitation through last 150,000 years and detailed temperature history of last 1000 years in polar ice sheets and in midtemperate latitude glaciers. This is shown in the Fig. 13.1.

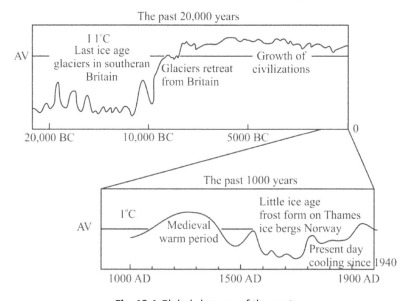

Fig. 13.1 Global changes of the past

Changes in the Norway snowline since 10, 000 BC and changes in average temperature in China since 3000 BC.

Ref. Meteorology and public safety WMO N.637

Table 13.1 Characteristics of Natural Archives

Archives	Temporal precision yr/season	Extent yrs	Derived Parameters
Tree rings	Yr/season	10^4 yrs	T, H, C_A, B, V, M, L, S
Lake sediment	Yr	$10^4 - 10^6$	T, B, M
Polar ice cores	Yr	10^5	T, H, C_A, B, V, M, S
Mid-lat ice cores	Yr	10^3	T, H, B, V, M, S
Coral deposits	Yr	10^5	T, C_W, L
Loess	10 Yr	10^6	H, C_S, B, M
Ocean cores	100 Yr	10^7	T, C_W, B, M
Pollen	1000 Yr	10^5	T, H, B
Paleosoils	100 Yr	10^5	T, H, C_S, V
Sedimentary rock	2 Yr	10^7	H, Cs, V, M, L
Historical records	Day/hr	10^3	T, H, B, V, M, L, S
Notation: T = Temperature H = Humidity C_A = Chemical composition of Air C_W = Chemical composition of water			

C_s = Chemical composition of soils

V = Volcanic eruptions

B = Information on Biomass, as in pollen samples

M = Geomagnetic field

L = Sea level

S = Solar activity

The natural archival environment data provides-

1. The general nature of climate change from glacial to interglacial
2. The identification of Milankovitch cycles and high frequency climate oscillation in both the ocean and continental archives
3. Changes in patterns of terrestrial vegetation corresponding to major climate variation
4. Changes in the basic nature of ocean circulation corresponding to changes in climate
5. Variation in the chemical composition of the atmosphere, particularly CO_2, CH_4
6. Significant temporal changes in the abundance of cosmic-ray produced isotopes likes 14C 10Bo, 36Cl
7. Fluctuations in a broad spectrum of ions deposited in polar ice
8. Major changes corresponding to climate change in particulate content of ice and transport of dust in the atmosphere

9. Significant variations in the extent of aridity, derived from lake sediments

The above findings document the internal and external forcing mechanisms, biogeochemical feedbacks and the general response of the geosphere-biosphere system

Table 13.1 gives the summary of the IGBP Report-6 which enumerates the present techniques for extracting information on global changes in the past.

Global Changes in the Holocene

The last 10, 000 years of the earth's history, reveals the present ecological condition were developed which covers the impact of human activities. This also provides us to anticipate the future.

The medieval climate optimum (1100-1200 AD) when a global warning of $1°C$ must have been accompanied by significant changes in the circulation of the atmosphere and oceans. From the present trend such conditions are developing from 2000 AD.

Global Change Through A Glacial-Interglacial Cycle

The study of past proxy data indicate changes in air and sea temperatures, distribution of ice sheets, glaciers, permafrost sea level, ocean circulation, land surface characteristics, continental vegetation and biomass and the chemical composition of atmosphere.

13.3 LAW OF THE SEA

The United Nations Convention on the law of the sea (UNCLOS) came into effect on 16 November 1949. This law provided for the first time a Universal legal frame work for the national management of marine resources and their conservation. The UNCLOS, oceans are divided into four domains.

1. Territorial Sea (coast to 12 nautical miles) where a country has exclusive right to exercise full Sovereignty
2. Continuous zone (coast to 24 nautical miles), where a country can exercise regulations relating to immigration, customs, fiscal, sanitation
3. Exclusive Economic Zone (coast to 200 nautical miles), where a country has the right to use both the living and non-living resources in the water column, Sea surface and sub surface
4. Legal Continental Shelf (coast to a maximum of 350 nautical miles), where country has the right to use only the non-living and sedentary resources (except water column) subject to satisfaction of certain geophysical conditions. CRZ: Coastal Regulation Zone.

Two important parameters that contribute the genesis of Tropical Cyclones are sea surface temperature (SST) and wind-field over the sea.

These fields contribute thermal and dynamic potential for the growth and sustain ability of Tropical Cyclone.

Sea surface temperature (SST), Outgoing Long Wave Radiation (OLR) and wind filed supports the prediction of Tropical Cyclone days in Southwest Indian Ocean, Pacific Ocean, El Nino, La Nina episodes clearly show effect on southwest monsoon over India rainfall activity over South-East Asia, Australia and South America.

13.3.1   THE ROLE OF OCEAN IN CLIMATE VARIABILITY

1. Heat, water, momentum, Greenhouse Gases (GHG) and many other substances cross the sea surface. Hence ocean is a central component of the climate system
2. Ocean releases heat and water vapor (80%) to the atmosphere
3. Ocean currents transport heat and salt around the global ocean
4. For global Hydrological Cycle maintains pole to equator, temperature gradient and fresh water transport
5. Buffering of atmospheric temperature changes by oceans huge heat capacity
6. In middle latitudes, atmospheric changes tend to precede ocean changes through air sea interaction
7. In tropical latitudes, changes in SST and tropical air temperature and winds are in phase with one another
8. In tropical oceans particularly in Pacific the Oceanic El Nino and the atmospheric Southern Oscillation phenomena are linked

CHAPTER - 14

Modern Aids of Communication and Detection

INTRODUCTION

In this chapter we shall discuss the presently available technology in communication, search and rescue operations. In order to understand the subject, the basics of radio, radar, satellites and transponders have been given, which are principally used in aviation. It is well known that the development of radio, TV, aircraft, satellites brought revolution in the field of communication, aviation and laid way for detection and rescue in hazards. These technologies have invaluable role in natural hazard/disaster detection and disaster management.

Heinrich Hertz conducted experiment in 1887 and proved the existence of electromagnetic waves and showed that these waves propagate in vacuum and can be stopped by a metallic screen. Radio waves are produced by changing fields in alternating current (AC). The electron flow alternately forwards and backwards (thus implies continuous changing of current) which produces fields along the wire.

An AC current can induce in an open circuit with a bare wire at the end. In this case electric and magnetic fields propagate outwards at right angles to the wire. If the wire is of the correct length, the fields will resonate and send continuous alternating waves of energy outwards. This outward propagation of fields form the transmitted radio waves. If a wire of same length is placed in the same direction in space as the transmitting aerial, the fields will affect the wire and induce and AC in it, so a receiver at a faraway distance can receive the transmitted signal exactly. This was exactly Marconi achieved in his experiments. The traditional aerial is called a halfwave dipole shown in Fig. 14.1.

The speed of propagation of electromagnetic wave is 3×10^8 m/s. The transmitted waves from a simple system with a single wire aerial travel in all directions around the antenna.

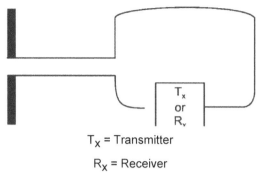

T_x = Transmitter

R_x = Receiver

Fig. 14.1 A half-wave dipole aerial

The magnetic field propagates at right angle to the electric field. The wave is polarised in the direction of the electric field.

The propagation of an alternating electromagnetic field in space constitutes electromagnetic waves. Electromagnetic waves are transverse waves because the electric and magnetic intensity vectors $\vec{E}$ and $\vec{H}$ of the wave fields are mutually perpendicular (shown in Fig 14.2) and lie in a plane perpendicular to the velocity vector $\vec{V}$ of wave propagation. The Vectors $\vec{V}$, $\vec{E}$ and $\vec{H}$ form a right handed system.

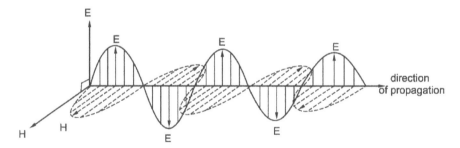

Fig. 14.2 Radio wave polarisation

According to classical electro dynamic theory electromagnetic waves originate by the accelerated electric charges. A frame or loop antenna is a closed AC circuit.

Radio communication includes transmission of any type of information by means of radio waves, that is electromagnetic waves of frequency less than 3×10^5 MHz. Radio broadcasting in transmission of speech, music, telegraphic signals by means of radio.

Television broadcsting in transmission of images by means of radio.

Radio communications are the transmission of modulated electromagnetic waves by the radio transmitter and their demodulation in a radio receiver.

The alteration or change of parameters of electromagnetic waves is called modulation. The wave which is modulated is called carrier wave and its frequency is called carrier frequency.

Depending upon the parameter of carrier wave that is altered in modulation we have:

Amplitude modulation see Fig. 14.3 (in which amplitude of the wave is changed).

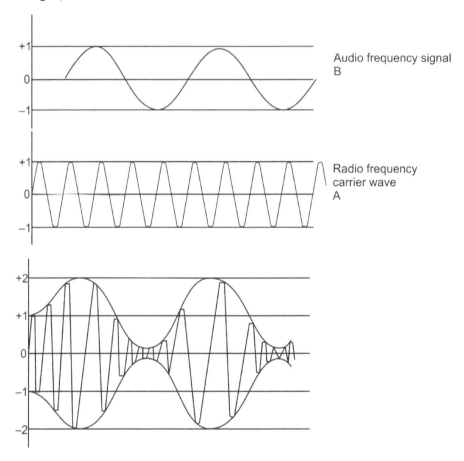

Fig. 14.3 Amplitude modulated signal

Keying

The receiver requires a beat frequency oscillator facility.

    (i) It consists of starting and stopping the continuous carrier wave, breaking it up into dots and dashes. This is generally called wireless telegraphy or interrupted carrier wave (ICW). The communication is by morse code.

(ii) Frequency modulation (in which only the frequency of the wave is changed). (See Fig. 14.4).

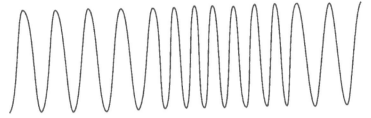

Fig. 14.4 Frequency modulated signal

(iii) Phase modulation (in which the initial phase of the wave is changed). (See Fig. 14.5).

Fig. 14.5 Phase modulation-reversing the phase

In radio broadcasting, the modulation frequency is low because the audible sounds frequency range is 16 to 20000 Hz, consequently there are no restriction to the choice of a carrier frquency. Thus the radio wave broadcast is accomplished on

Long radio waves $[\lambda = 10^3$ to 10^4m,   f = 30 to 300 kHz]

Medium audio waves $[\lambda = 10^2$ to 10^3 m,   f = 0.3 to 3 MHz]

and   Short radio waves $[\lambda = 10$ to 10^2 m,   f = 3 to 10 MHz]

$C = \lambda f$, or $f = \dfrac{C}{f}$ or $\lambda = \dfrac{C}{\rho}$ where C = speed of radio waves, λ = wave

length, f = frequency

An AC voltage in a wire reverses its direction in a number of times every second. The graph of radio wave is a sinusoidal curve as shown in Fig. 14.6.

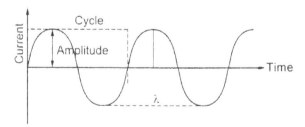

Fig. 14.6 Radio wave

Cycle: A cycle is one complete process or one complete series of values.

Hertz: One hertz is one cycle per second

Number of cycles per second is expressed in hertz (Hz).

f = frequency, of an AC or radio wave, is the number of cycles occurring in one second, expressed in hertz.

1 cycle per second = 1 Hz (hertz)

1000 Hz = 1 kHz (kilohertz)

1000 kHz = 1 MHz (mega hertz)

1000 MHz = 1 GHz (gigahertz)

λ = *wavelength:* The physical length travelled or covered by radio wave in one complete cycle of transmission (distance between two consecutive troughs or ridges)

Ex: Given $\lambda = 3$m, find f

$$f = \frac{C}{\lambda} = \frac{3 \times 10^8}{3} = 10^8 \text{ Hz or 100 MHz}$$

Ex: Given f = 100 kHz, find λ in meters

$$\lambda = \frac{C}{f} = \frac{3 \times 10^8 \text{m/s}}{100 \times 1000 \text{Hz}} = 3 \times 10^3 \text{ m} = 3000 \text{ m}$$

Ex: Given f = 325 kHz, find λ

$$\lambda = \frac{C}{f} = \frac{3 \times 10^8 \text{m/s}}{1000 \times 325 \text{ Hz}} \cong 923 \text{ m}$$

Ex: Given $\lambda = 3520$m, find f

$$f = \frac{C}{\lambda} = \frac{3 \times 10^8 \text{m/s}}{3250 \text{ m}} = 92307 \text{ Hz or 92.307 kHz}$$

Ex: Given $\lambda = 3.41$ cm, find f

$$f = \frac{C}{\lambda} = \frac{3 \times 10^8 \text{m/s}}{341 \text{m}} \times 10^4 \cong 90 \text{ GHz}$$
$$= 87.976 \times 10^9 \text{ Hz}$$
$$= 87976 \text{ MHz}$$

Ex: If a transmission is f = 100 MHz, find number of wavelengths in 60 ft. (given 1m = 3.28 ft)

$$\lambda = \frac{C}{f} = \frac{3 \times 10^8 \text{m/s}}{100 \times 10^6 \text{Hz}} = 3 \text{ m}$$

$$\therefore \text{ no. of wavelengths in 60 ft } = \frac{60 \text{ ft}}{3.28 \times 3} \cong 6\lambda$$

Modulation: Radio waves act as a vehicle (commonly called carrier waves) for the information. The waveform of information which is impressed upon the carrier wave is called modulating wave.

Oscillator: The radio frequency carrier wave is generated in the oscillator, whose frequency may be controlled by one or combination of several quartz crystals, a magnetron valve, or a semi conductor circuit incorporating varactor (or variable capacitor) diodes. These oscillations are of low frequency.

Transmitter: A radio transmitter consists of the following principal parts.

(i) A generator of sustained electromagnetic oscillations of the carrier frequency.

(ii) A modulator.

(iii) A transmitting antenna (which emits radio waves in the required direction).

Receiver: A receiver is a set consisting of :

(i) A receiving antenna.

(ii) A radio receiver.

Receiving antenna converts the energy of radio waves into the energy of high frequency electromagnetic oscillations. From these oscillations, the radio receiver chosenly separates those excited by the required transmitter and then amplifies and demodulates them. After amplification, the modulating oscillations are fed to the (reproducer) telephone/ loud speaker / TV kinescope etc.

A cathode ray tube is called kinescope, which is used to reproduce the picture in TV receiver.

In case of TV, the sending image is acomplished by modulating a carrier electromagnetic wave in proportion to the brightness of the various small spots (or portions) of the picture that is being transmitted. This is achieved by extrinsic or intrinsic photo-effect. The image (frame) is transmitted in continuous sequence line after line, and element after element in each line by scanning the image with the exploring spot. In Russian system, the Frame is divided into 625 lines with 833 picture elements per line. During each second 25 frames (distinct and separate pictures) are transmitted. The frequency of video signals (or modulation frequency) is about 6.5 MHz. To avoid distortion of video signals, the carrier frequency is used about ten times greater the ultrashort waves of meter band $\lambda = 1$ to 10 m and frequency more than 30 MHz are used.

Transmitter flow Diagram (See Fig. 14.7).

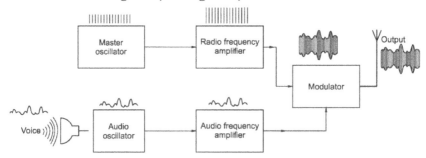

Fig. 14.7 Audio modulated transmitter

Radio Frequency Amplifier: When the signal is at the correct frequency, it is amplified to the extent that it is strong enough to pass through the remainder of the transmitter components. Such amplifiers are semiconductor circuits.

Microphone and Audio frequency amplifier: The audio frequency signal may be produced by the operators microphone and/or audio oscillator such as a recording device that again has to be amplified. After amplification any speech part of an audio signal will be processed in a "speech processor" before modulation takes place.

14.1   RADIO SPECTRUM

The whole of electromagnetic spectrum consists of radiation gamma rays ($\lambda < 10^{-10}$ m, f $\geq 10^{18}$ Hz), X-rays, UV-radiation, Visible light radiation, IR-radiation, Microwave (or very short radio waves), UHF (Ultra High Frequency), VHF (Very High Frequency), short and long radio waves ($\lambda \geq 10^{2}$ m, f $> 10^{5}$ Hz). Thus radio waves is only a part of this spectrum at the bottom. The spectrum of electromagnetic waves used for radio is divided into bands. Voice frequencies (audible range 16 Hz to 20 KHz). Sound waves are pressure waves and are propagated differently from electromagnetic waves. The internationally recognised radio frequency bands are given in Table 14.1.

Table 14.1

Frequency Band name	Abbreviation	f = frequency	λ = wavelength
Very low frequency	VLF	3 - 30 kHz	100 - 10 Km
Low frequency	LF	30 - 300 kHz	10 - 1 Km
Medium frequency	M	300 - 3000 kHz	1000 - 100 m
High frequency	HF	3 - 30 MHz	100 - 10 m
Very high frequency	VHF	30 - 300 MHz	10 - 1 m
Ultra high frequency	UHF	300 - 300 MHz	100 - 10 cm
Super high frequency	SHF	3 - 30 GHz	10 - 1 cm
Extremely high frequency EHF	EHF	30 - 300 GHz	100 - 10 mm

Radar Frequency Band names given in Table 14.2.

Table 14.2

Frequency Band name	Frequencies
L - band Radar	1 - 2 GHz
S - band Radar	2 - 4 GHz
C - band Radar	4 - 8 GHz
X - band Radar	8 - 12.5 GHz

Voice Communication: This is achieved by voice modulation of radio waves.

Long-Range Communication (on the basis of Global Distance): This lies in the frequency bands between VLF and HF. The frequency bands above HF is useful in direct wave or line of sight propagation (see Fig. 14.8). However these frequency bands are used in satellite technology. HF communication is superior for the following reasons.

- Aerials are shorter and less expensive to install.
- Static noise is less than MF and tolerable.
- For relatively less power very long range communication achieved by using sky waves during day and night.
- HF suffers less attenuation in the ionosphere.

Fig. 14.8 Ideal line of sight propagation

Increased efficiency obtained by beaming the radiation in the direction of the receiver.

Static Interference: Static interference is caused by electrical disturbances in the atmosphere. The word 'noise' is used for interference from electrical components in transmitters and receivers.

HF communication: HF communication depends on the ionospheric conditions (Efficiency lies on the ionosphere conditions that will produce the first return at the required skip distance from the transmitter).

The HF frequency band allocated for commercial avaiation purpose ranges 2 MHz to 22 MHz.

A low power is sufficient for transatlantic voice communication.

HF range depends on the following factors.

- Transmission power
- Time Day (this effects the electron density)

- Season of the year (this effects the electron density)
- Ionospheric disturbances (depends on solar activity)
- Geographical location
- Frequency in use (this determines the critical angle and the depth of ionospheric penetration)

HF Datalink: HF Datalink is a facility used in oceanic control to send and receive information over normal HF frequencies (using upper sideband of the selected frequency). The signal is phase modulated to send digital information. Modern instruments convert voice signals into digital information (like a digital cell phone) and vice versa, to provide digital/voice communications.

Short-range Communication: This will provide communication upto 80 nm range with an antenna altitude of 500 ft and 200 nm range at 20,000 ftm antenna short range communication frequency bands range from VLF (3-30 kHz) to HF (3-30 MHz). However these bands have shortcomings of complexity and static interference. The VHF band provides a working practical facility but aerial requirement is complicated.

The signal strength received by a simple antenna at a given range is proportional to the wavelength. Longer wavelength (lower frequency) will provide better reception.

VHF Communication: The VHF band used for Radio Telephoney (R/T) communication (at short range) lies between 117.975 MHz to 137.00 MHz. Within this band communication channels are available at 8 KHz seperation intervals. It requires transmitter producing 20 W power.

VHF is free from static but some background noise is picked up (due to vertically polarised receiver aerials). A frequency modulated UHF signal would provide absolute clarity of reception (however the equipment is complex and expensive). The VHF transmission range depends on the following factors.

- Transmission power
- Height of Transmitter
- Height of the Receiver
- Obstacles at or near the transmission site (that will block the signals or scatter them and cause attenuation).

14.2 SATELLITE COMMUNICATION

The most familiar Television pictures and sound is the gift of satellite communication. In aviation, the satellite communication is achieved through Inernational Maritime Organisation Satellite constellation (of four satellites) INMARSAT (see Fig. 14.9). The four satellites are positioned in geostationary equatorial orbit at a height of about 36000 km. The sensors based on these satellites provide communications by accepting transmissions of digital siganals in the 6 GHz band. The signals from the satellites cover almost all the area of the earth between latitudes 80 °N and 80 °S (see Fig. 14.9). The signals are not affected by the meteorological conditions (or static). This system requires special aerials for transmission and reception on these frequencies. It is noted that satellites do not reflect the signals. They receive the signals in 6 GHz band frequency andre-transmit them in different frequency of 4 GHz while those to the airborne aircraft in 1.5 GHz band. This reduces the attenuation of the signal.

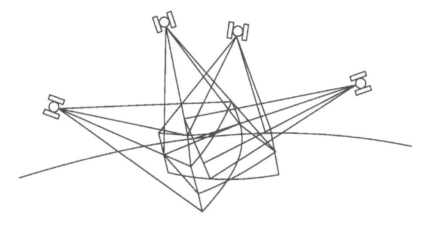

Fig. 14.9 Position fixing

Ground based receiving stations are located in a network, which serve each of the four satellite region (or segments) and connect to the conventional public and private telephone networks. In aviation, a pilot using satellite communication system in reality using an ordinary telephone, and airborne passengers as well use this system. The aircraft satellite communication (satcom) receivers operate on frequency between 1544 to 1555 MHz, while aircraft (satcom) transmitters use frequencies between 1626.5 MHz and 1660.5 MHz. Voice messages are digitised by the equipment on ICAO laid specific algorithms.

Search and Rescue Satellites

INMARSAT constellation of 4 satellites are used for search and rescue operations (Aircrafts, Ships, Fisheries Naval Boats etc). INMARSAT satellites receives the signals on international emergency frequencies (121.5 MHz and/or 243.0 MHz) and alerts Search and Rescue (SAR) centres to emergency beacons carried by survivors.

The international COSPAS - SARSAT (Satellite Constallation) system is dedicated to the provision of Search and Rescue facilities. This system uses four polar orbiting satellites to cover whole of the earth. It is a joint venture-COSPAS is the Russian name, SARSAT is the US name. One of these satellites (of constellation) can receive signals transmitted at 121.5 MHz, for example from a survivors Personal Locator Beacon (PLB). The satellite re-transmits the signal to a ground station, called a Local User Terminal (LUT), where the exact frequency measured and compared with the datum 121.5 MHz. The difference gives the Doppler shift. The variation of Doppler shift gives an indication of the lateral distance from the satellite path. From this the search area is determined.

Marine Beacon: Marine radio beacons transmit in the Low and Medium frequency bands (a signal consisting of several letters in code). Some marine radio beacons operate continuously during navigation season (commercial mechanised boats, ships, fisheries boats etc) and can be used by ships, boats etc., overseas and by aircraft for navigation purposes.

Kinescope: A cathode ray tube called kinescope, is used to reproduce the picture in TV receiver. The priciple of kinescope is based on cathode - luminiescence. A special device scans the kinescope screen horizontally and vertically in synchronism with transmission of the corresponding picture elements by the T.V broadcast station. The different brightness (intensities) at the various points of the viewing screen are obtained by modulating the intensity of the electron beam in proportion with the electro-magnetic waves received.

Ionosphere: The upper atmosphere (50 - 500 km altitude) is characterised by the presence of dense ions and free electrons is called Ionosphere. These ions cause reflection of radio-waves.

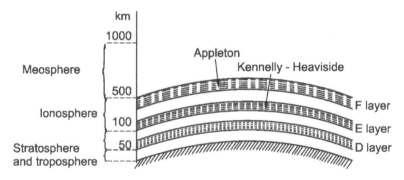

Fig. 14.10 Ionosphere layers

Ionosphere is divided into D-region 50-90 km, E = region 90- 50 km, F-region 150-500 km. F-region is further divided into F_1-region 150-250 km, F_2-region 250-500 km.

D-Region: Reflects low frequency radio-waves, absorbs medium and high frequency radio waves. D-region depends on solar radiation and disappears during nights.

D-region is well developed during solar flares, during which complete break-down of Medium and High frequency radio communication takes place. This is called sudden ionospheric distrubance (SID).

E-region: Strongly reflects Medium and High frequency radio waves. E-region begins to weaken after sunset but does not completely disappear. It disappears completely during polar nights.

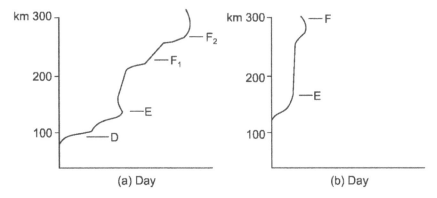

Fig. 14.11 Ionosphere diurnal effects

F_1-region: Found only during day time when the sun is fairly high. However when the sun is low and during night it merges with F_2 region. F_1-region is important for Medium and High frequency radio waves.

F_2-region: It is important in long distance radio communication. Upper F_2 region contains protons and electrons (mobile telephone) and vice versa, to provide digital voice communication.

Propagation of Radio Waves in the Atmosphere

The Principal features of propagation of radio waves in the atmosphere are given below.
- Diffraction of radio waves at the earth's surface.
- Absorption of Radio waves in the atmosphere and the surface of the earth.
- Reflection of radio waves from the surface of the earth
- Absorption, refraction and reflection by ionosphere. Inosphere is strongly charged by the ultraviolet rays, X-rays and corpuscualar radiation of the sun.

During normal conditions ultra short radio waves ($\lambda < 5$ m) are not reflected by the ionosphere. Direct waves propagated near the earth's surface are strongly absorbed by the earth. Consequently reliable reception of these waves (for example in T.V broadcasting) is possible only within the range of direct visibility, that is the distance within sight of the transmitting antenna. Long-distance T.V broad casting requires consecutives successive chain of relay stations called T.V. Station link; which receive, amplify and transmit signals along the chain work.

14.2.1 RADAR

The word Radar is derived from RAdio Detection And Ranging. The working principle of radar is based on reflection and scattering of radio waves by various objects. In this radio waves are used to detect the presence of an object, to find its location (distance and direction) and velocity.

A radar station consists of an ultra shortwave radio transmitter and a receiver with a single (i.e., common) transmitting and receiving antenna which produces a radio pencil beam (a narrow direction beam).

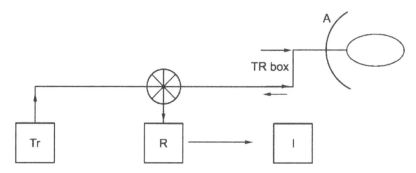

Fig. 14.12 Block diagram of a radar. T_r = Transmitter; R = Receiver; A = Antenna; T_R = Transmitter Receiver; I = Indicator

Transmission is in the form of short pulses of period of 10^{-6} sec (μs). During the interval between successive (consecutive) pulses the antenna is automatically switched over to reception of the echo-signal reflected from the target (object or obstacle). The distance of the target (radial distance from the antenna) is determined from the time interval between each transmitted pulse and the reception of the echo-signal. Radar uses ultra short waves (in the decimeter, centimeter and millimeter bands). The most successful operation (detection) depends on the size of the target (object) being detected, it should be many times larger than the wavelength (of radar or transmission wavelength λ). Radars are used efficiently and profitably in air navigation, detection of hazardous weather (like cyclones, tornadoes, thunderstorms, microbursts, clouding and precipitation etc.)

14.3   SATELLITE METEOROLOGY

Introduction

Satellite meteorology is the result of combination of science, satellite technology, computers and communications. Space age has began with the launching of sputlik-I by the Soviet Union on 4th October 1957. It may be regarded as communication satellite by virtue of radio beacon.

Observing the weather from space began with the launching of TIROS-I (Television Infrared Observational Satellite) by USA on 1st April 1960, since then 27 meteorological satellites have been launched by USA. The first weather satellite Cosmos-122 was launched by Soviet Union in June 1966 and 13 followed thereafter. INSAT-1, the first Indian national satellite was launched on 18th July 1980. INSAT-I series, is a multiple concept consisting of communication, television, radio broadcasting and meteorological services. METSAT launched by India in 2002, later renamed as Kalpana-1, is exclusively dedicated to meteorological services. INSAT-3D launched in 2007 contains a sounder.

All objects in the universe emit radiant energy as long as their temperature is more than 0 °K, but cease to radiate when it is o °K or less. The link between the distant stars and the earth is the electromagnetic radiation emitted by them. Radiant energy travels in space / vaccum in the form of electromagnetic waves, which travel in all directions away from the souce with speed of light, 3 × 10^{8} m/s.

The range of all possible radiation wavelengths constitutes electromagnetic spectrum. It consists of gamma rays, X-rays, ultraviolet radiation visible light radiation, infrared radiation, microwaves and short and long radio waves.

The wavelength (λ) and frequency (f) are related by the formula

$$C = f\lambda$$

where, C is velocity of light and associated energy (E) is given by the formula

$$E = h\nu,$$

where   h = Planksconstant,

ν = frequency.

The electromagnetic spectrum is given in Table 14.3.

Table 14.3 Electromagnetic spectrum

Wavelength		Wavelength	
10^{-6} nm	Gamma rays (MeV)	1mm	Millimeter waves (mm) Microwaves (cm, GHz)
10^{-5} nm		1 cm	
10^{-4} nm		10 cm	
10^{-3} nm		1 m	
10^{-2} nm		10 m	
10^{-1} nm	X-rays (A°)	100 m	
1 nm		1 km	
10 nm		10 km	
100 nm	Ultra-violet (nm)	100 km	Radio waves (MHz, kHz)
1 mm	Visible	1000 km	
	Near Infrared μm	1000 km	
	Thermal Infrared		
10 μm	(μm)	10000 km	
100 μm	Far Infrared (μm)	100000 km	

**nm = nanometer = -10^{-9} m, mm = micrometer = 10^{-6} m

Wavelength range of visible radiation, nomenlecture of microwave and radiowave frequencies, microwave bands are given in Tables 14.4, 14.5 and 14.6 respectively.

Table 14.4 Wavelength range of visible colours

Colour	nm	μm
Violet	380 – 430	0.38 - 0.43
Indigo	430 – 500	0.43 - 0.50
Blue	500 – 520	0.50 - 0.52
Green	520 – 565	0.52 - 0.565
Yellow	565 – 590	0.565 - 0.59
Orange	590 – 625	0.590 - 0.625
Red	625 – 740	0.625 - 0.740

Table 14.5 Nomenclature of microwave and radiowave frequencies

Observation	Full form	Frequency	Wavelength
EHF	Extremely high frequency(microwaves)	30-300 GHz	1 mm-1 cm
SHF	Super high frequency(microwaves)	30-3 GHz	1 cm-10 cm
UHF	Ultra-high frequency	3 GHz-300 MHz	10 cm-1 m
VHF	Very high frequency	300-30 MHz	1 m-10 m

Table 14.5 *Contd...*

Observation	Full form	Frequency	Wavelength
HF	High frequency	30-3 MHz	10 m-100 m
MF	Medium frequency	3 MHz-300 kHz	100 m-1 km
LF	Low frequency	300-30 kHz	1-10 km
VLF	Very low frequency	30-3 kHz	10-100 km
VF	Voice frequency	3 kHz-300 Hz	100-1000 km
ELF	Extremely low frequency	300-30 Hz	1000-10000 km

Table 14.6 Microwave bands

Band	(Wavelength)	f (frequency)
mm – band	1-7.5 mm	40-300 GHz
ku - k - ka band	0.75-2.5 cm	12-40 GHz
X – band	2.5-4 cm	8-12 GHz
C – band	4-8 cm	4-8 GHz
S – Band	8-15 cm	2-4 GHz
L – Band	15-30 cm	1-2 GHz

Satellite Orbits

They are two kinds

(i) Sun-synchronous or near polar orbiting (elliptical or circular orbits), Tropical orbit.

(ii) Geostationary or geosynchronus (circular orbits)

The *polar orbitting satellite* crosses the equator at the same local time and views any given place under comparable lighting conditions. They cover the whole earth and see every location of the earth twice a day at the same approximate time. Generally the altitude of the satellite varies 500-1500 km.

The satellite based sensors can provide :

(i) Atmospheric vertical temperature profile of the globe.

(ii) High resolution picture transmission (HR PT)

(iii) Automatic picture transmission (APT)

(iv) Collection and transmission of data from automatic weather stations to the ground stations.

Geostationary satellites move west to east in a circular orbit in the equatorial plane at an altitude of about 36000 km and have the circular velocity (464 m/s) same as the that of the earth. Consequently they appear sationary to a fixed ground object on equator. These satellites provide continuous weather data round the clock over fixed area from about 60° N to 60° S (but it will not see the polar regions). A constellation of 5 or 6 geostationary satellites spaced around the equator will provide the near global pictures. The satellite based sensors provide vertical temperature profiles, cloud motion, wind vectors and sea-surface temperature. They act as a platform for reception of data from automatic weather stations and transmit to the ground stations.

14.3.1 SATELLITE PAY LOADS, TV CAMERA

The payload carried by the TIROS-1 satellite in 1960 was a TV camera that relayed to the ground, whatever it observed of land, oceans and clouds.

Depending on the application besides TV cameras, the following are the common meteorological payloads.

(i)   A scanning radiometer (SR)

(ii)   Very high resolution radiometer (VHRR)

(iii)   Vertical temperature profile radiometer (VTPR)

(iv)   Scanning microwave radiometer (SMR)

(v)   Data collection transponder (DCT)

(vi)   Tape recorders

(vii)   Solar cells

SR : Scanning radiometer has two channels

(i)   Visible channel 0.5 to 0.17 μm

(ii)   Infrared channel (window) 0.7 to 10.5 μm

Scanning radiometer functions like TV camera, transmits strip image along the orbital track. It takes continuous images of earth's atmosphere and surface. The resolution of visible channel is 4-12 km and IR channel 25 km.

VHRR: It acts like SR with better resolution of visible channel 1 km and IR channel 8 km.

Note : Radiometer is an instrument used to measure electromagnetic radiation energy. Satellite based radiometers can provide radiation budget of surface – air system, sea surface temperature (SST), cloud cover distribution and cloud top temperatures.

Radiometer operates in visible, IR and microwave regions.

Spectrometer: A satellite based spectrometer provides:

(i)   Chemical composition of the surface layers in the atmosphere and ocean

(ii)   Measures the atmopheric concentrations of water vapour.

Resolution of an instrument is the smallest area on the ground that the given instrument is able to identify.

VTPR: It determines the temperature profile of the atmosphere on realtime operational basis over oceans and other regions (where the ground based upper air sounding station network is scanty or absent).

SMR: It operates on microwave bands and can provide global coverage of surface temperature data with an accuracy of $1^{O}C$.

DCT: It is used for collection of meteorogical and oceanographic data from remote uninhabited places on land and at sea by means of land based, ocean based data collection platforms (DCPs).

Tape Recorder: To store data before transmission to ground station. Such data storage on board makes provision for acquiring the global data at CDA (command data acquisition) station. CDA are managed by satellite launching countries.

CDA: A ground based station (on earth) at which various functions to control the satellite operations and to obtain data from the satellite are performed. The CDA transmits programming signals to the satellite, and commands transmission of data to the ground. Processing of data (by electronic machines and by hand) is accomplished at the CDA station. Raw and processed data are disseminated from CDA stations.

Solar Cells : Used for power. The sides and top of the satellite are studded with more than 9000 of these solar cells.

14.3.2 REMOTE SENSING

The process of observing an object with the help of electromagnetic spectra (band of radiation wavelengths) that an human eye cannot see is called remote sensing. In meteorology this term is used for observing meteorological parameters with the aid of radiometers on board satellites. Infact remote sensing is measurement of returned (or back) radiation from the earth-atmosphere system. The satellite based sensors are used for the detection of earth observation (both atmosphere and sea), detection of severe events like cyclones, tornadoes, tsunamises etc, delivary of warnings directly to the people of the affected area and for data relay and communication capabilities.

Microwave remote sensing is a powerful tool in application of meteorology and oceanography. It has the ability to measure water vapour and liquid water even in the presence of most clouds. Land surface emissivity in microwave region is high and is variable. It must be noted that microwave sensors reception signal strength on board geostationary satellites is poor and hence not useful, while its reception on polar low orbit altitudes is fairly good and hence mostly used.

TV Carmeras: TIROS-1 was operational for 78 days but it clearly demonstrated beyond doubt the potential use of satellites for surveying global weather conditions from space.

TIROS-1 has two TV cameras one with low resolution and the other one high resolution. The cameras were operating only in the portion of the earth which was sunlit.

TIROS-8 was launched on 21st December 1963, equipped with APT (Atutomatic picture transmission) capability, which provided satellite pictures world over to meteorological agencies in real time as the satellite passed over their region. There were 50 ground stations including one at Mumbai, India.

The successor TIROS series was ESSA (Environmental Science Services Administration) series initiated in 1966 by U.S. ESSA series transmitted thousands of space based weather system images which were received by more than 300 APT reception stations in 45 countries by 1970.

NASA (National Aeronautical and Space Administration) US launched Nimbus series satellites which had on board very advanced instruments for mapping ozone, sea ice, radiation budget components, coastal zone properties and sea surface temperature. The series carried improved Vidicon Camera system for daylight coverage, which were transmitted through the APT system.

Another APT service was provided parallelly by the Russian METEOR series of satellites.

NOAA, AVHRR: TIROS next generation satellite programme, name TIROS-N launched on 13th October 1978 followed by many satellites in the series. This followed by NOAA (National Oceanic and Atmospheric Administration) series by US. This series till 2005 had NOAA - 18 (series).

All NOAA satellites placed in near circular polar orbit. It carried on board Advanced Very High Resolution Radiometer (AVHRR) which had 4-channel radiometer initially, subsequently by 6 channel radiometer. One in visible (VIS), two in near infrared (NIR), one in mid-wave infrared (MIR) and two in thermal infrared (TIR) wavelength regions (See Table 14.7). All channels had resolution 1.09 km at the sub-satellite point.

Table 14.7 NOAA AVHRR channel specifications

Channel Number	Spectral Band	Wave length Range (um)	Uses
1	VIS	0.58 - 0.68	Daytime cloud and surface mapping
2	NIR	0.725 - 1.00	Land-water boundary delineation
3A	NIR	1.58 - 1.64	Snow and ice detection
3B	MIR	3.55 - 3.93	Night time cloud mappin SST retrieval
4	TIR	10.3 - 11.3	Night cloud mapping, SST retrieval
5	TIR	11.5 - 12.5	SST retrieval

14.3.3 GEOSTATIONARY SATELLITES

The first US Geostationary Operational Environment Satellite (GOES) was launched on 16th October 1975, the first new generation (GOES-8) launched on 13 April 1994. GOES-12 or GOES-East is located on equatorial orbit at 75°W and GOES-10 or GOES-west is located at 135 °W.

The current GOES satellites have 5 channel imaging radiometers whose specification are given in the Table 14.8.

Table 14.8 GOES channel specifications

Channel Number	Spectral Band	Wavelength range (um)	Resolution
1	VIS	0.55-0.75	1
2	SWIR	3.8-4.0	4
3	WV	6.5-7.0	8
4	TIR	10.2-11.2	4
5	TIR	11.5-12.5	4

(SWIR : shortwave infrared)

INSAT: (Indian National Satellite)

INSAT 1 : was launched on 18th July 1980, (India became the seventh nation after Soviet Union, USA, France, Japan, China and UK in satellite launching).

INSAT was a multipurpose satellite used for long distance telecommunication, meterorological earth observation and data relay, and direct TV broadcasting.

INSAT-1 A launched on 10th April 1982 was deactivated on 6th September 1982 due to malfunctioning. INSAT - IB launched on 30 [th] August 1983 was a geostationary satellite, located at 74 °E.

India launched METSAT on 12th September 2002, which was renamed Kalpana - I, is an exclusive meteorological satellite.

INSAT - 3D launched 2007 carried a sounder.

The other geostationary satellites are:

METEOSAT of Europe locate at 0^0 meridian

INSAT - ID of India located at 85^0E

GMS of Japan located at 140^0E

14.3.4 IMDPS (INSAT METEOROLOGICAL DATA PROCESSING SYSTEM)

IMDPS is receiving and processing Meteorological data from AWS (automatic weather stations). An AWS station records, stores hourly data and transmits the same to Kalpana - 1 through INSAT Data Relay Transponder (DRT). Cloud imageries are supplied and transmited through INSAT-3C to various IMD forecasting offices and also to other users through MDD (Meteorological Data Dissemination) scheme.

IMDPS generates cloud imageries and thereby derives the following meteorological products.

1. Cloud Motion Winds (CMW), using 3 hourly consecutive half-hourly images from the operational Kalpana-1 satellite at 0000 UTC and 0730 UTC and sends to users through GTS, IMD website.

2. SSTs are computed from INSAT - IR imagery from 0000 and 1200 UTC. SSTs also computed from NOAA satellite data using multichannel algorithm.

3. OLR (Outgoing Longwave Radiation) and quantitative precipitation estimates at $1.0^0 \times 1.0$ grid is computed from INSAT - IR data on 3 hourly/daily/weekly/monthly basis.

4. Atmospheric soundings are generated from US polar orbiting (satellite) NOAA series. The soundings include temperature and humidity profiles and standard level geopotentials.

5. METEOSAT - 5, satellite is located in equatorial orbit at long 63°E. and covers India and adjoining area.

Primary Data Utilization Centre was installed in 2000 AD for reception of data from METEOSAT-5. The data is received in VIS, IR and WV-channels at 30 minute intervals.

14.4 GLOBAL POSITIONING SYSTEM

Introduction

Satellite navigation is the concept of instrument navigation. It was developed as military system by the US and the Soviet Union. This system further developed with the computer system and provides more accurate position fixing for aviation purposes. ICAO calls this system as "Global Navigation Satellite System" (GNSS). The US calls this system "Global positioning System" (GPS). While the Russians call it "Glonass" system which is commercially available with suitable receiver equipment. The concept of satellite navigation is basically the principle of distance measuring equipment. The receivers measure the time that it takes radio signal to travel from satellite based transmitter at a known point in space. Knowing the speed of radio wave propagation and time taken for its journey from transmitter to the receiver the distance can be calculated.

Space Segment

The GPS consists of constellation of satellites (upto 29) that transmit L-Band radio signals. The satellites are all moving in a circular orbit of 26,500 km radius with inclination of 55° to the earth's spin axis and period of revolution of 12 hours around the earth. The satellites move in six different orbital planes with identical orbital characteristics. Each GPS satellite broadcasts signals in two L-band frequencies L1 at 1.57542 GHz (or 19 cm wavelength) and L2 at 1.22760 GHz (or 24.4 cm wavelength). The L1 carries the coarse acquisition or CA code, which can be received and interpreted by all receivers. The transmission from every satellite includes the orbit that each satellite is following and its position in the orbit as an almanac. The receiving computer uses that almanac to calculate where each satellite should be in its orbit. This provides which one will be in view and approximate range from each one. The ranges received at the receivers are not exact and hence called "pseudo-ranges". The satellite transmits its own exact position and path, called the "ephemeris". Each satellite carries a very accurate atomic clock which is in agreement with those in the constellation of satellites. This helps in measuring the true ranges (from the satellites instead of pseudo-ranges.)

The satellites (constellation) are supported by ground stations, which receive signals from all the satellites in turn as they pass overhead. The ground stations monitor the signals, and amend the almanac and ephemeris messages transmitted from the satellites.

Control Segment

The Control Segment consists of the ground stations. The Master station does all the calculations and the other ground stations provide communications between the satellites and the Master station.

User-Segment (Aircraft Equipment)

The GPS receiver is a simple device consisting of a small screened aerial mounted on the skin of the aircraft. Hand-held devices are also capable of showing position accurately. The computers and their associated software are the complex parts of the aircraft equipment. Computers can calculate actual track and groundspeed made good and required track and ETA (Expected time of arrival) to the intended waypoints. Given inputs from flight instrument or an air data computer, the GPS computers can make all navigational calculations.

The receiver can be any one of the three categories; multi-channel, multiplex or sequential. Sequential receivers lockon to one satellite at a time, measure the pseudo range, then reach and lockon to the next. This system is not useful in aviation. Multi channel receivers are very expensive but they receive the signals from all the satellites in view independently and simultaneously (preferred type for aviation). Multiplex receivers 'time share' their receiving and computing time between all the satellites in view. A multiplex receiver can store the information from each while it is receiving information from others.

Pilot interface with some GPS instruments is by means of a numeric keypad, but those designated for light aircraft tend to use a menu system. A full computer or alpha-numeric keyboard controls the flight management system of airlines, and is ued to interface with the GPS equipment.

GPS Accuracy

The course acquisition (CA) signal is assumed to be accurate in the horizontal plane to within 30 m for 95% of the time. It is assumed accurate within 300 m for 99.99% of the time. The computed vertical position is however less accurate. The 95% position should be within 156 m (500 ft) vertically and 99.99% position within 500 m (1600 ft).

GLONASS

The Russian satellite navigation is called GLONASS. It uses separate frequencies for each of its active satellites which have orbital incliation at $60°$, to transmit navigational data. All GLONASS satellites orbit the earth every 12 hours at an altitude of around 19000 km.

Some Applications of Global Positioning System (GPS)

Electromagnetic-surveillance technology is used in USA to monitor habitual / repeated offenders. Coast to coast authorities are using electronic monitoring devices to fight crime (this include gang members who were released on bail or probation, people accused of committing repeated violence against women etc).

i-Secure track corporation, based in Omaha, Neberska is manufacturing these devices and leasing them to Police and courts. It is said in Massachusetts, about 700 offenders were fitted with electronic bracelets which send signals via satellite to computer services to know their where abouts or if they go to the places which were debared (exclusion zones). The local law allows judges to impose electronic monitoring offenders. This system is cost-effective alternative to prison. GPS is also used in air navigation in search and rescue operations. A similar device is prepared by Indian Railways to locate railway accident sites promptly.

14.4.1 INDIAN REGIONAL NAVIGATION SATELLITE SYSTEM (IRNSS)

IRNSS designed to provide accurate position information service to users in India and extending up to 1500 Km from its boundary which its primary service area.

IRNSS will provide two types of services.

1. Standard positioning service (SPS),
2. Restricted service (RS), to provide a positioning of better than 20 meters in primary service area.

   IRSS-1B Navigation Satellite of India.

   IRNSS-1A, the first of the seven satellites of the IRNSS constellation.

   Seventh GPS Navigation Satellite, IRNSS was launched into sub-geosynchronous transfer orbit with a perigee (nearest point to earth) of 284 Km and apogee (farthest point earth) of 20657 Km.

   The Indian Regional GPS (global positioning system) constellations consist of seven satellites. The constellation is similar to the GPS operated by US with 24 satellites, Glonass 21 satellites of Russia, Galihio of Europe and Beldou system of China.

1. The GPS is a network of 30 satellites orbiting the earth at an altitude of 20,000 Km.
2. India has seven satellites orbiting the earth, of while 4 (four) satellites always be visible from India. So these will be able to tell precise location up to 100 meters.
3. The GPS system will be used for terrestrial, aerial and marine navigation vehicle tracking, fleet management disaster management, mapping and geodetic data capture, Navigation for drive.

   Also use in Defense. The 7-satellites monitor the entire country (India) and also potential threats by missiles, fighter planes. Another advantage is that signals are not jammed.

   Drawbacks: It does not offer global coverage and covers only India and its neighborhood. Commercial air lines that fly over India will not install it because they already have GPS tracking device on board.

   Launching and some details of (last 3 IRNSS) given below.

   IRNSS - 1A, July 2013

   IRNSS – 1B, April 2014

   IRNSS – 1C, Oct 2014

   IRNSS – 1D, March 2015

   IRNSS – 1E, Jan 2016

IRNSS – 1E carried payload – a Rubidium atomic clock, c-band Transponder. It provides accurate position information service to users in India and the region extending up to 1500 Km.

IRNSS – 1H, Aug 2017. It is called Now IC. It is the Indian Regional Version similar to US GPS. It will drive all positional based activates on ground, sea and air by giving near accurate location details of persons/abject. It will be put to military commercial and common every day uses.

14.4.2 GAGAN (GPS AIDED GEO AUGMENTED NAVIGATION)

GAGAN is a satellite based Augmentation system (SBAS) implemented jointly with Airport Authority of Indi a (AAI) and ISRO (India Space Research Organization) to provided navigation services with accuracy and integrity required for civil Aviation applications and to provide better Air Traffic Management over Indian Air Space. The first GAGAN was flown on GSAT-8 in 2011 and second on GSAT-10 launched on Sept 29, 2012.

INSAT Applications

The telephone circuit devices through INSAT connect remote in accessible areas to major cities in India. 85% of Television services in India's populations is via INSAT, over 200 AIR stations are linked through INSAT-g-network VSAT (Very Small Aperture Terminals) used for Telecommunications sector. In additions INSAT supports over 29000 VSATS for e-commerce and e-governance. National stock Exchange use VSAT Technology across the country for instantaneous transactions. Interactive training and developmental communication, distance learning achieved through exctuusive channels.

Kalpana I Satellite exclusively used for meteorological purpose. The imaging (instruments is achieved through) Very High Resolution Radiometer (VHRR) and CCD (charge coupled Device) service used for collection of meteorological data and to Provide timely warnings on impending cyclones on real time and hydro-meteorological data for monitoring river stage, inflows, outflows across barrages.

EDUSAT (launched in 2004) used for distance education. About 35000 class rooms are connected through EDUSAT network provide services at Primary, Secondary and University levels.

The satellite based telemedicine network connected 375 hospitals (305 remote and rural hospital including those in Jammu and Kashmir, North eastern region and Andaman and Nicobar Islands, 13 mobile units and 57 super speciality hospitals in major cities)

14.4.3 IRS APPLICATIONS

Imagery taken by *Indian Remote Sensing* satellite system has found application in (i)Agriculture (ii) Urban planning (iii) crop health monitoring (iv) crop yield estimation and (v) drought assessment and allied fields (vi) soil mapping (viii) ground water potential zone mapping (viii) mineral targeting tasks (ix) ocean fishing zone identification (x) coastal zone napping (xi) forest cover mapping (xii) Biodiversity characterization and (xiii) monitoring forest fire is now carried out using IRS imagery)

IRS space craft also provides timely inputs to flood and earthquake damage assessment necessary for giving support for disaster management. For Archaeological Survey IRS imagery are useful.

1.4.4 GEOSTATIONARY OPERATIONAL ENVIRONMENTAL SATELLITE (GOES)

The first US GOES was launched on 16 October 1975. Since then GOES 2-12 have been launched. Current GOES-12 or GOES-East positioned on equatorial orbit (altitude 36000 km) at 75^0 W longitude and GOES 10 or GOES-west is positioned at 135^0 W longitude. The GOES imager is a 5-channel imaging radiometer, whose spectral band, wavelength range and resolution are given below.

Channel Number	Spectral Band	Wavelength range (µm)	Resolution km
1	VIS	0.55-0.75	1
2	SWIR	3.8-4.0	4
3	WV	6.5-7.0	8
4	TIR	10.2-11.2	4
5	TIR	11.5-12.5	4

VIS = Visible, SWIR = short wave infrared, WIV = water Vapour, TIR = Thermal infra-red

GOES provide half hourly observations. The instruments on board satellite (GOES) measure earth emitted and reflected radiation from which: (i) atmospheric temperature (ii) winds (iii) moisture and (iv) cloud cover data can be derived.

GOES serves satellites owned and operated by NOAA (National Oceanic and Atmospheric Administration).

NASA (National Aeronautics and Space Administration) manages the design, development and launch of the spacecraft. NOAA owns responsibility for the: (i) command and control, data receipt and (ii) product generation and (iii) distribution. Each satellite (in the series) carries two main interments – an Imager and a Sounder. These instruments (have) acquire high resolution visible and IR data and temperature and moisture profiles of the atmosphere. On board instruments continuously transmit data to ground terminals, where the data are processed and re-broadcast to primary weather services in US and the world over received by Global Research Community.

The functions of GOES (1-M)

1. Acquisition, processing and dissemination of imaging and sounding data.

2. Acquisition and dissemination Space Environment Monitor (SEM) data

3. Reception and relay of data from ground-based Data Collection Platforms (DCPs) that are installed at carefully selected urban and remote areas, to NOAA-CDA (Command and Data Acquisition) station.

4. Continuous relay of weather facsimile (WE FAX) and other data users, independent of all other functions.

5. Relay of distress signals from people aircraft or marine vessels to the Search and Rescue Ground Stations of the Search and Rescue Satellite Aided Tracking (SARSAT) system,

GOES provides the instantaneous relay functions for he SARSAT system. A dedicated Search and Rescue Transponder on board GOES designed to detect emergency distress signals originating from earth based sources. These unique identification signals are normally combined with signals received by a low Earth Orbiting Satellite Systems and relayed to Search and Rescue Ground Terminal. The combined data are used to perform effective Search and Rescue operation.

As noted earlier, GOES West located at 135^0 West longitude and GOES East at 75^0 West longitude. A common ground station, the CDA (Command and Data Acquisition) station located at Wallops, Virginia supports the interface to both satellites.

Delivery of products involves ground procession of the raw instrument data for radiometric calibration and Earth Location information and retransmission to the satellite for relay to the data are received at Control Centre and disseminated to the National Weather Services (NWS) National Meteorological Centre Camp Springs Maryland and NWS forecast offices (including the National Hurricane Centre Miami Florida and National severe Storms Forecast Centre, Kansas city Missouri. Processed Data also received by Department of Defence installations universities and a number of private commercial users).

14.4.5 SATELLITE MEASUREMENTS OF OCEAN PARAMETERS ROUTINELY

1. Sea surface temperature (SST)

2. Wave height
3. (Sea) Surface height
4. Sea ice distribution
5. Radar/back scatter
6. Ocean colour

Meteorological parameters that are important for ocean forcing are:
1. Wind speed
2. Sea level pressure (atmospheric pressure)
3. Cloudiness or cloud coverage
4. Water vapour content in the atmosphere.

14.4.6 GSAT-17 COMMUNICATION SATELLITE

17 satellites working orbit. GSAT-17 launched on 29 tunes 2017 from the European Space part of Kourou in French Guiana.

GSAT-17 designed to provide continuity of services of operational satellites in C, extended C and S bands. The spacecraft support various users across India.

GSAT-17 Main functions

(i). Broadcasting (ii). Telecommunication and (iii). VSA services carrier over 40 transponders. GSAT-17 will add to the services they provide for broadcasting, telecommunication, VSAT Services meteorology, Search and Rescue, among others.

14.4.7 CARTOSTAT–SERIES

Cartostat–2: The first satellite in the services was launched in 2007. The satellites on board camera had a spatial resolution of less than one meter. It can cover a swath of 9.6 Km in the image. The camera can be tilted up to 45^0

Cartostat–2A: It was launched exclusively for defense surveillance and its panchromatic camera can be fitted up to 45^0 to capture images. The satellite was launched in April 2008.

Cart state 2B: caret stat – 2Bs panchromatic camera can be moved to 26^0 to take hodographs

Cartostat–2C: It was launched in June 2016. Resolution 25 cm in September 2016, the Indian Army carried out a "surgical strike" on terror camps situated across the India Pakistan line of control in Jammu Kashmir.

Cartostat–2D: Launched in February 2017 with 103 satellites as part of PSLVC 37 mission. Used mostly for cartographic observations including defence survellians.

Cartosat–2E: launched on 23 June 2017. Functions same as in Cartostat-2D

Cartosat–3: Launched on 27 Nov 2019. A high resolution (one foot or 25 cm) earth observation satellite. U.S. Satellite World view-3 has maximum resolution 31 cm. Cartosat helps large scale urban planning and infrastructure development.

14.4.8 SATELLITE NAVIGATION (SAT NAV)

Location based information data via satellite are important in our daily life if data is available through compact hand held devices like mobile phones. According to a Military Scientist, in modern day war, Sat Nav is very useful to decide where to hit and where to dodge.

Satellite aided Navigation is "the art knowing where you are and moving towards where you want to go in the shortest possible time. Satellite aided Navigation will certainly change our life in future. It will particularly influence in the areas, (i) agriculture (ii) ocean fleet monitoring (iii) weather prediction and (iv) locating people in distress (v) that is in disaster mitigation.

ISRO has a two pronged Sat Nav plan.
1. GAGAN (The GPS aided Geo Augmented Navigation and
2. Indian Regional Navigation Satellite System (IRNSS)

Now a day's man has been using various navigation techniques, to this Sat Nav will help the world to revolutionize with innovative solutions using position information and time. Government planners, emergency service providers, infrastructure companies transport companies, travelers use (Sat Nav) satellites to know exact position information.

Sat Nav primary uses

(i)   Sailing ships to know their position information for docking and harbor operations

(ii)   Railways use to avoid collision and safety at unmanned level crossing

(iii) It is useful in power grids and banks to know exact time of power transfer and money transactions.

(iv)   As per ISRO scientist, a farmer can use fertilizers optimally and plant multiple crops.

(v)   Flow of rivers can be tracked and people be warned in case of impending flood situation with Sat Nov powered devices tike a receiver on buoy.

(vi)   During disasters, Sat Nov can be used as an additional communication device, which will help in disaster mitigation.

(vii) Sat Nov applications can be made in several programmes Govt. India, like Digital India, Smart cities, AMRUT (Atal Mission for Rejuvenation and Urban Transformation)

(viii) Stakeholders and people to use when they require advice on phone with Sat Nov Chip and receiver
       GAGAN has been jointly implemented by ISRO and AAI (Airport Authority of India, to smoothen air traffic management across the Indian Skies, which improves the safety of aircrafts.

Special features of Sat Nav

Sat. Nav provides spatial positioning and enable hand held devices to determine accurate position using time signals.

Present Global Satellite Navigation Systems

  (i) GPS (Global Positioning System). It consists of upto 32 medium orbit satellites operational since 1994

  (ii) GLONASS: It consists of 24 satellites, operated by Russia Aerospace Defence Forces.

  (iii) DORIS (Doppler Orbitography and Radio positioning Integrated by satellite).

French precision navigation system. Based on static emitting stations around the world. Limited in usage and coverage, used with other traditional QNNS system.

Satellite Based Instrumental Data

Satellite based meteorological data is proved to be a boon in observing weather over vast oceans, mountains, dense forests and inaccessible areas. Large weather systems, like tropical cyclones Tornadoes, Mid-latitude cyclones, Anticyclones, Jet Streams etc are clearly identified.

The capability of weather satellites to observe and monitor weather systems depend on resolution of spectral channels of the radiometer, the orbital period of satellite, height of satellite above earth etc.

  (i) Satellite remote sensing is essentially the measurement of returned radiation.

  (ii) The absorption spectra of atoms that occur in UV-region of electro-magnetic spectrum is a transition from one energy state to another. Absorption or emission of radiation takes place when there is a transition one energy state to another.

  (iii) In the visible region, gasses in the atmosphere account very little absorption. Atoms and molecules in a gas have electronic rotational vibration energy.

The main absorption bands are those of three atmospheric gases. (i) water vapour at 6.7 μm (ii) CO_2 at 15μm and (iii) ozone at 9.6 μm. In addition to these, there are minor absorption bands pertaining to Methane, Nitrous-oxides and other gases.

Geostationary satellite remains fixed over one location on equatorial orbit (36000 Km. altitude). It scans one-Third of earth's surface under the satellite.

GOES- East and GOES-west together provide full coverage of the American continents and most part of the Pacific and Atlantic Oceans.

ESA (European Space Agency) first satellite was Meteosat, stationed over Western Africa (intersection of the equator and the Greenwich meridian. Resolution 1 km in visible light. Metsat (or kalpana -1) launched by India on 12 Sept 2002 is a first exclusive India's meteorological satellite, located at 74^0 E on equatorial orbit and INSAT-3A launched in April 2003 located 93.5^0 E. INSAT-3D launched in 2007, which carried a sounder.

Note:

1. All satellite sensing is "remote sessing", using electromagnetic radiation at different wavelengths.

2. Radiation used in satellite system-
   (i)   Microwave – 0.1 to 30 cm
   (ii) IR 0.7 to 100 µm
   (iii) Visible – 400 to 700 nm
   (iv) $c = \lambda v$
         Where $C = 3 \times 10^8$ m/sec = speed of light
                  λ = wavelength
                  v = wave frequency

3. **Satellite radiation sensors (a) Active (b) passive:** In Active system the satellite radiates energy down-ward and then senses reflected or re-radiated from earth (like that of radar & slider) system that emit radiation to sense surface properties.

   In passive system, the satellite sensors detect radiation from earth's surface or reflected/re-radiated solar energy.

4. Atmosphere is transparent, from micro waves $\lambda > 5$ mm. Electromagnetic radiation spectra nearly passes. Visible and near IR completely passer though atmosphere.

   Solar emission is maximum in visible range and most of short wave solar energy reaching the earth. Thermal IR (10-12µm) transmission is greater in atmosphere window.

   The best long wave thermal channel is at 10 µm and another at 5 µm and also 60-80µm.

   INSAT-3DR An advanced weather satellite put into orbit, 09 Sept 2016. It carried a multispectoral imager (MSI), which images the earth from geostationary altitude of 36000 km every 26 minutes. It provides information on parameters such as sea surface temperature (SST), snow cover, cloud motion winds (CMW) etc. Second payload, a 19-channal sounder that provide information on vertical profile of temperature (VPT), Humidity and integrated ozone while Data Relay Transponder will provide Service Continuity to ISROS previous meteorological missions.

   The third payload had the Search and Rescue pick up and relay alert signals originating from the distress beacons of marine aviation and land based users to Indian Mission Control Centre (INMCC).

The fourth payload had the main users of the serivce-the Indian Coast Guards, Airport Authority of India (AAI), Directorate General of Shipping, Defence Services and Fisheries.

The Indian Service Region will cover a large part of the Indian ocean and also include Bangladesh, Bhutan, Maldives, Nepal, Seychelles, Srilanka and Tanzania for providing Distress Alert Services.

14.4.9 UN INDIRECT SATELLITE IMAGERY FOR DISASTER RISK REDUCTION (UNISDR) MODEL

A UN model which uses satellite imagery and on ground assessments to estimate direct and indirect damages. This will help the centre to estimate scientifically, the relief and reconstruction packages for disaster hit regions.

National Institute of Disaster Management (NIDM) prepared PDNA (Post Disaster Need Assessment) model for estimating disaster losses.

14.5 GEOPHYSICAL INFORMATION SYSTEM (GIS)

In India National Spatial Data Infrastructure (NSDI) forging digital links to construct information Highway, that could place data secreted in official vaults to be available on a personal computer. The data may be from diverse fields (like street maps, forest areas, hospitals, tax collection etc).

NSDI aims to use geophysical information system to mold satellite imagery and Survey of India Toposheets, superimposing data on water resources, flooding, rainfall, crop patterns, civic layouts etc., to produce three dimensional (3-D) digital maps. These maps will be of great use in disaster management besides in agriculture, irrigation, townplanning etc. As a first step major cites in India will be mapped on a scale 1:1000 and subsequently entire country will be mapped in a phased manner (similar to the google-earth).

GIS enables us to use (any one with a personal computer) to face several layers of data such as forest cover, taxes collected from a region or area affected by epidemic (like cholara, bird flue etc) on to a base map of the region and then carryout geographic and cartographic analysis. NSDI will act as online database to maintain such data layers and base maps in easily retrievable form.

Defence Ministry cleared 1600 Survey of India maps (given no objection certificate for use). GIS unshakling information technology (IT) and space technology for everyday use. Some of the classical Survey of India maps will be in public domain on a designated website. Work on 3-D digital "live" maps of major cities will progress as given below.

(i)   NSDI will use GIS to fuse satellite imagery (photographs) to create database of urban and rural areas.

(ii)   Satellite images will help in preparation of maps of major cities on scale 1 : 1000. Data on municipal plans, forests, ground water, roads will be superimposed. 4800 Survey of India maps on scale 1 : 50,000 will be open to public (many for the first time).

(iii) Maps will be updated using satellite aereal photography. Public private partnership will be encouraged. Commercial use will include vehicle navigation and data integration. User shall have any of the four controls. (a) Hand-held cellphone (or PADS), (b) Computers (laptop and desktop), (c) Internet kiosks in rural areas, (d) Printouts and Television.

14.5.1 SOME APPLICATIONS OF GIS

GIS is a poweful tool to disclose what actually lies on the ground.

High Resolution GIS Surveillance helps in tracking terror activities. Recently (2007 - 08) in Birmingham anti-terror arrests made by the use of High Resolution Surveillance. Indian army uses GIS to track terrorist infiltration across Indo-Pakistan border.

Service providers can deliver updates on road blocks and traffic congestions. On a rainy day planning a driveout of town can be simpler.

GIS helps accurately in assessing standing crops. It can provide changes in cultivated areas and also quality of gram.

GIS based 3-D maps will deliver the civic bodies authorities the truth on the ground, helps in tracking car thieves, it gives existence of a road or street or any area has been encroached, water logged, fields inundated.

GIS softwares can be applied to the tasks of locating sites for tubewells, percolation tanks in accordance with a set of parameters. Example : No well should be more than 500 m from a habitation.

Groundwater in a specific area, say north of Secunderabad adjacent to National Highway-7, can be mapped accurately using GIS.

Using GIS, medical team/health professionals can track geographic links of a disease outbreak (like birdflue) and act to control it.

GIS is of great help in disaster management, for example in flooding. It can provide information on roads, hospitals, railway tracks etc., and diret the rescue team in a pinpointed manner.

Using GIS and personal computer (PC) by a click of mouse one can superimpose physical and geographical data. GIS scheme will be profitably rolled out through public private partnerships. Any citizen, including academics, can use the data by paying fee. Besides use in disaster management, digital maps loaded with data are useful in various economic activities. The pilot project on Chandini Chowk (Delhi) had clearly shown that even crowded places can be accurately reconstructed with the help of GIS. Some map making companies are developing detailed urban maps useful in tourism and transport, entertainment, health, education and business.

Government of India considering a mechanism through GIS to enable investigation of certain types of terror incidents and other offences having inter-state and international linkages by a central agency. At present such matters are probed by CBI (Central Bureau of Investigation) only when they are referred to it by States concerned or by courts. Since "law and Order" comes under the control of States within the federal set up, the CBI or any Central Agency cannot take-up the cases of investigation on its own.

Use of Satellite (VSAT hub) by Indian Railways in Disaster Management

Indian Railways run about 12000 trains everyday, that carry more than 125 lakh passengers across the country of about 65000 km railway tracks. Available statistical records show the highest rate of railway accidents 300 per year.

In order to help the disaster victims an efficienct communication has been planned through VSAT hub in Delhi to receive updated reports from the accident sites. The Railway VSAT is connected to INSAT 4CR launched by ISRO (Indian Sapce Research Organisation). The VSAT hub was designed and commissioned by Hughes communication Ltd through its satellite broadband services. Indian Railways will install mobile satellite terminals in about 100 Accident Relief Trains (ART) which will be stationed across the country. Earlier ARTs would reach the accident site with sole purpose of rescue operations. With this satellite terminal in ARTs there will be instant Vedio and voice data connectively with bigger Railway Stations on route which will facilitate quick rescue, relief operation. The satellite terminal in ART will be in contact with the VSAT hub in Delhi via INSAT 4CR within minutes. The satellite terminal also allows connectivity with normal BSNL (Bharat Sanchar Nigam Ltd) phones and Railways Administrative lines near the accident site to other stations. With this conncectivity standard passengers can get in contact with their homes/relatives.

Arrangements are also being made for using the VSAT hub to provide internet connectivity on trains by installing portable antenna as roof of outstation trains. The antenna will continuously track signals with INSAT 4 CR and the VSAT hub at Delhi. This will ensure wireless connectivity to passengers with laptop in the train.

Some uses of Cartosat 2 Satellite High Resolution Imageries

Indian Space Research Organisation (ISRO) launched Cartosat-2 in January 2007. The resolution of imageries is one meter, but trying to improve to half meter resolution. With the cartosa-2 imageries it is planned to prepare cartographic map of India. The data will be used in urban development. The imageries could provide information required for infrastructure development, disaster management and watershed development. The imageris can help building a national database which will be very useful for a long term disaster management. The life of Cartosat-2 is five years that is upto 2012.

India has the world's largest constellation remote sensing satellites. Seven satellites in orbit provide imagery of the earth in a variety of spectral bands with a resolution of one meter (even less). The satellites provide remote sensing data, which are received at about 20 ground stations across various parts of the globe including US and Europe. Village Resource Centres established in 2004 are providing a variety of space based products and services including tele-education, tele-medicine and information on natural resources. There are about 400 Village Resource Centres in the country.

Consists of starting and stopping the continuous carrier wave, breaking up into dots and diesels. This is generally called wireless telegraphy or interupted carrier wave (I.C.W). The communication is by Morse code. The receiver requires a beat frequency oscilltor facility.

CHAPTER - 15

Energy

An object is said to possess energy if it is capable of doing work.

Ability of an object to do work is its energy.

Or

Energy is the capacity to do work.

Newton's First Law of motion

Statement: Everybody (material points) remains or persists in its state of rest or uniform motion in a straight line until it is acted on or compelled by external force to change that state. This is called Law of Inertia.

Force: Force is a vector quantity, it has both magnitude and direction.

Force is a measure of the mechanical motion exerted on a body (or particle) by other bodies or field. Force is considered to be a physical entity, which tends to accelerate any physical body to which it is applied.

Newton's Second Law of Motion

The rate of change of momentum of a particle (or body) is directly proportional to the magnitude of force acting on it and inversely proportional to the mass of the particle (or body).

Or

The first time derivative of the momentum of a particle (body) is equal to the force acting on it.

$$\frac{dp_i}{dt} = F_i \text{ or } \frac{d}{dt}(m_i v_i) = F_i [\text{where } P = m_i v_i]$$

Newton's Third Law of Motion: Action and Reaction are equal and opposite

Or

The actions of two bodies (particles) on each other are equal in magnitude and opposite in directions/sense

$$F_{ij} = -F_{ji}, \text{ where } (i \neq j)$$

F_{ij} = Force acting on the ith particle (body) and exerted by the jth particle (body)

F_{ji} = Force acting on the jth particle (body) and exerted by ι^{th} particle (body)

Work: Work is a scalar quantity.

Work is done when the point of application of force is moved in the direction of the force. An element of work δw done by force F in the displacement d of a particle (body) is equal to the scalar product of vector $(\vec{F})$ and $d_s^{\rightarrow}$.

$$\delta w = \left(\overline{F}.ds^{\rightarrow}\right) = Fds\cos\alpha = F_x ds$$

Where $\quad\vec{r}$ = radius vector of the particle (body)

α = angle between F and dr^s

$ds = \left|dr^{\rightarrow}\right|$, F_x = F cos α

(projection of $\vec{F}$ on x axis or on the tangent of the path

W = Fs, where W = Work done (in Joule) by force F (Newton) S = Point of application moved through a distance 'S'.

The unit of work is Joule

As stated earlier, a body is capable of doing work is said to possess energy. The quantity of work is Q a measure of its energy. Hence energy is measured in the same units as work its Joules.

Power: Power is the rate at which work is done

$$P = \frac{W}{t} \text{ or } \frac{\delta w}{\delta t}, \frac{s}{t} = V$$

$$P = \frac{FS}{t} = FV \text{ where } F = \left|\vec{F}\right|$$

Momentum: It is the Product of mass and velocity

Movement = $m\upsilon$, where m = mass υ= velocity

Impulse = (magnitude of force) × (time)

Time is during which the force acts.

15.1 DIFFERENT FORMS OF ENERGY

For performing work flow (transfer) of energy is required. This is achieved by changing one type of energy into another.

Energy occurs in many forms, like potential energy (PE), Kinetic energy (KE), Internal energy, heat energy, nuclear energy, electric energy, wind energy, solar energy etc.

Potential Energy: The energy possessed by a body due to its position or state is called potential energy

PE = mgh (Joule) where

m = Mass of particle energy

h = Height of particle above ground

g = Gravity of the place

Kinetic Energy: The energy possessed by a body (object, or particle) because of its motion is called kinetic energy.

$$KE = \frac{1}{2}mv^2 \text{ (Joule)}$$

v= Velocity of the particle

M = mass of the particle

Electric charges possess electric (potential) energy in an electric field

Fuels possess stored energy.

Energy is stored in food is converted to body energy (for activities) by chemical process.

Magnetic poles possess magnetic (potential) energy in a magnetic field.

Heat is a form of energy (Fuel energy).

Radiant energy: In universe every body emits radiant energy as long as its (body's) temperature is above zero degree kelvins.

Radiant energy is transmitted in the electromagnetic (wave) radiation. This form of energy exists even in the absence of matter.

Geothermal energy: The source of geothermal energy is mainly due to radioactive decay taking place deep in the earth's core (which is responsible for volcanic eruptions, earthquakes). Geothermal energy generated deep in the earth's core is very diffuse with low temperature. Geothermal energy application classified as low temperature (less than 90 °C), moderate temperature (90 to 150 °C) and high temperature (more than 150 °C).

Low and moderate temperature resources are:

(i) Direct use, (ii) Ground source heat pumps and (iii) power generation.

Heating water directly in buildings, industrial processes, green houses, aqua-culture (temperature source 38-140 °C)

Resource temperature 4-38°C, used through pump device, one place to another.

Power plant taps accumulated energy inside the earth.

WIND ENERGY

INTRODUCTION

Wind Energy has been in use in some form or the other as a source of power from the dawn of civilization. This is an ideal example of renewable natural energy source without causing any pollution and interfering with nature. As early as 5000 BC, Egyptians used wind energy to drive boats and ships across the river Nile. Several centuries BC, Chinese developed wind mills for pumping water. In Persia, in 200-100 BC, vertical axis wind mill developed with woven reed sails and used for grinding grain. With the passage of time wind energy use was spread to Mediterranean region and from there it spread over Europe in 11th century, mainly by merchants. By the beginning of 20th century, the Dutch perfected the wind energy machines and then spread all over the world. However at that time fossil-fueled steam engine ships came into existence and slowly took over the wind driven wooden sailing ships because of improved technology. Similar replacement observed in America. Thousands of American farmers who had steel-towered wind mills for pumping water were replaced by rural electrification by fossil fuels. Present day research and development programmes aim at producing electricity (of the order 3 MW or more) by using wind mills. For this purpose wind speeds of the order 6 mps or more required, which is a restriction in selecting wind mill site. According to WMO estimates the world wide wind energy potential of generation of electricity is about 20 million MW.

WIND GENERATED ELECTRICITY

Towards the end of 19th century wind mills were the principal sources of power in agriculture in Denmark. The power generated was of the order 200 MW. In about 1890 -1908, Prof. P. La Cour erected a windmill at Ashov (Denmark) and pioneeringly generated electricity. The wind will had four blades of diameter 23 meters, erected on a steel tower of height 24.4 m. This wind mill was used to drive two 9 KW D.C. generators. This was the first recorded wind generated electricity. By 1944 wind generated electricity reached the peak with about 88 wind mills. This electricity was converted to AC in 1955, when it generated about 2000 kwh/per year. In 1941, the world's largest wind mill (Smith-Putnam) was erected at Grandpa's Knob in central Vermont (America). This wind mill had two blades of diameter about 53.5 m, erected on a tower of 33.54 m high. It generated electricity about 1.25 MW. However it failed after about 18 months due to breakage of main bearing, which could not be replaced. Technical problems of converting wind energy into electricity gradually overcome and at present any country with suitable wind climate can establish wind power electricity.

In 1931, the Russians erected their first wind mill to generate electricity at Yalta near the Black Sea. It had three blades of diameter 30.5m, mounted on a tower of 30.5 m high but was provided with an inclined strut to bear the high wind thrust. It was connected to an A.C generator and used as a supplementary power source. The site had annual mean wind speed of 15 kmph and its annual output was 279000 kwh.

Wind Energy Estimates

Air in motion is called wind. It has mass and its source of energy is the sun. The mass of the atmosphere is about 5.6×10^{18} kg. Horizontal wind speed (V_h) in general is much greater than the vertical wind speed (V_v). $V_h >> V_v$. The total energy of the atmosphere is present as potential energy (PE) and kinetic energy (KE). According to one estimate the total atmospheric PE = 2242.3×10^6 J/m^2 and KE = 1153.4×10^3 J/m^2. This shows that the average

amount of atmospheric KE = $\dfrac{1153.4 \times 10^3}{2242.3 \times 10^6} \square \dfrac{1}{2000}$ of the PE. The winds are

the result of conversion of atmospheric PE into KE. This conversion is mainly carried out by the work of atmospheric pressure force.

Atmospheric engine is mainly driven by insolation. We have studied that there is an excess of insolation as against the outgoing terrestrial radiation in equatorial region (equator to 38° N or S). Whereas there is a net loss of radiation from the ground between lat 38 and poles. As a result there is a transfer of energy from equator to poles at the surface level of the earth. This transfer of energy is effected by the atmospheric winds and ocean currents. It is estimated that ocean currents transfer about 30% of excess heat from equator to pole wards and the remaining is carried out by latent heat of vaporisation (cyclones, thunderstorms or convective cells), that is through different scales of weather and winds.

The KE of a moving air stream of unit mass $= \dfrac{1}{2} V^2$ .....(15.1)

where V is the speed of air.

The mass of air that passes through cross section area A with a speed V, r density in unit time ρ A V .....(15.2)

($\therefore$ mass = volume × density = AV × ρ)

$\therefore$ The wind power (P) of a air stream that passes through a cross section area A, speed V, density ρ in unit time is equal to the product of (15.1) and (15.2).

i.e., $\qquad P = \dfrac{1}{2} V^2 \times \rho \, A \, V = \dfrac{\rho A V^3}{2}$ .....(15.3)

Let A be the circular area swept by a rotor of diameter D

then $\qquad A = \pi \dfrac{D^2}{4}$ .....(15.4)

substituting 15.4 in 15.3 we have

$$P = \frac{1}{2}\rho\pi\frac{D^2}{4}V^3 = \frac{\pi}{8}pD^2V^3 \qquad .....(15.5)$$

Power available (P_A)conventionally expressed as

$$P_A = K_r\, D^2\, V^3 \qquad .....(15.6)$$

Where K_r = Constant, depends on wind dynamics and efficiency of rotor.

Table 15.1 The empirical value of constant K_r is given below

Unit Power	Unit of area	Unit of wind velocity or speed	Value of Kr
K W (kilowatts)	m^2	mps	0.00064
K W	m^2	Kmph	0.000014

In 1927, a German engineer A.Betz of Gottingen showed that maximum energy that can be extracted is 0.59259 P. This efficiency can be achieved by suitable blade design. In practice this factor may not exceed by 0.4

i.e., $\qquad\qquad P \simeq 0.4\, KAV^3$

Taking air density $\rho = 1.201\text{Kg/m}^3$ at normal atmospheric pressure 1000 hPa, and temperature T = 290° K. and assuming a rotor conversion efficiency of 75%, $K_r \simeq = \frac{\pi}{8} \times \frac{1.201 \times 0.593 \times 0.75}{1000}$

$$K_r \simeq 0.0002$$

The effect of the height of the wind mill tower on the performance can be significant. Empirical power law of indices has established which will be discussed subsequently.

The actual mechanical or electrical output would be less than 0.0002 $D^2\, V^3$. The energy produced per annum by a wind mill is given by

$$Ey = K_r\, D^2\, V^3 \times K_s\, t\; (KWh) \qquad ......(15.7)$$

where t = average number of hours in a year

K_s = Semi empirical factor associated with the

statistical nature of wind energy recovery.

The value of K_s = 2.06, was suggested by Pontin (1975) after computer analysis.

Thus $K_r\, K_s = 0.0002 \times 2.06 \simeq 0.0004$.

$Ey = \Sigma\; Em$,

where Em monthly power available

$$Em = K_r\, K_s\, D^2\, V^3 \times t\;\; Kwh \qquad .....(15.8)$$

t = no. of hours in a month.

G. Hellman derived an empirical formula relating wind speed (V_z) at an altitude Z and wind speed (V_{10}) at 10 m agl (above ground level) is given below.

$$V_z = V_{10} [\, 0.2337 + 0.656 \log_{10} (Z + 4.75)]\qquad\qquad.....(15.9)$$

The above formula was further modified by N. Carruthers as

$$V_z = K\, Z^{0.17}$$

where K is proportionality constant

Much work done by using formula

$$V_z = V_o\, Z^{\alpha}$$

where V_o = wind speed at standard height

α = exponent changes with height, temperature, time, nature of ground

From the last formula we have $= \dfrac{V_2}{V_1} = \left(\dfrac{Z_2}{Z_1}\right)^{\alpha}\qquad\qquad.....(15.10)$

In this $\propto$ is called power law index.

where V_2 = mean wind speed at height Z_2

V_1 = mean wind speed at height Z_1

Taking logarithms of (15.10) we have

$$\ln V_2 - \ln V_1 = \mu\, [\ln Z_2 - \ln Z_1]$$

or $\alpha = \dfrac{\ln V_2 - \ln V_1}{\ln Z_2 - \ln Z_1}\qquad\qquad.....(15.11)$

Power Law or Wind Field in Urban Areas

Davenport gave a similar empirical formula called power law relating surface wind ($\bar{u}$) and gradient wind (V_g) as follows.

$$\bar{u} = V_g \left(\dfrac{Z}{Z_G}\right)^{\alpha}\qquad\qquad.....(15.12)$$

where $\bar{u}$ = surface mean wind speed

V_g = gradient wind speed

Z = height of surface wind measurement

Z_G = height of gradient wind measurement

Davenports values of , $\propto$ Z_G in different environments are given below in Table 15.2.

Table 15.2

Environment	$\propto$	Z_G meters
Flat open country	0.16	270
Suburban area	0.28	390
Urban centers	0.40	420

$\propto$ - changes with pressure, temperature, height, time and nature of ground. $\propto$ is determined empirically from the observational data.

The power law $\dfrac{V_2}{V_1} = \left(\dfrac{Z_2}{Z_1}\right)^{\alpha}$ is valid upto 100-150 m agl. This law is used to estimate mean wind speed at any height (Z_2), if the mean wind speed at another height (Z_1) is known.

Density

Mass per unit volume is called density. Mathematically density of a body is defined as the limit of the ratio of the mass Δ m of an element of the body to its volume Δ v as $\Delta \upsilon \rightarrow 0$.

i.e., $\qquad \rho = \lim_{\Delta v \to 0} \dfrac{\Delta m}{\Delta \upsilon} = \dfrac{dm}{d\upsilon}$

or $\qquad d\,m = \rho \; d\upsilon.$

Integrating we have $m = \int\limits_{0}^{v} \rho \; dv$

The units of density: 1kg/m^3 in SI units

and 1 gm/cm^3 CGS system.

Typical values of densities at room temperature are given below

Mercury 13.6 gm/cm^3, water 1.00 gm/cm^3, sea water 1.03 gm/cm^3, Dry air at 0 °C and 1000 hPa pressure has density $1.277 \times 10^{-3} \text{ gm/cm}^3$ or 1.277 kg/m^3.

The density of air changes with height and atmospheric conditions (pressure, temperature, RH etc).

For dry air $(\rho \text{ gm/m}^3) = \dfrac{348.8 \times \text{Atmospheric Pressure in hPa}}{\text{Temperature (in}^\circ K)}$

Put $T = 273^\circ$ K, $\quad$ P = 1000 h Pa we have

$$\rho \text{ (gm/m}^3) = \dfrac{348.8 \times 1000}{273} \approx 1277 \text{gm} / \text{m}^3$$

or $\qquad \rho = 1.277 \text{ kg/m}^3$

The density of air changes in space and time and at a given location with temperature, pressure, humidity etc. We can find the changes in ambient air density due to atmospheric changes using ideal gas equation.

$$\rho = \frac{P}{R\,T_v}$$

where

ρ = density in gm/cm^3

P = Pressure 10^3 hPa

p = atmospheric pressure dynes/cm^2

T_v = Virtual temperature of air in $^\circ$K

R = gas constant,

for dry air

= 2.8703 × 10^6 ergs/gm $^\circ$K

$$T_\upsilon = \frac{T}{\left(1 - \frac{3e}{8p}\right)} \; \square \; T\left(1 + \frac{3e}{8p}\right)$$

Where   e = partial pressure of water vapour

T = temperature of moist air in $^\circ$K

For practical use, air density may be assumed to decrease by 1% of the surface value for every 100 m of height from ground level. According to Putnam, the density of air at 300 m asl 600 m asl and 3 km asl are 1300 gm/m^3, 1200 gm/m^3 and 900 gm/m^3 respectively.

DPTA

The principle of measurement of wind direction and velocity by Dines pressure tube anemometer. The instrument devised by W.H. Dines mostly used as standard. DPTA provides a continuous record of wind direction and velocity. The records serve as a basis in the search for windy districts and for locating winds power sites.

Wind Energy Estimates

Examples

Wind energy that can be produced by a wind mill in year is given by the formula

$$E_y = K_r\, K_s\, D^2\, V^3\, t \ (kwh/year)$$

Where K$_r$ $\square$ 0.00020, constant; K$_s$ $\simeq$ 2.06 semi empirical factor, suggested by Pontin in 1975. The product K$_r$ K$_s$ $\simeq$ 0.0004, t = number of hours in a year.

The following examples Table 32.3, 32.4, 32.5 give the wind energy estimates of a wind mill based on average wind speed for various months. Data extracted from IMD climatological tables (based on surface observation data at 0300 UTC and 1200 UTC for the period 1931 - 60).

Note: Em (Kwh/month)

Table 15.3 Station: Begum pet (Hyderabad-India) Lat 17° 27N, Long 78° 28E Height amsl 545 m

Months	Jan	Feb	Mar	Apr	May	June	July	Aug	Sep	Oct	Nov	Dec	Annual kwh/year Ey
Total no. of hrs	744	672	744	720	744	720	744	744	720	744	720	744	
Mean wind speed $\bar{V}$ = Kmps	8.1	8.9	9.6	10.9	12.4	23.8	22.1	18.3	12.6	8.9	8.0	7.4	
$\bar{V}$ = mph	2.25	2.47	2.67	30.3	3.44	6.61	6.14	5.08	3.50	2.47	2.22	2.05	
V^{-3}	11.39	15.07	19.03	27.82	40.71	288.8	231.47	131.10	42.87	15.07	10.94	8.61	
E_m for D = 3.65 m	46.5	55.6	77.7	109.9	166.3	1141.4	945.3	535.4	169.4	61.5	39.5	35.2	3343.7
D = 5 m	87.3	104.3	145.8	206.3	312.0	2141.7	1773.8	1004.6	317.9	115.5	81.1	66.0	6306.3
D = 7 m	171.1	204.4	285.8	404.4	611.5	4197.8	3476.6	1969.1	623.1	226.3	159.0	129.3	12458.4

Table 15.4 Station: Puri, India Lat 19° 48'N, Long 85° 49'E, Height amsl 6m

Months	Jan	Feb	Mar	Apr	May	June	July	Aug	Sep	Oct	Nov	Dec	Annual kwh/year Ey
Total no. of hrs	744	672	744	720	744	720	744	744	720	744	720	744	
Mean wind speed $\bar{V}$ = Kmps	11.9	15.9	20.5	24.0	26.2	23.2	23.3	19.7	15.9	12.3	10.2	0.5	
$\bar{V}$ = mph	3.31	4.42	5.69	6.67	7.28	6.44	6.47	5.47	4.42	3.42	2.83	0.14	
V^{-3}	36.26	86.35	184.22	296.74	385.83	267.09	270.84	163.67	86.35	40.0	22.67	0.003	
E_m for D = 3.65 m	143.7	309.2	730.3	1138.3	1529.4	1024.6	1073.6	648.8	331.3	158.6	87.0	0.01	6895.8
D = 5 m	269.7	580.3	1370.6	2136.5	2870.6	1923.0	2015.0	1217.7	621.7	297.6	163.2	0.02	13465.9
D = 7 m	528.8	1137.3	2686.4	4187.6	5626.3	3769.2	3949.5	2386.7	1218.6	583.3	319.9	0.0	26393.6

Table 15.5 Dwaraka-India, Lat 22° 22′N, Long 69° 0.5 E, Height amsl 11m

Months	Jan	Feb	Mar	Apr	May	June	July	Aug	Sep	Oct	Nov	Dec	Annual kwh/ year Ey
Total no. of hrs	744	672	744	720	744	720	744	744	720	744	720	744	
Mean wind speed $\bar{V} = Kmph$	13.5	14.8	16.1	16.7	20.0	23.5	27.1	21.9	15.0	11.8	12.4	12.3	
$\bar{V} = mps$	3.75	4.11	4.47	4.64	5.56	6.53	7.53	6.08	4.17	3.28	3.44	3.42	
V^3	52.73	69.43	89.31	99.90	171.88	278.44	426.96	224.76	72.51	35.29	40.71	40.0	
E_m for D = 3.65 m	215.3	256.1	364.7	394.7	701.8	1100.2	1743.3	918.5	286.5	144.1	160.9	163.3	6449.3
D = 5 m	404.1	480.6	684.4	7409	1317.1	20649	3271.9	723.9	537.7	270.4	301.9	306.5	12104.3
D = 7 m	492	941.9	1341.4	1452.1	2851.6	4047.2–	6412.9	3375.9	1054.0	530.1	591.7	600.8	23721.5

SOLAR ENERGY

The fusion furnace of the sun converts about 4 million tons (4×10^9 kg) of hydrogen into helium and radiates an amount of energy 3.8×10^{23} Kw/s as a by product. The solar energy incident on $1m^2$ area held normal to the sun's rays at the outer boundary of the atmosphere in one second is 1.38×10^3 W/m². It is virtually constant and called solar constant (accepted by IGY).

A house with peak load requirement of 2.5 KW would require a solar collector of 3.6 m² with 100% efficiency. The total solar energy received at the earth in one day is equivalent to 35 lakh nuclear explosions or 10000 hurricanes or 10^8 thunderstorms or 10^{11} Tornadoes. If this energy were stored it will satisfy all the world needs of domestic and industrial use for about 100 years.

According to one estimate, the solar radiation received at the outer boundary of the atmosphere is about 1.74×10^{14} KW. Of this about (30%) 5.2×10^{13} KW is reflected back to the space as shortwave radiation. 8.2×10^{13} KW (about 47%) is directly absorbed by the land, oceans and atmospheric system and the remaining 4.0×10^{13} KW (about 23%) is utilised in driving the hydrologic cycle through evaporation, convection and precipitation. About 3.7×10^{11} KW ($\sim 0.21\%$) of the incident solar energy is utilised (as input into Biosphere) for driving winds, waves, convection and sea currents. 4×10^9 KW (0.023% of the incident solar energy) is used in photosynthesis process on earth. The green canopy of plant kingdom consumes solar energy about (0.01 ly/min) 7 watt m². The total solar energy intercepted by the earth $= 2.55 \times 10^{18}$ Cal/min or 3.67×10^{21} Cal/day.

Solar energy per unit area expressed in langley (ly) or Kilo-langley (kly).

$$1 \text{ ly} = 1 \text{ Cal/cm}^2,$$

It is estimated that sun radiates each minute about 56×10^{26} Cal of energy in all.

SOLAR BATTERY

Photoelectric cell or solar battery is a device for generating electric power from sunlight Fig. 15.5. It generates power at the rate of 100 watts / m^2 of illuminated surface area. The basic unit of a typical solar battery is a thin wafer of extra pure silicon, containing less than 1ppm of impurity. For solar battery a tiny amount of arsenic is added. This silicon-arsenic wafer is called n-type silicon. Similarly if the silicon wafer is doped with tiny amount of boron then it is called p-type silicon. The combination of these two wafers is called p-n junction.

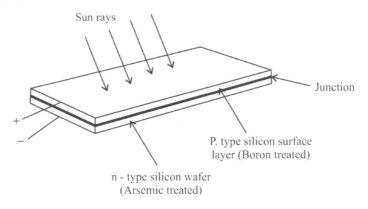

Fig. 15.1 Solar battery cell

In the solar battery, one electrical lead is attached to the surface and the other to the body of wafer. The device is a battery with positive terminal at the p-contact and the negative terminal at the n-contact. A series of such wafers (or so called solar battery cells) are used side-by-side.

Note: The sun is the primary source of heat, while the earth is the secondary source of heat.

15.2 ENERGY RESOURCES

Fuel: Fuels are substances defined as which on combustion produce heat energy without creating undesired byproducts. The production of heat mainly depends on the carbon content or hydrogen content or both.

The caloric value of fuel: Heat generated by fuels is measured in their caloric value, which is the maximum amount of heat liberated on complete combustion of 1kg of solid or liquid fuel or one ml of gaseous fuel.

Fuels are classified as three types-solid, liquid and gas and each may be natural or artificial. Example of solid, liquid gaseous natural fuels are (a) wood, lignite, peat, coal, anthracite, (b) petroleum, diesel oil, (c) natural gas, coal gas, while artificial fuels are (a) charcoal, coke, (b) artificial petrol Kerosene, spirit (c) producer gas, coal gas, water gas etc respectively.

Gaseous fuels are more economical for use:

Gaseous fuels can flow through pipes and can be lighted quickly at any place using burners. Most of the gaseous fuels are derived from coal or oil. All fuels contain carbon (in free state or combined state). The popular gaseous fuels are (i) Producer gas, (ii) Water gas, (iii) Semi-water gas, (iv) Coal gas, (v) Oil gas, (iv) Natural gas and (vii) Town gas.

1. **Producer gas:** It is obtained by the incomplete combustion of coal, coke or charcoal in furnaces called gas producers.

   It is a mixture of gases containing about 25% CO, 70% N_2 and 4% CO_2 and in small qualities of H_2, CH_4 and O_2. Still heat remains in the gas form which is released.

   The gas is poisonous, combustible but non-supporter of combustion. It is the cheapest of all the fuel gases.

   It is mainly used in metallurgical and other industrial operation. Also used to heat ovens (in the manufacture of coal gases), to run gas engines etc.

2. **Water gas:** It is obtained by passing steam over red hot coke.

   $$C + H_2O \longrightarrow CO + H_2 - 2900 \text{ cal}$$

   The reaction is endothermic and red hot coke cools down after sometime.

   Water gas is a mixture of CO (44%) and H_2 (49%), and traces of CO_2 N_2. The calorific value of water gas if high 2800-3100 $Kcal/m^3$. Water gas is used as a fuel and a source for the industrial preparation of hydrogen. It is used for lighting purposes with the help of Welsbach mantles (becomes incandescent on strong heating). Its flame is short and hot, it is used for Welding purpose.

3. **Coal gas:** When ordinary coal is subjected to destructive distillations (in closed retorts out of contact with air) at about $1000°C$, the following products are obtained.

   (i) Coal gas, (ii) Coke (iii) Coal tar, (iv) Ammoniacal liquor.

   One ton of bituminous coal gives about 13000 cft of coal gas.

   Coal gas is a mixture of H_2 (48%), CH_4 (30%), ethane (30%), ethylene (4.5%), CO (7%), N_2 (8.5%), O_2 (6.5%) and CO_2 (1.5%).

Coal gas calorific value 4500-5900 Kcal/m^3. It contains about 95% combustible material and mixed with water gas.

It is used as a fuel both in the household and metallurgical operations to create on inert atmosphere in certain chemical processes as a reducing agent and as an illuminating gas.

4. **Oil gas:** It is prepared by cracking a mineral (Kerosene oil) by letting it fall in the form of a thin stream on a retort heated to redness over a furnace which is broken upto gaseous hydrocarbons of ethylene and methane series.

Oil gas is used as a fuel gas in laboratories. Gaseous fuels are more economical compared to solid fuels (like coal, wood etc.)

15.3 FOSSIL FUELS

Fossil fuels are rich substances of energy, formed from the remains of plants and animals buried underground some 200 to 500 millions of years ago.

Petroleum (or mineral oil): It is complex mixture of hydrocarbons (mostly alkanes and cycloalkanes). Petrol occurs deep below earth crust, entrapped under rocky strata. It is a viscous black liquid.

In crude form the natural oil is called petroleum or crude oil. In Latin, Petra means rock, oleum means oil. In contact with petroleum deposits, there exists a gas which flows out from oil wells is called Natural gas. Natural gas in a mixture of methane, ethane and a small amount of propane. Petroleum and Natural gas can be regarded as stored form of solar energy.

Petroleum and natural gas can be regarded as stored form of Solar energy. These considered are formed by the decomposition of plants and microscopic animals. The plants and microscopic animals buried under pressure of earth crust are converted to petroleum natural gas under geological condition of high pressure, high temperature in the absence oxygen.

In modern society petroleum is a major source of energy, used in industry, agriculture, communication, transportation and used to prepare synthetic fibers, plastics, films, drugs and pesticides, insecticides, detergents etc.

Composition of petrol

Element	Percentage
Carbon	83-87
Hydrogen	10-14
Nitrogen	0.1-2.0
Sulphur	0.05-6.0
Oxygen	0.05-1.5

15.4 PETROLEUM REFINING

After locating oil field an oil well is dug by drilling bore (down) into the rock. Under pressure of gas above, the crude oil rushes out to the surface. The wells are also called gushers. The oil obtained from the mine is conveyed (by a system of pipelines) to a distant place and distilled.

Crude oil is a mixture of alkanes boiling in the range 0-400 °C. The oil is subjected to fractional distillation. The process of dividing petrol into fractions at different temperatures by freeing impurities is called refining.

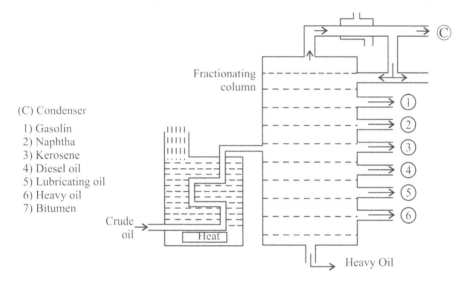

(C) Condenser

1) Gasolin
2) Naphtha
3) Kerosene
4) Diesel oil
5) Lubricating oil
6) Heavy oil
7) Bitumen

Fig. 15.2 Flow sheet of petroleum refining

Sl. No	Fraction or Distillate	Boiling temperature range	Number of carbons
	Refinery gases at top	Up to room temperature	$C_1 - C_4$
1	Gasoline (petrol or motor spirit)	30-150°C	$C_5 - C_9$
2	Naphtha		$C_7 - C_{10}$
3	Kerosine or Paraffin oil	150-240°C	$C_{10} - C_{12}$
4	Diesel or gas oil	240-350°C	$C_{13} - C_{14}$
5	Lubricating oil	Above 270°C	$C_{15} - C_{18}$
6	Paraffin wax	Above 400°C	$C_{18} - C_{30}$
7	Heavy fuel oil & Bitumens	From residue $C_{30} \longrightarrow$	Paints, road surfacing
	General Hydro carbon formula C_nH_{2n+2}		

In petroleum refining crude oil is continuously pumped through heated pipes at the bottom as shown in the figure. The highest boiling fraction condenses at the bottom and the lowest fraction at the top.

15.4.1 PETROLEUM REFINING PRODUCTS

Petroleum refining products in the fractionating column at various stages in order from bottom are described and some other important byproducts also given.

1. **Asphalt:** It is composed of almost of bitumen. Asphalt is a sticky, black, viscous liquid or semi-solid. It is used in concrete pavements, roofing, water proofing for fabrics.

2. **Diesel fuel:** It is composed of about 75% saturated hydrocarbons and 25% aromatic hydrocarbons. Diesel oil is 18% denser than gasoline (density of diesel 850 gm/lt, density of gasoline 720 gm/lt). It gives 18 % more energy per unit volume than gasoline, hence used in trains, lorries and buses.

3. **Heavy fuel oil:** It is a residue. It is heavier than gasoline and naphtha. Carbon chain C_{20} to C_{70}. It is used in ships, paints.

4. **Gasoline or petrol:** It consists of carbon chain C_5 to C_{12}. It is a mixture of Alkanes (paraffin's), cycloalkanes (napthanes) aromatics, alkanes (olefins). Mainly used as motor fuel and aeration spirit.

5. **Kerosene:** It has carbon chain $C_{10} - C_{15}$. It is used as stove fuel, burning lamps, mainly used as aviation fuel for jet engines.

6. **Liquid Petroleum gas (LPG):** It emerges from the lowest level (ground) in oil refinery. It is a mixture of propane and butane. It is used as fuel for cooking and driving vehicles.

7. **Lubricant:** It is a thick dark liquid with carbon chain $C_{13} - C_{18}$. Mainly used as lubricant oil (to reduce friction). It contains 90% base oil (mineral oil)

8. **Paraffin:** It has carbon chain $C_{20} +$ (above). It is also called paraffin wax. Used for making candles. Ointments, Vaseline.

9. **Mineral Oil:** It is a byproduct of distillation of petroleum to produce gasoline. It is colorless oil, composed of alkanes and cyclic paraffin. It is used as transformer oil. Used in baby lotions, cold creams, ointments, cosmetics, coolant etc.

10. **Tar:** It is a black viscous liquid produced in destructive distillation of organic matter. Tar is mostly produced from coal as a byproduct in coke production. Tar is a disinfectant and used as disinfectant.

11. **Bitumen:** Asphalt and tar are common forms of bitumen. Bitumen is an organic liquid. It is viscous, black sticky. Bitumen is mainly used for road surfacing and paint. It is viscous, black sticky. Bitumen is mainly used as road surfacing and paint. In the past it was used for waterproofing of boats and coating for buildings.

12. **Pitch or resin:** Pitch is a name of any bunch of high viscous liquid, which appear as solid. It is made of petroleum products or plants. The former is called bitumen while the latter is called resin or rosin.

15.4.2 PETROLEUM RESOUCES

Oil producing countries of the world. USA, Russia, Venezuela, Iran, Mexico, Romania, Iraq, Myanmar, India and Pakistan.

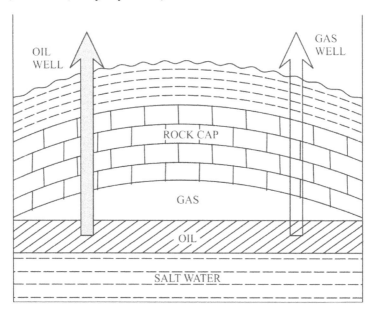

Fig. 15.3 Occurrence of petroleum or crude oil

Petroleum Resources of India

Oil Field are located in Assam and Gujarat. In Gujarat oil fields located at (a) Cambay, (b) Ankleshwar, (c) Kalol

In 1978 ONGC discovered gas at Sisodra, Gajera, West Metwam and Walner in Gujarat, oil and gas at Viraj West sobhasan in Gujarat.

In 1978-79 at Uran to Trombay

Name of Refinery	Year of production started	Capacity in Lakh tonnes 1.4-82	Actual crude run Lakh tonnes	
			1980-81	1981-82
Barauni IOC	1964	0.5	30.6	4.5
BPCL, Bombay	1955	48.7	42.12	49.9
BR & PL, Bongaigaon	1979	10.0	0.5	4.3
CRL Cochin	1966	33.0	29.2	31.2
Digboi	1901	5.0	5.0	5.0
Ganchali (IOC)	1962	8.5	6.4	7.5
Haldia	1975	25.0	23.1	22.8
HPCL, Bombay	1954	35.0	31.1	34.8
HPCL, Visakhapatnam	1957	15.0	13.2	11.8
Koyali (IOC)	1965	73.0	69.7	70.4
MRL Madras	1969	28.0	261	28.0
IOC Mathura	1982	48.6	-	5.2

15.5 MINERAL WEALTH OF INDIA

Minerals: Some metals occur in nature in free state and most others in the combined state, as compounds with other elements. Metal that found in free state are said to be 'native'. Ex: copper, silver, gold, mercury, platinum.

Def: The natural materials found in the earths crust in which metals found in a combined state are called minerals.

Ores: The minerals from which metals are extracted easily and economically are called ores. All ores are minerals but all minerals are not ores. The process of taking out the ores from the earth (mines) is called mining.

Native minerals have no affinity or little effinity for oxygen, moisture, others are chemical reagents. Eg. Ag, Au, Pt. In combined states (i) as oxides, examples bauxite Al_2O_3, tin stone SnO_2, (ii) as carbonate-Magnesite $Mg\ CO_3$ lime stone $CaCO_3$, (iii) as sulphate-gypsum $CaSO_4$, $2H_2O$, calestine S_rSO_4, barytes $Ba\ SO_4$, (iv) as silicates-spodumene $Li\ Al\ (Si\ O_3)_2$, (v) as Phosphate-amblygonite $Li\ F\ AlPO_4$, triphylite $(Li\ Na)PO_4$, $(Fe,\ Mn)_3$ $(Po_4)_2$, (vi) as sulphide-galena PbS, zinc blend ZnS, cinnabar HgS, (vii) as halide rock salt $Na\ Cl$, carnallite $KCl\ Mg\ Cl_2$, H_2O, horn silver $AgCl$.

India has rich mineral resources. The following table gives some common minerals and their occurrences in India.

Table 15.6

S. No	Metal	Mineral found in India	Places in India
1	Lithium (Li)	Lithium mica (Li, Na, K)$_2$, Al$_2$ (SiO$_3$) (FOH)$_2$	Bihar, Rajasthan, Karnataka, Kashmir
2	Sodium (Na)	Native borax Na$_2$ B$_4$O$_7$ 10H$_2$O	Laddakh & Kashmir
3	Potassium (K)	Nitre (Indian saltpetre) KNO$_3$	Haryana, Bengal
4	Beryllium (Be)	Beryl 3BeO, Al$_2$O$_3$, 6SiO$_2$	Bihar odisha, coastal AP (Nellore) Rajasthan (Ajmer) Tamil Nadu (salem)
5	Magnesium (Mg)	(i) Dolomite MgCO$_3$, CaCO$_3$ Magnesite Mg CO$_3$	Tamil Nadu (Salem)
6	Calcium (Ca)	(i) Gypsum Ca SO$_4$ 2H$_2$O (ii) Limestone (Chalk or Marble) CaCO$_3$ (iii) Flurospar CaF$_2$	MP Rajasthan, Tamil Nadu, Maharashtra, J & K, Rajasthan, MP MP (Jabalpur)
7	Barium (Ba)	Heavyspar BaSO$_4$	Rajasthan (Alwar), Rayalaseema (Kurnool)
8	Boron (B)	Borax Na$_2$B$_4$O$_7$, 10H$_2$O	Ladakh, Kashmir
9	Aluminium (Al)	Bauxite Al$_2$O$_3$, 2H$_2$O	MP, Odisha, Chhattisgarh (Balaghat) , Jammu, Tamil Nadu
10	Tin (Sn)	Tinstone SnO$_2$	Bihar (Hazaribagh), Odisha
11	Antimony (Sb)	Stibnite Sb$_2$S$_3$	Punjab, Karnataka
12	Copper (Cu)	Copper pyrites CuFeS$_2$	Bihar, Odisha, MP, Rajasthan
13	Silver (Ag)	Native-Argentite Ag$_2$S Horn Silver AgCl	Kolar fields, Hatti mines (Karnataka) Rajasthan
14	Gold (Au)	Native-Bismuth Aurite Au$_2$ Bi	Alluvial sands of Ganga & Brahmaputra, Irrawaddy
15	Zinc (Zn)	Zinc blende of ZnS	Rajasthan (Zewar mines)
16	Chromium (Cr)	Chromite FeO Cr$_2$ O$_3$	karnataka, Bihar, Odisha, Tamil Nadu, Andaman Islands
17	Manganese (Mn)	Pyrolusite MnO$_2$, Braunite Mn$_2$O$_3$	Maharashtra (Panchmahal) Karnataka, Bihar
18	Iron (Fe)	Haematite Fe$_2$O$_3$ Magnetite Fe$_3$ O$_4$	Bihar (singh bhum) Karnataka (Mysore)
19	Cobalt (Co)	Cobalt glance (CoFe)AsS	Khetri, mixed with copper
20	Nickel (Ni)	Smalltite (Fe, Ca, Ni)As$_2$	Ores at Khetri & Kolan and Tamil Nadu

15.6 METALLURGY

The Process of extracting the metal from their ores and refining them is called metallurgy

There is no single method of extraction of all the metals. various steps of metallurgy given below.

1. Grinding (or crushing) the ore

2. Pulverising the ore-crushed ores taken to stamp mills, or pulverisers to make fine powder

3. Ore-dressing (concentrating the ore); separating impurities of earthy matter (like sand, clay, lime). The aggregate of impurities called gangue, Removal gangue is called ore dressing. This is achieved by the process (a) Hand picking, (b) Hydraulic classifier, (c) Froth flotation process, (d) electromagnetic separation, (e) leaching

4. Conversion of the concentrated ore into metallic oxide. This is done by (a) Calcinations, (b) Roasting

5. Extraction of the metal from metallic oxide. This is achieved by the methods (a) Smelting, in this process oxide ore in the fused state reduced by carbon to free metal (b) Carbon monoxide reduction (c) Hydrogen reduction method, (d) Magnesium reduction method, (e) Aluminum reduction method, (f) self reduction method (g) Amalgamation process and (h) Hydro metallurgy process.

6. Purification or refining metals. The common refining methods are (i) liquation, (ii) Distillation, (iii) Oxidation, (iv) Poling, (v) Electrolytic method, (vi) Vapour phase method, (vii) Zone refining method, (viii) Chromatography method, and (ix) Solvent extraction.

15.7 METAL EXTRACTION FURNACES

The main furnaces are: (i) Kilns, (ii) Blast furnace, (iii) Reverberatory furnace, (iv) open hearth to furnace, (v) Electric furnace and (vi) Bessemer's converter.

1. **Kilns:** Kilns are large enclosed spaces in which the materials are mixed before they are put into a furnace. Here no chemical reaction or fusion takes place. Here the ore is only dried.

2. **Blast furnace:** Blast furnace is used for the manufacture of cast iron, for smelting roasted copper pyrites ore in the extraction of copper metal and smelting of Pbo in the extraction of lead metal.

3. **Reverberatory furnace:** The furnace consists of three main parts-Fire place, Hearth (or bed) and Chimney. The charge is placed on the hearth and the fuel is placed on the fire place. Air enters from the fire place

through iron gates and waste gases escape out chimney. There are doors on one side of the furnace for introducing charge and tap holes on the other side to remove the charge.

4. **Open hearth furnace:** This furnace is used for the manufacture of steel from cast iron.

5. **Electric furnace:** It is very useful where very high temperatures are required, which can be easily controlled by power supply regulation. Heroult's furnace is commonly used (as in the case of manufacture of steel).

6. **Bessemer's Converter:** It is a pear shaped vessel and is used for the manufacture of steel from cast iron.

Processes Functioning

1. **Calcinations:** In this process ore is subjected to the action of heat, but is confined to the operation of expelling. Ex-lime stone heated to expel CO_2.

   Bauxite heated to high temperature to expel water

2. **Roasting:** ore alone or with the addition of other material heated to below their melting point to effect chemical changes in them, which will help in the process

3. **Smelting:** The separation of metal from chemical combination is called reduction. Smelting is an operation, whereby the metal is separated by fusion from the ore. Various reducing agents used like carbon, CO, H, Al.

4. **Blast Furnace:** It is a tall structure with gates at the bottom and flare opening at the top. An air blast is supplied to the furnace by means of bellows, fans or blowing machines through the nozzles at the bottom (the nozzles are called tuyeres). The materials to be treated are charged into the furnace mixed with the fuel and as the substances melt, they rundown to the bottom and accumulated below in crucible or hearth, which is removed (and further treated).

5. **Regenerative furnace:** The heat carried away to the flues by the escaping gases at top is again utilized in this furnace. A flowing column of air is heated by the hot flue gases, the air is then brought back to the fire and returned to the furnace.

6. **Electrolytic refining:** The impure metal is made the anode of an electrolytic cell, the cathode is a thin plate made of any pure metal,

upon which the deposit can be stripped. The impurities after electrolysis remain in solution while the pure metal gets deposited on the cathode.

Important Steel Plants in India

Bhilai (MP), Durgapur (WB), Bokaro (Bihar) and Rourkela (Odisha).

Most abundant elements of the earth's crust.

O_2 (about 50%), Si (26%), Al (7.3%), Iron (Fe, 4.1%), Ca (3.2%), Na (2.3%).

Index

A

Absorbed radiation dose 167
Abyss 20
Abyssal benthic zone 21
Acid rain 157, 159, 178, 179
Acid soils 48
Agriculture 39, 40, 48, 51, 52
Agro climatology 317, 319
Air pollution 180, 181, 182, 194, 197
Algae 13, 216, 222, 241, 303
Alluvial soil 47, 48, 126, 212
Anabolism 13, 18, 327, 431
Anaerobic 221
Aquatic Ecosystem 12, 15, 16, 255
Aquatic plants 12, 433
Arable land 39, 40, 47, 52, 270
Aridity Index 128, 152
Arcthi- benthic zone 2
Arobic 221
Atmosphere 154, 157, 158, 159, 162
Atmospheric heat process 92
Atmospheric nitrogen 32, 47, 173, 174
Atmospheric structure 93

B

Bank 55, 72, 119, 122, 125
Bapmon 179, 180
Basin 63, 65, 66, 226, 289
Beach 26, 120, 193
Benthic biotic environment 20
Biotic units 22
Benthic Division 20
Biodegradable 17, 198, 243, 257, 436
Biodiversity 1, 3, 4, 7, 8
Bio-divesity Act-2002 240
Biogeographic Zones in India 260

Biological diversity in India 245
Biological environment 1, 2, 81, 294
Biological factors 4, 28
Biomass 158, 241, 300, 303, 316
Biomes 16
Biosphere 33, 36, 74, 75, 89
Biota 17, 20, 198
Black soil 47, 48, 93, 127
Blue-green algae 24, 171, 303
Brown algae 24
Buffer solution 159

C

Canyon 119
Capillary water 44
Carbon and its compounds 171, 201
Carnivores 13, 14, 15, 18, 21
Cartostat 368
Catabolism 13, 18, 431
Chemical weapons 83, 84, 85, 88, 89
Chemical weathering 41, 126, 271, 304, 305
Chemotropism 5
Chloroplasts 9
Chromosphere 97, 98
Climate 12, 13, 16, 17, 33
Climate change 1, 3, 6, 7, 17
Climatic controls 275
Climate disasters 76
Climate over earth history 305
Climate system 33, 198, 286, 287, 289
Climatology 76, 274, 275, 293
Climax importance 12
Climax species 10
Climax-types 12
Component of climate system 291

Composition of
Atmosphere 178, 195, 324
Consumer-micro 10
Contour-bounding 125
Corona 97
CRBN 71
Contour strip-cropping 125
Current fallow 51, 53
CWR 316

D

Darwin's theory 6
Deep-sea Benthos 30
Density of soil 43
Desert soil 47, 48, 127, 128, 152
Desertification 54, 127, 128, 129, 152
Deserts 5, 16, 53, 76, 77
Deserts-geographical 129
Disaster 71, 72, 73, 74, 75
Disaster Management 72, 73, 82, 89, 147
Dispersal of air pollutants 183
Drinking water 216, 220
Dust fall Jar 186, 188

E

Earth's atmosphere 176, 294, 306, 319, 357
Earthquake 55, 72, 73, 75, 82
Earthquake intensity 142, 143
Earthquake parameters 140, 147
Earthquake safety 151
Earthquake waves 137, 141
Earth-structure 130
Ecological pyramids 15, 18
Ecological succession 10, 17
Ecology 2, 10, 29, 37, 75
Ecosystem-Aquatic 12
Effects of Particulate matter 164
Environ Automatic
weather station 296, 331
Environment 1, 2, 3, 4, 5
Environment Marine issues 7
Environment protection 78

Environmental
degradation 71, 79, 262, 267
Environmental hazards 81
Environmental issues 6, 7, 17, 78, 89
Environmental
pollution 7, 74, 78, 79, 81
Erosion Accelerated soil 151
Erosion control-mechanical
measures 125
Erosion normal 54

F

Fault 64, 134, 135, 136, 137
Fertilizers 55, 56, 70, 88, 158
Field capacity 44, 45, 70
Fire services 199, 202
Food chain 2, 13, 14, 15, 18
Food web types 14
Food webs 14, 15, 18
Forest Conservation Act 1980 3
Forest fire 72, 154, 157, 172, 180
Forest policy 226, 230, 232
Forest products 225, 227, 228, 230, 234
Forests 1, 3, 5, 12, 16
Fungus 13, 14, 434

G

GAGAN 365, 369, 376
Gene 8
Geological time scale 310, 311, 319
Geophysics 114
Geostationary Satellites 358, 360
Geotropism 5
GFD 114
GIS 51, 376
Global warming effects 188
Global water 62, 68, 69, 116, 202
GLONASS 363, 364, 370
GOES 360, 366, 367
GPS 361, 362, 363, 364, 365
Gravitational water 44
Gravity water 44, 58, 64, 215
Ground water 57, 58, 60, 62, 63

H

Habitat 239, 243, 244, 245
Hardness of water 202, 217
Hazard 72, 73, 74, 75
Hazards of nuclear fallout 168
Henrich events 304, 305, 309
Herbivores 13, 14, 15, 18
Heterotrophic bacteria 31
Heterotrophic succession 11
Heterotrophs 9
High volume sampler 160, 186, 187
Human population
growth 246, 248, 249
Humus 33, 41, 42, 46
Hydrocarbons 154, 158, 160
Hydrography 114
Hydrological cycle 196, 202, 212
Hydrology 58, 60, 212
Hydroscopic water 44
Hydrotropism 5, 8, 11

I

IGBP-1989 321
Igneous rocks 41
Impact of climate change on Agriculture 300, 315
Important Ocean current 286, 301
Indian Forest Act 3, 232, 238
Indian Forest organization 227
INSAT 354, 365, 371, 374
In-situ conservation 259
Ionosphere 95, 96, 348
IRNSS 364, 365
IRS Application 366

K

Kharif 52
kyoto protocol 193, 435

L

Lamarck's theories 5
Land resources 39, 51, 126
Land use statistics 39, 51, 227

Land utilization 51, 53, 272
Landslide or Slip Erosion 122
Laterite soil 127
Law of the sea 256, 324
Lichen 12, 13, 136
Littoral system 21
LPI 250, 265

M

Man and environment 3, 19
Marine biodiversity 242, 244, 268
Marine biosphere 36, 37
Marine carbon cycle 32, 38
Marine environment 19, 20, 21, 22
Marine environment classification 20
Marine organism 30, 37
Marine phosphorus cycle 32
Marine sulphate cycle 33, 38
Mechanism of erosion 122
Mesosphere 95
Metabolism 33, 57, 300
Metamorphic rocks 41, 48, 132
Micro consumers 10
Microfauna 46
Microflora 46
Milankovitch cycles 313, 319, 323
Mineral matter 45
Minerals 11, 14, 31, 39
Mitigation 72, 74, 75, 81
Monsoon season 52, 53

N

NAAQS – 1970 160, 161
National Authority 85, 88
National Biodiversity Authority 242, 265, 266
Natural resources 1, 2, 7, 39
Nekton 22, 23, 27, 31
Nitrogen cycle 32, 37, 174
Nitrogen Fixation 32, 38
NOAA 351, 366, 367
Noise pollution 194, 247, 257
Nuclear energy 7, 170, 202

Nuclear fission 169, 170, 202
Nuclear fusion 170, 202

O

Ocean Carbon cycle 37, 38
Ocean currents 109, 110, 111
Ocean tranches 117, 118
Oceanographic 358
Oceans 13, 20, 32, 60
Omnivores organism 8
Organic production in the sea 34, 36
Oxygen production 35
Ozone 91, 94, 95

P

Paleoclimate 105, 314
Parasite 8, 16, 31, 46
Pelagic division 21, 22, 27
Pelagic environment 21, 22
PETN 87
pH value 159, 178, 179
Photo tropism 5
Photosphere 97, 98
Photosynthesis 2, 8, 9, 13
Phylum 8, 23, 24, 25
Phytoplankton 7, 11, 27
Plant–phylum or group 241
Plate boundaries 134, 135
Plate tectonics 132, 133
Pollution control 80,177, 267
Pollutions standard index 160, 200
Population density 28, 246, 247
Populations of the sea 23
Predator 8, 197
Primary succession 11, 12
Producers 9, 10, 12, 13
Productivity 17, 46, 55
Public liability
Insurance Act 1991 89, 154, 261
Pyramid biomass 16
Pyramid number 15
Pyramid of energy 16

R

Rabi 53
Racial development theories 5
Radar 341, 348, 353
Radio nuclides 166
Radio spectrum 347
Radioactivity 74, 163, 166
Rain water 56, 57, 64
Red soil 47, 48, 127
Reef 30, 119
Remote sensing 180, 358, 366, 370
Reproduction 9, 27, 28, 31
Reservoir 12, 33, 36, 47
Resident time 165
Reverse osmoses 202, 223, 224

S

Saline soil 49
Satellite communication 147, 349, 350
Satellite meteorology 354
Satellite orbits 356
SATNAV 376
Sea icing 107, 120
Sea life 23, 243
Sea mountains 118
Search and rescue satellites 350
Sedimentary rocks 41, 130
Seismic waves 135, 139, 142
Seismic zoning of India 148
Sensitivity 5, 318, 337
Sewage treatment 202, 222, 224
Shoal 119
Slip erosion 55, 122
Smog 164, 176, 179
Soil 1, 40
Soil and water conservation 123, 124
Soil erosion 54, 56, 70
Soil fertility 47, 55
Soil formation 41, 42
Soil profile 42, 67, 70
Soil testing 49, 50
Soil types 63, 126
Soil water 44

Solar activity 97, 98, 304
Solar energy 2, 5, 13, 16
Stream Bank Erosion 122
Suspended particulate
matter (spm) 163,177
Sustainable development 267, 256
Sustainable use 9, 256

T

Tectonics 136, 286
Terrestrial ecosystem 12, 16
Thallophyta 24
Thermal comfort 327
Thermosphere 95
Thermotropism 5
TIROS 354, 359
TNT 87, 88
Topoclimatology 318
Toxic chemicals 86, 87
Toxic pollutants 243, 257
Trench 117, 118
Tropism 5,8
Troposphere 93, 307
Trough 118, 278
Type of food chains 13
Types of erosion 54

V

Vertebrates 24, 26
Vulnerability 273, 299

W

Water - global 68
Water and its pollution 205
Water conservation measures 123, 151
Water Erosion 122
Water impurities 216
Water pollution act (1974) 207, 224
Water resources 57, 58
Water uses 62, 69
Weapons of mass destruction 83, 84
Weathering 41, 48
Wind Erosion 122
WMO 306, 380
World environmental day 6, 78
World human population 249, 271
World population explosion 7, 78
World's land utilization 270
World's water 60, 439

X

Xerosere 12

Z

Zooplankton production 34, 36